Strontium Metabolism

Strontium Metabolism

*Proceedings of the International Symposium
on Some Aspects of Strontium Metabolism
held at Chapelcross, Glasgow and Strontian,
5–7 May, 1966*

Edited by

J. M. A. LENIHAN

(Western Regional Hospital Board, Glasgow)

J. F. LOUTIT

(M.R.C. Radiobiological Research Unit, Harwell)

J. H. MARTIN

(United Kingdom Atomic Energy Authority, Chapelcross)

ACADEMIC PRESS
London and New York
1967

ACADEMIC PRESS INC. (LONDON) LTD
Berkeley Square House
Berkeley Square
London, W.1.

U.S. Edition published by
ACADEMIC PRESS INC.
111 Fifth Avenue
New York, New York 10003

Library of Congress Catalog Card Number: 67–24318

PRINTED IN GREAT BRITAIN BY
ROBERT MACLEHOSE & CO. LTD
THE UNIVERSITY PRESS, GLASGOW

237885

LIST OF CONTRIBUTORS

B. ÅBERG, *Department of Clinical Biochemistry, Royal Veterinary College, Stockholm 50, Sweden.*

G. C. ARNEIL, *Department of Child Health, Royal Hospital for Sick Children, Oakbank, Glasgow, C.4.*

H. AUERBACH, *Argonne National Laboratory, 9700 South Cass Avenue, Argonne, Illinois 60440.*

B. O. BARTLETT, *Agricultural Research Council, Radiobiological Laboratory, Letcombe Regis, Wantage, Berks.*

J. L. BEAMER, *Biology Department, Pacific Northwest Laboratory, Battelle Memorial Institute, Richland, Washington.*

M. C. BELL, *Agricultural Research Laboratory, University of Tennessee, Oak Ridge, Tennessee.*

R. S. BRUCE, *Agricultural Research Council, Radiobiological Laboratory, Letcombe Regis, Wantage, Berks.*

A. M. BRUES, *Argonne National Laboratory, 9700 South Cass Avenue, Argonne, Illinois 60440.*

L. BURKINSHAW, *M.R.C. Environmental Radiation Research Unit, Leeds General Infirmary, Leeds.*

L. N. BURYKINA, *Department of Public Health, U.S.S.R. Department of External Affairs, Rachmanovsky Lane, 3, Moscow.*

T. E. F. CARR, *M.R.C. Radiobiological Research Unit, Harwell.*

A. CATSCH, *Lehrstuhl für Strahlenbiologie, Technische Hochschule, Karlsruhe, Germany.*

C. L. COMAR, *Department of Physical Biology, New York State Veterinary College, Cornell University, Ithaca, New York, U.S.A.*

A. H. DAHL, *Colorado State University, Fort Collins, Colorado, U.S.A.*

R. J. DELLA ROSA, *Radiobiology Laboratory, University of California, Davis, California, U.S.A.*

D. DEPCZYK, *Institute of Occupational Medicine, Lodz, Poland.*

G. DEROCHE, *Argonne National Laboratory, 9700 South Cass Avenue, Argonne, Illinois 60440.*

R. DOMANSKI, *Institute of Occupational Medicine, Lodz, Poland.*

G. C. FARRIS, *Colorado State University, Fort Collins, Colorado, U.S.A.*

T. M. FERGUSON, *Agricultural Research Laboratory, University of Tennessee, Oak Ridge, Tennessee.*

M. GILLNER, *Department of Clinical Biochemistry, Royal Veterinary College, Stockholm 50, Sweden.*

M. GOLDMAN, *Radiobiology Laboratory, University of California, Davis, California, U.S.A.*

D. GRUBE, *Argonne National Laboratory, 9700 South Cass Avenue, Argonne, Illinois 60440.*

R. Hesp, *Health and Safety Department, U.K.A.E.A., Windscale Works, Sellafield, Seascale, Cumberland.*

P. L. Hackett, *Biology Department, Battelle Memorial Institute, Pacific Northwest Laboratory, Richland, Washington, U.S.A.*

G. E. Harrison, *M.R.C. Radiobiological Research Unit, Harwell.*

J. M. Judd, *Atomic Energy of Canada Limited, Chalk River, Ontario, Canada.*

A. Kent, *Overwood, Main Road, Elderslie, Renfrewshire.*

R. Kirchmann, *Department of Radiobiology, Belgian Nuclear Centre, Mol, Belgium.*

V. A. Knizhnikov, *Institute of Biophysics, Ministry of Public Health of the U.S.S.R., Moscow.*

J. M. A. Lenihan, *Western Regional Hospital Board, Regional Physics Department, 9–13 West Graham Street, Glasgow, C.4.*

I. Lewin, *Veterans Administration, Edward Hines, Jr., Hospital, Hines, Illinois 60141.*

J. Liniecki, *Institute of Occupational Medicine, Lodz, Poland.*

E. Lloyd, *Argonne National Laboratory, Argonne, U.S.A.*

J. F. Loutit, *M.R.C. Radiobiological Research Unit, Harwell.*

R. S. Lowrey, *Agricultural Research Laboratory, University of Tennessee, Oak Ridge, Tennessee.*

A. N. Marei, *Institute of Biophysics, Ministry of Public Health of the U.S.S.R., Moscow.*

W. V. Mayneord, *7 Downs Way Close, Tadworth, Surrey.*

R. O. McClellan, *Fission Product Inhalation Program, Lovelace Foundation, 5200 Gibson Boulevard, S.E. Albuquerque, New Mexico 87108.*

F. R. Mraz, *Agricultural Research Laboratory, University of Tennessee, Oak Ridge, Tennessee.*

B. E. C. Nordin, *M.R.C. Mineral Metabolism Research Unit, Leeds General Infirmary, Leeds.*

I. L. Ophel, *Atomic Energy of Canada Limited, Chalk River, Ontario, Canada.*

T. M. Paul, *McGill University, Montreal, Canada.*

B. Ramsbottom, *Health and Safety Department, U.K.A.E.A., Windscale Works, Sellafield, Seascale, Cumberland.*

J. Rivera, *Health and Safety Laboratory, U.S. Atomic Energy Commission, New York, U.S.A.*

J. Rundo, *United Kingdom Atomic Energy Authority, Harwell.*

J. Samachson, *Veterans Administration, Edward Hines, Jr., Hospital, Hines, Illinois 60141.*

R. Scott Russell, *Agricultural Research Council, Radiobiological Laboratory, Letcombe Regis, Wantage, Berks.*

P. L. Sheldon, *Fission Product Inhalation Program, Lovelace Foundation, 5200 Gibson Boulevard, S. E. Albuquerque, New Mexico 87108.*

J. G. Shimmins, *Western Regional Hospital Board, Regional Physics Department, 9–13 West Graham Street, Glasgow, C.4.*

S. C. Skoryna, *McGill University, Montreal, Canada.*

D. A. Smith, *Gardiner Institute of Medicine, Western Infirmary, Glasgow W.1.*

H. SMITH, *U.K.A.E.A.*, *Chapelcross Works, Annan, Dumfriesshire.*

H. SPENCER, *Veterans Administration, Edward Hines, Jr., Hospital, Hines, Illinois 60141.*

A. SUTTON, *M.R.C. Radiobiological Research Unit, Harwell.*

D. M. TAYLOR, *Department of Biophysics, Institute of Cancer Research, Belmont, Sutton, Surrey.*

R. C. THOMPSON, *Biology Department, Battelle Memorial Institute, Pacific Northwest Laboratory, Richland, Washington, U.S.A.*

O. VAN DER BORGHT, *Department of Radiobiology, Belgian Nuclear Center, Mol, Belgium.*

L. M. VAN PUTTEN, *Radiobiological Institute TNO, Rijswijk, The Netherlands.*

S. VAN PUYMBROECK, *Department of Radiobiology, Belgian Nuclear Center, Mol, Belgium.*

J. VAUGHAN, *M.R.C. Bone-Seeking Isotopes Research Unit, Churchill Hospital, Oxford.*

V. VOLF, *Institute of Radiation Hygiene, Prague, Czechoslovakia.*

D. WALDRON-EDWARD, *McGill University, Montreal, Canada.*

J. M. WARREN, *Western Regional Hospital Board, Regional Physics Department, 9–13 West Graham Street, Glasgow, C.4.*

F. A. WHICKER, *Colorado State University, Fort Collins, Colorado, U.S.A.*

M. WILLIAMSON, *M.R.C. Bone-Seeking Isotopes Research Unit, Churchill Hospital, Oxford.*

G. WITHROW, *Agricultural Research Laboratory, University of Tennessee, Oak Ridge, Tennessee.*

P. L. WRIGHT, *Agricultural Research Laboratory, University of Tennessee, Oak Ridge, Tennessee.*

G. A. P. WYLLIE, *Department of Natural Ph losophy, University of Glasgow, Glasgow, W.2.*

Preface

This volume contains a representative collection of reports describing current research on the physiological behaviour of strontium. The problems with which it deals are of concern to scientists involved in assessment of possible hazards from radioactive fallout in man and other animals as well as to clinical investigators interested in bone growth and other aspects of calcium metabolism.

Scotland has a particular interest in strontium, for the element is named after the remote Argyllshire village of Strontian. More recently two teams of research workers have made modest contributions to the current programme of work which, though inspired by the unexpected and curiously difficult problems associated with fallout from nuclear weapons, has extended into many fascinating areas of biological and clinical science. The Glasgow group, drawing resources from the university Department of Child Health, the Royal Hospital for Sick Children and the Physics Department of the Western Regional Hospital Board, have been concerned mainly with the estimation of ^{90}Sr in the bones of children and have recently extended the scope of their project to include analytical studies of diet and urine. The Chapelcross group, forming part of the health and safety organization in a large nuclear establishment, have been concerned mainly with the possibility of modifying the uptake and metabolism of Sr by the administration of drugs. The investigations are being conducted both in animals and in human subjects.

The common interests of the two groups initiated the original proposal for an international symposium that would bring together research workers from many parts of the world to discuss current activity and future prospects. The organization of the meeting was facilitated by the patronage and active encouragement of the Medical Research Council and the United Kingdom Atomic Energy Authority.

The scientific sessions, held at Chapelcross on 5th and 6th May, 1966, embraced four main topics. The first session dealt with the passage of ^{90}Sr through food chains and its incorporation into the tissues of man and other animals. In this matter the widespread distribution of ^{90}Sr gives the opportunity for isotopic tracer investigations exciting the imagination and challenging the ingenuity of the experimenter. Secondly, there was a group of papers on work with tracer substances, such as ^{85}Sr, used to investigate metabolic processes in man. Among the topics of major interest here are the incorporation of Sr into bone and the usefulness of radioactive Sr as a tracer for Ca in the study of bone formation and growth. Thirdly, there were reports on investigations with animals, where the more rapid growth and less exacting assessment of radiation hazards allows many kinds of experiment to be made in a more precise and systematic way. Lastly came reports on methods of modifying Sr metabolism by inhibiting uptake or by enhancing excretion in the hope of diminishing the hazard associated with acute or chronic exposure of an abnormal kind.

After the conclusion of this rigorous diet of study, the participants repaired to Glasgow, to examine some historic relics from the early history of Sr and, on 7th May, to Strontian itself. There, on the sunlit hillside where strontianite was first discovered, an informal and evocative colloquium reviewed the history and geology of the element. Fortified by expert knowledge and guidance, members of the symposium departed with an impressive collection of specimens, of varying authenticity, gathered from the rocky slope or from the spoil-heaps of the long abandoned workings.

This volume is a record of the proceedings of the symposium. The 38 papers are, in the main, reproduced as the authors presented them. In a number of instances, however, the original text has been shortened or otherwise modified in the interest of clarity and to avoid overlapping. The discussion that followed each paper was often lively and provocative, but, as so often happens, did not survive very happily the translation into cold print. Many of the questions answer themselves after the text has been studied. Other exchanges, illustrating different approaches to the interpretation of experimental observations, illuminated aspects of intellect and personality that are probably of limited interest to the wider circle of readers to whom the present volume is offered. The record of the discussion has therefore been severely curtailed.

For a century or more, Sr had humdrum uses and made a mildly spectacular contribution to the manufacture of fireworks. A great and terrible firework has made the name of the once obscure element an epitome of mankind's anxiety over the impact of nuclear science on military technology. The pages that follow give an account of the ways in which scientists and doctors throughout the world are trying patiently, honestly and with modest success to come to terms with the challenge of strontium.

Contents

List of Contributors   v
Preface   ix

Historical

Opening remarks—W. V. MAYNEORD   1
Scotland and Strontium—ANDREW KENT   7
The Strontian Mines—G. A. P. WYLLIE   13

Fallout Studies

Some Principles of Strontium Metabolism: Implications, Applications, Limitations—C. L. COMAR   17
Time Course of the Transfer of Strontium-90 Through Food Chains to Man—R. S. BRUCE, B. O. BARTLETT and R. SCOTT RUSSELL   33
Strontium-90 from Fallout in Human Bone—J. F. LOUTIT   41
Predicting Strontium-90 Concentrations in Human Bone—JOSEPH RIVERA   47
Studies in Strontium Metabolism–1. Correlation of Strontium-90 Levels in Foetal Bone and Maternal Diet—J. M. A. LENIHAN   57
Studies in Strontium Metabolism–2. Strontium-90 in Urine: Correlation with Diet and Age—JANET M. WARREN   63
Studies in Strontium Metabolism–3. Preferential Calcium/Strontium Renal Tubular Absorption Measured by Activation Analysis—GAVIN C. ARNEIL   67
Strontium Metabolism in Man—V. A. KNIZHNIKOV and A. N. MAREI   71
Chronic Accumulation of Strontium-90 in the Body as Predicted from the Retention of a Single Tracer Dose—JULIAN LINIECKI   83
Effect of Age on Radioactive and Stable Strontium Accumulation in Mule Deer Bone—G. C. FARRIS, F. A. WHICKER and A. H. DAHL   93
Skeletal Distribution of Strontium and Calcium and Strontium/Calcium Ratios in Several Species of Fish—I. L. OPHEL and J. M. JUDD   103

CONTENTS

Tracer Studies in Man

Influence of Dietary and Hormonal Factors on Radiostrontium Metabolism in Man—HERTA SPENCER, ISAAC LEWIN and JOSEPH SAMACHSON   111

Kinetics of Strontium-85 Deposition in the Skeleton During Chronic Exposure—J. RUNDO   131

An Attempt to Quantitate the Short-Term Movement of Strontium in the Human Adult—T. E. F. CARR   139

A Comparison Between Calcium-45 and Strontium-85 Absorption, Excretion and Skeletal Uptake—J. G. SHIMMINS, D. A. SMITH, B. E. C. NORDIN and L. BURKINSHAW   149

Ratio of the Faecal to Urinary Clearance of Strontium in Man—G. E. HARRISON and ALICE SUTTON   161

Tracer Studies in Animals

A Comparison of the Metabolism of Calcium and Strontium in Rabbit and Man—ELIZABETH LLOYD   167

The Role of Oxidative Phosphorylation in Calcium and Strontium Absorption from the Gastro-intestinal Tract—D. M. TAYLOR   175

Studies on the Dynamics of Strontium Metabolism under Conditions of Continual Ingestion to Maturity—MARVIN GOLDMAN and R. J. DELLA ROSA   181

Variation in "Turnover Rates" in Different Parts of the Skeleton in Relation to Tumour Incidence due to Strontium-90 Deposition—JANET VAUGHAN and MARGARET WILLIAMSON   195

Studies on Soft-tissue Dosage from Strontium-90—A. M. BRUES, H. AUERBACH, D. GRUBE and G. DEROCHE   207

Some Aspects of Strontium-90 Metabolism in Minature Swine Ingesting Strontium-90 Daily—R. O. MCCLELLAN, J. L. BEAMER and P.L. SHELDON   213

Metabolism of Strontium-89 and Calcium-45 Given Orally to Pigs 1, 3, or 7 Days after Whole-body Irradiation—M. C. BELL, R. S. LOWREY and G. WITHROW   223

Influence of Absorption of Strontium in Food, on the Strontium-85 Body Burden in Guinea Pigs—O. VAN DER BORGHT, R. KIRCHMANN and S. VAN PUYMBROECK   229

Strontium-90 Transfer to Progeny Via Material Placenta and Milk—L. N. BURYKINA   237

Effect of Dietary Strontium on Reproductive Performance of the Laying Hen—F. R. MRAZ, P. L. WRIGHT and T. M. FERGUSON   247

Urinary/Fecal Excretion Ratios of Strontium and Calcium in the Rat—R. C. THOMPSON and P. L. HACKETT   255

(*xii*)

CONTENTS

Radiostrontium, Stable Strontium, Stable Calcium and Phosphorus in Sperm DNA—BERTIL ÅBERG and MARGOT GILLNER   261

Modification of Strontium Metabolism

Removal of Radiostrontium from the Mammalian Body—A. CATSCH   265

High Strontium Diet and Radiostrontium Retention—DANUTA DEPCZYK, TOMINSLAW DOMANSKI and JULIAN LINIECKI   283

Ingested or Injected Strontium as Influenced by Oral Treatment Shortly Before or After Exposure—V. VOLF   297

Effect of a Low Phosphorus Diet on Strontium-90 Toxicity in Mice—L. M. VAN PUTTEN   307

Studies on the Inhibition of Radiostrontium Uptake from the Human Gastro-intestinal Tract with Sodium Alginate—R. HESP and B. RAMSBOTTOM   313

The Mechanism of Citrate in Influencing the Excretion of Radioactive Strontium—H. SMITH   323

Inhibition of Absorption of Radioactive Strontium by Alginic Acid Derivatives—DEIRDRE WALDRON-EDWARD, T. M. PAUL and STANLEY C. SKORYNA   329

Opening Remarks

W. V. MAYNEORD

Though I certainly have no official brief to speak on behalf of the Medical Research Council or the United Kingdom Atomic Energy Authority, I feel sure that they as sponsors of the Symposium would wish me first to welcome you all to this Conference, extending a particular welcome to those of you who have travelled long distances overseas.

If, in a land of renowned religious eloquence, I might presume to choose a text for the Symposium, I suppose it might be from the Prophecy of Daniel (Daniel, Chapter 12, verse 4): "Many shall go to and fro, and knowledge shall be increased." At least we hope so.

Certainly, no one can deny the fascination or utility of knowledge of strontium metabolism. Acquired at first under some social pressure and dominated by the urgent need to deduce radiation hazards, this knowledge shows signs of being increasingly useful in a wider field of human metabolic studies. We all hope that this Symposium may accelerate this transition as well as improving our estimates of radiation risks. It is still true that the physics is relatively easy, but the physiology complex.

You will have deduced from your preliminary programme that there is particular and justifiable local pride in the fact that the element strontium was named after the village of Strontian in Argyllshire, and that a pious pilgrimage is to be made to that place. I hope, therefore, that some account of the discovery of the element might be interesting as a preliminary item of your programme.

It seems (Partington, 1962, p. 656) that a peculiar mineral from a lead mine in Strontian was brought to Edinburgh about the year 1787 and examined by Crawford and Cruickshank, who showed that it contained a peculiar "earth" different from baryta. Crawford's paper "On the medicinal properties of muriated barytes" was published in 1790 in the Medical Communications of the Society for promoting Medical Knowledge.

It seems that the new substances were to be associated with medicine from the start, for Adair Crawford was a Physician to St. Thomas's Hospital, London, and the Lecturer in Chemistry at Woolwich Arsenal (*Ibid.*, p. 156). He was a pioneer in the study of problems now associated with specific and latent heat. William Cruickshank, a diplomate of the Royal College of Surgeons, later became Lecturer in Chemistry at the Royal Artillery Academy, Wool-wich, Ordinance Chemist and Surgeon to the Ordinance Medical Department (*Ibid.*, p. 273). He was a leading authority on gases, or "airs" as they were then designated.

History has strange ironies, but what better parents could we have chosen for ^{90}Sr than Crawford, learned physician and chemist, working with Cruick-shank, a Surgeon of Artillery; medicine and ballistics already in conjunction?

The material was independently examined by Hope in 1791, who showed in a careful series of experiments that the mineral was the carbonate of a new "earth" which he called Strontites (*Ibid.*, p. 656). Thomas Charles Hope was Lecturer in Chemistry and Professor of Medicine in Glasgow, and succeeded the great Professor Black as Professor of Chemistry in Edinburgh in 1799. Hope is best known perhaps for his work on the maximum density of water. Some of Hope's strontium preparations are still preserved in Edinburgh. He was certainly a skilled and learned experimenter who went far to establishing the presence of a new element in these materials.

In 1793–94 the specimens were examined by a famous chemist of the period, Klaproth, who in 1789 had discovered uranium, and who came to the same conclusions about the "earth" as Hope, whose paper was not then published. Klaproth called the earth "Strontian". These results were verified among others by Kirwan, who had received specimens from Crawford and who published his results in the Transactions of the Royal Irish Academy in 1795.

These were exciting years in which the foundations of modern chemistry were being laid. The redoubtable Samuel Johnson, then near the end of his life but still interested in "Chymistry", was "knitting his brows in a stern manner" and demanding (Boswell, 1791) "Why do we hear so much of Dr. Priestley?", of whom, it need hardly be said, he heartily disapproved.

In 1793 Thomas Young entered St. Bartholomew's Hospital as a student and in the same year published his theory of the accommodation of the eye in the Transactions of the Royal Society and was elected a Fellow at the age of 21 (Oldham, 1933).

From 1798–1801 Count Rumford was ardently devoting himself to the development of the Royal Institution (*Ibid.*, p. 39) with Young as Professor of Natural Philosophy from January 1802 to July 1803, while Davy appears as lecturing in 1802.

Those of you interested in other fields will think of these years in terms of George III and the Prince Regent, of the newly won independence of the American Colonies, the mounting struggle with Republican France, of Pitt, Fox, Sheridan and Mrs. Siddons. Reynolds was just dead and old Haydn paying his visits to England in 1790–92 and 1794–95, producing his last great symphonies and complaining of London fog "so thick you could spread it on bread" (Hughes, 1950). They were exciting times in literature too. From 1799 to 1805 Wordsworth was engaged upon "The Prelude", 1798 saw the publication with Coleridge of the Lyrical Ballads. Indeed, in the year 1808, in which as we shall see Davy finally separated the element strontium, Coleridge delivered, also at the Royal Institution, a famous course of lectures on poetry and the fine arts. It would of course be unforgivable not to mention that Robert Burns was living near here in Dumfries from 1792 until his death there in 1796; as Carlyle put it, a noble and gentle soul "wasting itself away in a hopeless struggle with base entanglements".

With all this in mind we return to London and in 1806 we see young Humphry Davy beginning his researches on the "electrization" of metals. In his Bakerian Lecture to the Royal Society on 20th November 1806 "On some chemical agencies of Electricity", he described (Davy, 1807) his techniques and some preliminary results explaining them with a theory based upon "the repellant and attractive energies communicated from one particle to another so as to establish a conducting chain in the fluid. The oxygene of a portion of water is attracted by the positive surface at the same time that the other conductive part, the hydrogene is repelled by it." This theory was to be roughly handled by Michael Faraday 30 years later. Davy was already experimenting with the electrical transfer of materials in solutions of strontium compounds (*Ibid.*, pp. 14, 25). However, in his second Bakerian Lecture (Davy, 1808a) on "Some new phenomena of chemical changes produced by Electricity, particularly the Decomposition of the fixed Alkalies, and the Exhibition of the new substances which constitute their bases; and on the general nature of alkaline bodies", he describes (*Ibid.*, p. 1) the simple and beautiful experiments which enabled him to separate the "base of potash" and of soda as small globules of metal.

The source of power was a combination of Voltaic batteries belonging to the Royal Institution "containing 24 plates of Copper and Zinc 12 inches square, 100 plates of 6 inches and 150 of 4 inches square, charged with solutions of alum and nitrous acid" (*Ibid.*, p. 3).

There is a delightful description by Dumas in his "Eloge Historique' of an interview with Faraday in which he was shown Davy's original notebook. I quote from Silvanus Thompson (1898): "Faraday took out a notebook, opened it and pointed out with his finger the words written by Davy, at the very moment when by means of the battery he had just decomposed potash, and had seen the first globule of potassium ever isolated by the hand of man. Davy had traced with a feverish hand a circle which separates them from the rest of the page; the words 'Capital Experiment' which he wrote below, cannot be read without emotion by any true chemist."

"But should the bases of potash and soda be called metals?" Davy (1808a, p. 31) asks and concludes they should. The names present problems. "I have consulted with many of the most eminent scientific persons in this country upon the method of derivation, and the one I have adopted has been most generally approved. It is perhaps more significant than elegant" (*Ibid.*, p. 33). The metals were called Potassium and Sodium.

However, in a paper which even more closely concerns us (Davy, 1808b), read on 30th June 1808 on "Electrochemical Researches, or the Decomposition of the Earths; with observations on the metals obtained from the Alkaline Earths, and on the Amalgam procured from Ammonia", Davy took the next step (*Ibid.*, p. 333). "Barytes, Strontites and lime, slightly moistened were electrified by iron wires under naphtha, by the same methods and with the same powers as those employed for the decomposition of the fixed alkalies" (*Ibid.*, p. 334). At first "Mixtures of lime, strontites, magnesia and red oxide of mercury treated in the same way gave similar amalgams" but "the quantity of metallic substances obtained were exceedingly minute" (*Ibid.*, p. 338).

Davy's battery evidently needed refurbishing, but in May 1808 having "500 pairs of double plates of six inches square" he tells us (*Ibid.*, p. 339) that "A globule of mercury, electrified by the power of the battery of 500, weakly charged, was made to act upon a surface of slightly moistened barytes, fixed upon a plate of platina. The mercury gradually became less fluid and after a few minutes was found covered with a white film of barytes and when the amalgam was thrown into water, hydrogene was disengaged, the mercury remained free, and a solution of barytes was formed."

Having been successful with barytes, Davy attacks its analogues. "With strontites I obtained a very rapid result" he says. "All these amalgams I found might be preserved for a considerable period under naphtha" (*Ibid.*, p. 340). Again, there were questions of names (*Ibid.*, p. 346). "The new substances will demand names; and on the same principles as I have named the bases of the fixed alkalies potassium and sodium, I shall venture to denominate the metals from the alkaline earths barium, strontium, calcium and magnium; the last of these words is undoubtedly objectionable but magnesium has been already applied to metallic manganese and would consequently have been an equivocal term." Later, in deference to "the candid criticisms of some philosophical friends" Davy (1812) was induced to "apply the termination in the usual manner".

Davy gave a shorter account of these findings in his "Elements of Chemical Philosophy" published in 1812. He there refers to "carbonate of strontia, or strontianite, a mineral found at Strontian in Scotland". "I first procured this metal in 1808 but in quantities too small to make an accurate examination of its properties" (*Ibid.*, p. 343). Davy records in his last paragraph "None of the compounds of this body have as yet been applied to any of the purposes of the arts, and its combinations are rare in nature" (*Ibid.*, p. 345). One fascinating thing, however, he did do, at a time when even the laws of multiple proportions were under fire and the assessment of numbers surely very difficult. Speaking of "Strontia or strontites, the substance procured by burning strontium", "From indirect experiments" says Davy (*Ibid.*, p. 344) "I am disposed to regard it as composed of about 86 of strontium and 14 oxygene; and supposing it to contain one proportion of metal and one of oxygene, the number representing strontium will be 90 and that representing the earth 105."

REFERENCES

Boswell, J. (1791). "The Life of Samuel Johnson, II, 464" (Everyman edition, 1906, London: Dent, Vol. 2, p. 481).

Davy, H. (1807). *Phil. Trans. R. Soc.*, **97**, 1–5.

Davy, H. (1808a). *Phil. Trans. R. Soc.*, **98**, 1–44.

Davy, H. (1808b). *Phil. Trans. R. Soc.*, **98**, 333–370.

Davy, H. (1812). "Elements of Chemical Philosophy", Part 1, Vol. 1, London.

Hughes, R. (1950) "Haydn", p. 76. Dent, London.

Oldham, F. (1933). "Thomas Young", p. 16. Arnold, London.

Partington, J. R. (1962). "A History of Chemistry", Vol. III. Macmillan, London.

Thompson, S. P. (1898). "Michael Faraday", p. 59. Cassell, London.

Scotland and Strontium

ANDREW KENT

*Department of Chemistry, The University,
Glasgow, W.2*

Britain is an apposite centre for a symposium on Strontium-90 This island provided the first source of the element (a carbonate from Scotland), and a sulphate mine at Yate in Gloucester, providing a preponderance of present world supplies, affords our single remaining precedence of this kind. Strontium is the only element named after a British locality (Day, 1963).

To the chemist, element 38 ranks among the less rare components of the lithosphere. Its relatively high atomic weight of 87·63 is deceptively average for the "triad" completed by calcium and barium. There are four naturally occurring isotopes: ^{87}Sr (^{87}Rb, e) is at least in part a product of an earlier radioactivity than the parvenu ^{90}Sr. Although the "fellow traveller" of calcium in biological processes, this member of the triad accompanies barium in mineralogy. Its history has acquired a patina of romance, roughened in places by acerbities of priority (Partington, 1942, 1951).

Argyll, on Scotland's west coast, represents an ancient Celtic kingdom of Dalriada. Its northern peninsula of Ardnamurchan was long conceded to the Vikings before Somerled, a Lord of the Isles and chief of the clan Macdonald, campaigned against these Norwegian overlords of the West. Ardnamurchan was liberated by the MacIans, sons of John (Ian) of Islay, in the 12th century. Their castle of Mingary still stands in ruined splendour, and names like Strontian in Sunart and other parts are memorials to their ancient fame.

Such a border district was exposed to bitter clan warfare, and had become a Cameron country before the Glorious Revolution of 1688 brought a relative stability, despite incidents like the Campbell massacre of Macdonalds in nearby Glencoe and the forays of the 'Fifteen. Even pacific prospecting became possible among the hydrothermal intrusions of galena (PbS) that occur with iron, barium and other minerals in the secondary rock. Sir Archibald Murray had organized lead mining by 1722 at Strontian, whence sea transport by Loch Sunart was readily available to Glasgow and beyond. Here the associated crystalline "heavy spars" were, exceptionally, carbonates whose showy character must have attracted attention from early days. All such progressive interests of this period in Scotland had benefit from the resurgence of interest in Newtonian science and philosophy. Here the chemical highlight was the career of Joseph Black whose coterie of brilliant students in Edinburgh and Glasgow included Adair Crawford, an Ulster Scot, and Thomas Charles Hope, the son of a professor at Edinburgh (Kent, 1950).

Murray soon disposed of his mineral rights to the York Building Company whose president, the Duke of Norfolk, attempted to expand this Scottish interest and built the settlement of New York to house their English miners and metallurgists in a staff of some 500 men and women. Later, when the Camerons had been outlawed following the 'Forty-five rebellion, Riddell of Linlithgow took over the mines. In 1798, forty miners and thirty labourers still produced some £4,000 worth of metal per annum, until this prosperity was ruined by the collapse of prices after the Napoleonic wars. In 1872 production of lead ore had fallen to 12 tons per annum. An attempted revival of the industry failed in 1904. At present a Canadian concern is re-examining the

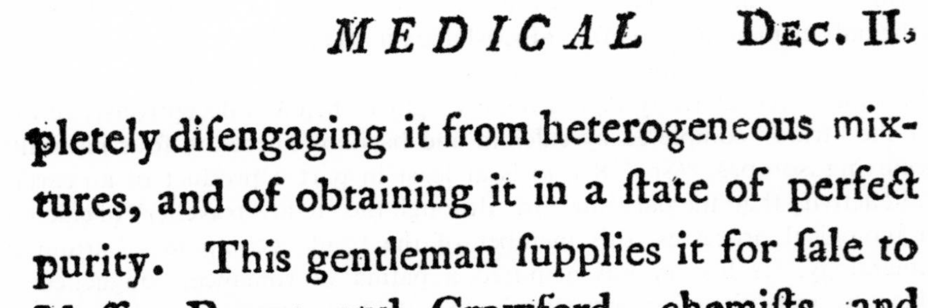

M E D I C A L   Dɛc. II,

pletely difengaging it from heterogeneous mixtures, and of obtaining it in a ſtate of perfect purity. This gentleman ſupplies it for ſale to Meſſrs Payne and Crawford, chemiſts and druggiſts, Nº 66. Leadenhall Street, London. It may not be improper to mention, that there is a mineral ſold for the aërated Barytes at Strontian in Scotland, which I have found to be a different ſubſtance from the true Terra ponderoſa. It appears to be a new earth, the properties of which have not yet been ſufficiently examined."

FIG. 1. The first published comment on the distinction of Strontian mineral ($SrCO_3$) from "aerated Barytes" ($BaCO_3$). "This gentleman" is William Cruickshank, laboratory assistant to Adair Crawford in 1789. (Crawford, 1789, p. 436.)

area for lead and silver, so the Strontian history may yet again deny the old saying that "the manufactures of Argyllshire are very limited, consisting solely of whisky and gunpowder".

The eighteenth was the century of Horace Walpole and Strawberry Hill, the age of collectors and private museums. One notable event, in Edinburgh about 1787, was a dealer's display of curios from the Sunart mines with the Strontian spar or mineral as the most exciting of these crystalline rarities.

A P P E N D I X.

POSTSCRIPT TO THE HISTORY.

ON Monday, the 4th of November 1793, Dr HOPE, Pro-feſſor of Medicine in the Univerſity of Glaſgow, read a paper, entitled, *An Account of a Mineral from Strontian, and of a peculiar Species of Earth which it contains.* Want of room, and the length of the diſſertation, prevent its appearance in the preſent volume. But as the diſcovery of a new earth cannot fail to be intereſting, it has been thought proper to treſpaſs a little on the order of time, and to inſert here the following abſtract.

THE mineral is found in the lead-mine of Strontian in Argyle-ſhire. It was brought to Edinburgh about ſix years ago in conſi-derable quantity. It was generally received as the aërated barytes. At that time, Dr HOPE had ſome doubts of its being the barytic ſpar, and uſed, in his prelections, when he filled the chemical chair in the Univerſity of Glaſgow, to mention ſuch of its diſtin-guiſhing characters as he had then diſcovered. The Strontian ſpar ſometimes is colourleſs, oftener it has a greeniſh or yel-lowiſh hue. Its texture is fibrous, and it frequently ſhoots in-to cryſtals, which are ſlender ſpiculæ or hexagonal columns. The ſpecific gravity of it goes from 3.650 to 3.726.

THIS mineral is inſipid, and requires nearly 800 times its weight of water to diſſolve it. It effervesces with acids, and

FIG. 2. From the forward notice of T. C. Hope's paper on "A Mineral from Strontian" in 1794. Hope's "chemical chair" was in fact a Lectureship he held from 1787 to 1791, when he was made Professor of Medicine. (Kent, 1950.)

Although young T. C. Hope—soon to take up appointment as Glasgow College Lecturer in Chemistry (1787–91)—had known other specimens in private collections, it was only now that a Mr. Ash displayed to him the brilliant red flame displayed by such crystals especially in their "muriated" (chloride) derivatives. He took some further Strontian "fossil" and its problems to Glasgow, where a furnished laboratory was at his disposal; in his lectures he expressed his doubts on the derivation of the Strontian spar from "barytes". In 1791, during his final summer in the Glasgow appointment, he concentrated his time and energy on this conundrum (Hope, 1798).

At this period the crust of our planet was believed to be composed of "primitive earths" with phlogisticated or saline variations derived from the three basic types: vitreous (SiO_2), argillaceous (Al_2O_3) and absorbent or calcareous (CaO). There was also, in our terms, a lasting confusion between these oxides and their carbonates, the "aerated earths". However in a thesis published in 1754, at Edinburgh, Black established "magnesia usta" (MgO) as a "peculiar" earth with a character and a chemistry distinct from those of quicklime, at the same time explaining the distinction between the simple earths (oxides) and their carbonates. Twenty years later, Scheele provided a similar distinction for a third alkaline-earth, barytes (BaO) derived from "terra ponderosa" or "heavy spar" ($BaSO_4$). There were now five "primitive earths", and Lavoisier's first table of his elements in 1789 admitted them all, (though reluctantly). Clearly a new "earth" or "element" of this kind could not be conclusively established without isolation of the new base (oxide) and evidence of its distinctive chemical behaviour.

While Hope pursued these objectives Adair Crawford, renowned for his study of Animal Heat (forming his Glasgow M.D. thesis of 1780), expressed his views in two announcements. Firstly in a letter, sometimes overlooked, where he tells Dr. Andrew Duncan that "a mineral sold for the aerated Barytes at Strontian in Scotland . . . appears to be a new earth" (Crawford, 1789): secondly, in his published address of 10th November, 1789, on "Muriated Barytes", in which he records that he and his notable assistant, William Cruickshank, in attempting a synthesis of their medicament ($BaCl_2$) from the Scottish mineral had obtained an unexpectedly different substance. Crawford, who acknowledges a similar independent conclusion by his colleague Babington, considers this material as one which has not "hitherto been sufficiently examined": and he forwarded specimens to Dr. Kirwan for further investigation. It is interesting that one review of the published paper makes no mention of the incidental comments on strontianite (Crawford, 1790).

Hope first presented his finding to a public audience, the College Literary Society of Glasgow, in an address (unpublished) in March, 1792 (Hope, 1798, p. 1). Repeated to the Royal Philosophical Society of Edinburgh in November, 1793, this produced such interest that an abstract was "rushed" into the Transactions then in course of publication. The full paper was incorporated in the Society's next volume. Hope had isolated the earth (our strontia, SrO) named it "strontites" (*Ibid.*, p. 8) and distinguished the new base—an "element" to him as a known disciple of Lavoisier—from both lime and barytes (*Ibid.*, p. 36). He recorded flame tests and colours for all three alkaline earths (by candlelight) and compared the properties of about a dozen salts before concluding, after an appreciation of Crawford's report, that his own work has established the "sixth simple earth" (*Ibid.*, p. 147). He had at one time considered its derivatives as calcium–barium mixtures; and in confuting this he had unwittingly provided the first of those "triads" which would lead to a far off revolution in chemical theory. Hope's paper, which earned then, and later, some incomprehensibly aggressive reviews (Anon, 1796), was followed almost at once—and independently—by the delivery of Kirwan's similar findings in his address to the Royal Irish Academy in January, 1794.

Meantime, on the Continent, the "Schottische Fossil" underwent other scrutinies. F. G. Sulzer considered in 1791 that it "ebenfalls eine neue Grunderde zu erhalten scheint". His findings were available to Kirwan, and he appears to have isolated the earth (oxide) from the carbonate before Hope made any similar claim; but it is Klaproth whose extensive chemical investigations are held to have again, in 1793, provided independent evidence of a new earth distinct from Scheele's barytes (BaO) (Partington, 1942, 1951).

The final steps that established strontium as a new alkaline-earth metal had to await Davy's electrolytic analyses in 1808. It is perhaps at this point fair to say that systematic study of strontium compounds was first attempted in the chemistry department at Glasgow, and that the first published suggestion of a distinction from barium was offered by Adair Crawford in 1790, before Hope's conclusive study detailed for the first time the characterization of this earth *and* its chemical distinction from both calcium and barium.

It is certainly evident that the only element first found on Scottish soil received early and adequate attention from some brilliant young Scottish chemists; Crawford, Cruickshank and Hope. In naming the oxide Strontites Hope brought unwittingly a foretaste of the violence inseparable from strontium-90; for "Strontium" carries yet a far, thin sound of battle where the sons of Ian fought it out with the Norsemen by the shores of Loch Sunart in Argyll.

REFERENCES

Anon (1796). *Mon. Rev.*, **19**, 242: Thomson, T. (1816). *Ann. Philosophy*, **8**, 133, is reproduced with commendation by Partington (1942), 164; but see also Thomson, T. (1831). "History of Chemistry", vol. II, p. 209, and Partington (1951), 96.

Crawford, A. (1789). *Med. Commentaries*, **4**, 436.

Crawford, A. (1790). *Medical Communications*, **2**, 356. (A notice in *Mon. Rev.* 1791, **6**, 243, overlooks the brief reference to the strontian mineral).

Day, F. H. (1963). "The Chemical Elements in Nature", p. 174.

Davy, H. (1808). *Phil. Trans. R. Soc.*, **98**, 333. (See also Mayneord, W.V., this volume, p. 1.)

Hope, T. C. (1798). *Trans. R. Soc. Edin.*, **4**, part II, 1–39. An advance abstract appeared in the same periodical in 1794, **3**, part I, 141 (*recte* 143). For information on Crawford and Hope, reference may be made to Dictionary of National Biography; for Cruickshank, see Coutts, A. (1959), *Ann. Sci.* **15**, 121.

Kent, A. (1950). "An Eighteenth Century Lectureship in Chemistry", p. 157. Glasgow.

Lavoisier, A. (1789). "Traité Elémentaire de Chimie", Paris.

Partington, J. R. (1942). *Ann. Sci.* **5**, 157, and (1951), **7**, 95.

(The Hunterian Museum of Glasgow University provided an extensive series of strontium and barium minerals for examination, and for comparison with those later found by members of the Symposium at the original minehead.)

The Strontian Mines

G. A. P. WYLLIE

Natural Philosophy Department, The University,
Glasgow, W.2

Strontium has its name, via the mineral strontianite, from the village of Strontian in the North West Highlands, where lead ore and barytes were formerly mined. The mines were discovered in 1722 by Sir Alexander Murray of Stanhope, the proprietor of Ardnamurchan at that time.

In the troubled political conditions of the period, when prospecting and mine management were occupations freely interchangeable with secret-service work for several employers, the mines had a complicated business history. Successive changes in management or ownership were as often directed by the risk of political imprisonment as by the chances of an essentially risky trade.

Of the miners' hamlet of New York, named from the York Buildings Company (a London waterworks company that invested widely in Scotland and did much to develop the Strontian mines) there remains no visible evidence. They left the proof of their labour in the workings that extend for several miles to east and west. The Bellsgrove Opencast, which is bridged by the road to Glenhurich Forest, reaches 12 metres in width: it is 300 metres long and goes to a depth of over 30 metres. The Fee Donald mines, about 2 miles to the east, seem to have been the main source of strontianite.

The geological history of the area has been in no way less complex than the economic or political history.

Molten material from deep in the earth's crust, making its way up to regions of lower temperature and pressure, differentiates in a complicated way and slowly solidifies, rejecting at each stage its more fusible and volatile components. Regions of weakness in the country rock surrounding the intrusion become tracks for the passage of large amounts of water, carbon dioxide and hydrogen sulphide, at temperatures and pressures still high enough for silica and metallic sulphides and carbonates to be freely mobilized both from the intrusive material and the country rock.

The Strontian mineral veins lie in a region that has been subject to at least three intrusions. The earliest of these, before or early in the Caledonian mountain-building period, involved the thorough permeation of earlier sediments by igneous material, followed by a kneading of the plastic mixture, during the mountain folding, which left the country rock as an injection gneiss. About the Old Red Sandstone period, another igneous mass was intruded to the south. This is highly differentiated, containing regions of ultrabasic rock with a high content of iron and magnesium, and shading to a clinkstone—a fairly fine-grained grey granite—towards its northern edge where it meets the injection gneiss. Along part of the Bellsgrove Opencast working, where the Strontian Main Vein reached its maximum thickness, the north wall of the vein is gneiss and the south (hanging) wall is granite.

Further to the east, there is a group of veins more closely associated with the third intrusion, a set of basalt dykes that are considerably crushed and, particularly in the Smiddy Vein, extremely decomposed. These dykes were probably intruded within a few tens of millions of years of the main wrench

FIG. 1. Geological Survey map of Strontian (Crown Copyright reserved: published by permission of the Controller, H.M. Stationery Office).

of the Great Glen Fault, in which Strontian appears to have slid about 100 km south-west with respect to the southern part of Britain. Heating due to this distortion may have contributed to the Strontian mineralization.

At the Smiddy Vein, which is most interesting in its mineral assemblage, the country rock is red gneiss, relatively rich in potassium, and the vein rock is mainly a mixture of calcite with decomposed basalt, barium sulphate and silica, with lead and iron sulphides and the rarer strontium minerals, strontianite, the carbonate, in which a new substance was first recognized, and celestine, the sulphate. The workings are no longer easily accessible, but the spoil heaps along the Fee Donald Burn contain moderately weathered samples of all these minerals.

Some Principles of Strontium Metabolism: Implications, Applications, Limitations

C. L. COMAR

*Department of Physical Biology, New York State Veterinary College,
Cornell University, Ithaca, New York*

INTRODUCTION

Interest in strontium can be visualized in terms of a chain of events that may involve release to the environment, incorporation into the food chain and man's diet, deposition and retention in the body, delivery of radiation dosage and finally production of harm. Information on aspects of metabolism that govern the body burden to be attained has come from two approaches: controlled metabolic studies with limited numbers of individuals, and assessment of data from surveys. Each approach has advantages and limitations, as well as pitfalls even for the wary. Controversy has frequently occurred when survey results have been improperly used to generate physiological knowledge and when arbitrary relationships useful for predictive purposes have been regarded and misused as fundamental biological properties. These disagreements have largely been resolved, and it is fair to say that field data appropriately taken and interpreted have conformed to concepts developed from controlled and theoretical studies.

The behavior of Sr in biological systems can be treated either in terms of the movement of Sr itself, or of the comparative movement of Sr and Ca. Emphasis to date has been largely on Sr–Ca relationships. This has been objected to in many quarters with the suggestion that strontium deserves the dignity of being granted independence. By all means, let Sr be granted independence, providing that the interests of the investigator and that the purposes of the investigation are better served. It is quite true that we can now obtain good data on stable Sr and that increasing variations in Sr–Ca relationships show up with increasing experimental refinements. Nevertheless, it still appears advantageous for many purposes and essential for some to consider the simultaneous behavior of both elements. The basic reasons are that homeostatic control of Ca leads to a remarkable constancy of Ca concentrations in many important tissues and fluids (e.g., bone, plasma and milk) and that Sr metabolism is not regulated by normal amounts of stable Sr, but is regulated to a large extent by levels of Ca.

There would appear ample justification for a strong recommendation that data always be obtained for both Sr and Ca in simultaneous or at least parallel experiments. The investigator and others can then use either the data for both elements or just those for Sr as they see fit.

The kinetics of the individual nuclides have been studied intensively and will be discussed by others. This paper is concerned more with the principles of the comparative behavior of the two nuclides to provide a basis for evaluating and coping with food chain contamination as well as understanding the governing processes.

STRONTIUM–CALCIUM RELATIONSHIPS BETWEEN PRECURSORS AND PRODUCTS

Knowledge of Sr–Ca relationships has been developing through what might be thought of as four stages: (*a*) establishment of precursor–product ratios; (*b*) quantitation of contribution of physiological processes to differential behavior; (*c*) consideration in terms of rate constants; and (*d*) mechanism of movement across membranes. Emphasis is given primarily to discussion of rate constants, because stages (*a*) and (*b*) have been well worked over whereas the mechanisms (stage *d*) are yet to be understood.

Precursor–product ratios

At the outset, some 10 years ago, it appeared essential to be able to answer questions such as: If ^{90}Sr is introduced into the food chain, what will be the levels at various steps and how will the various foods contribute to dietary intake? If an individual chronically ingests a diet contaminated with ^{90}Sr, what will be the levels in the body?

$$OR_{product/precursor} = \frac{Sr/Ca \text{ of product}}{Sr/Ca \text{ of precursor}}$$

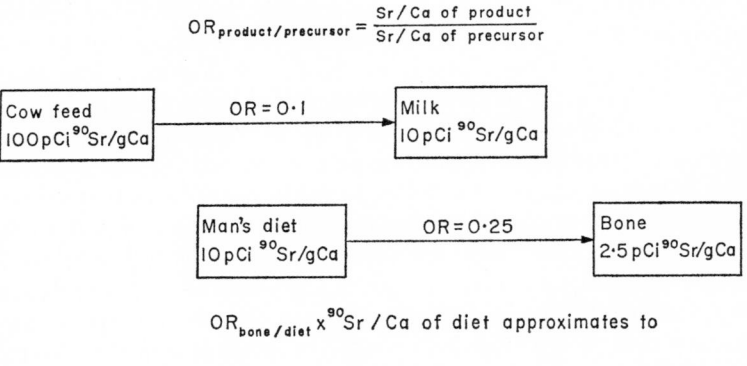

$$OR_{bone/diet} \times {}^{90}Sr/Ca \text{ of diet approximates to}$$

A   Uniform concentration in bone from constant intake

or

B   Maximum concentration in new bone from short-term intake

FIG. 1. Terminology for Sr–Ca relationships between precursor and product.

To aid in systematizing the data, the terminology of the Observed Ratio (OR) was proposed (Comar and Wasserman, 1964) as illustrated in Fig. 1. Over the years much information (and mis-information) has been published on gross relationships in numerous systems. The essentially steady-state relationships in the food chain through man are reasonably established from analyses of stable Sr and Ca, and the extent of approach to steady state after environmental contamination is apparent from analyses of ^{90}Sr and Ca. These relationships, illustrated in Figs. 2 and 3 (compiled from HASL-165, 1966),

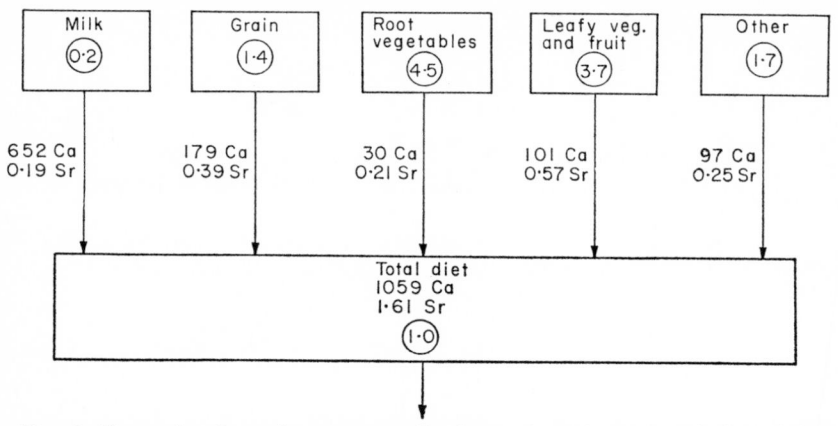

FIG. 2. Example of steady-state contributions of stable Sr (mg/day) and Ca to total diet by various dietary constituents, New York City. Circled values are Sr/Ca; total diet set at ①.

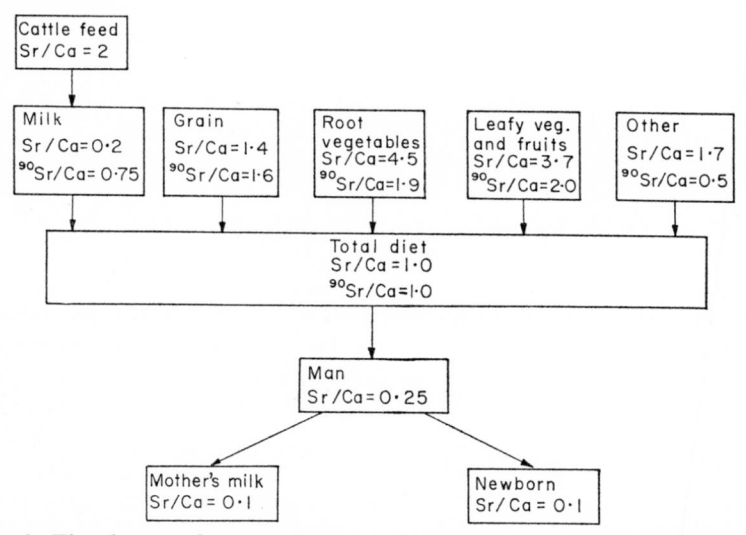

FIG. 3. The degree of approach to steady-state conditions illustrated by comparison of Sr/Ca and ^{90}Sr/Ca ratios in dietary constituents, New York City, 1965.

have been most useful in assessing the role of various dietary constituents and the effects of dietary modifications on body burdens of ^{90}Sr.

Discrimination factors

Particularly in regard to animals and man it appeared useful to quantitate the contribution of various physiological processes to the differential behavior of the two elements. As shown in Fig. 4, this was done by defining the

$$OR = (DF_1)(DF_2) \cdots (DF_n)$$

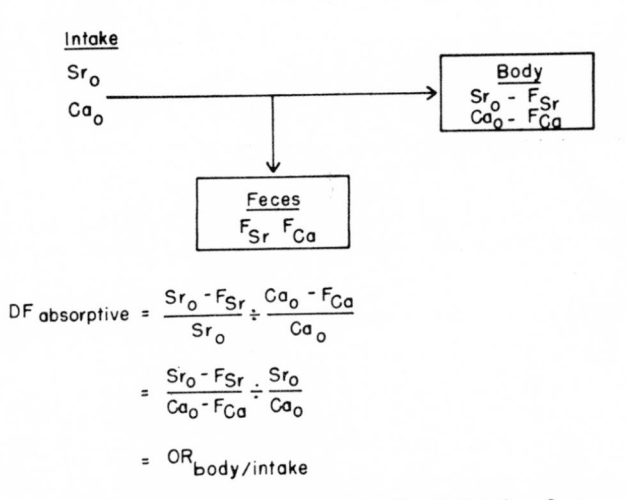

DF = Discrimination Factor

   = Ratio of fractional retentions in body as a result of process

$$DF \text{ absorptive} = \frac{Sr_0 - F_{Sr}}{Sr_0} \div \frac{Ca_0 - F_{Ca}}{Ca_0}$$

$$= \frac{Sr_0 - F_{Sr}}{Ca_0 - F_{Ca}} \div \frac{Sr_0}{Ca_0}$$

$$= OR_{body/intake}$$

FIG. 4. Definition of the Sr–Ca discrimination factor.

OR value as the product of the Discrimination Factors (DF) that are operative. In terms of physical meaning, the DF value is the ratio of fractional retentions that result from the process. For example, in a one-step process (absorption) if the intake is Sr_0 and the fecal excretion is F_{Sr}, then the fractional retention of Sr is $\dfrac{Sr_0 - F_{Sr}}{Sr_0}$ and that of Ca is similarly $\dfrac{Ca_0 - F_{Ca}}{Ca_0}$. The DF$_{absorptive}$ is then the ratio of these two quantities; by rearrangement of terms it can be seen that the DF$_{absorptive}$ is equal to OR$_{body/intake}$, as it must be in a one-step process.

If two or more discriminatory steps occur in sequence, as illustrated for absorption and urinary excretion in Fig. 5, then the calculation is straightforward. The $DF_{absorptive}$ is the same as in Fig. 4 and the $DF_{urinary}$ is the ratio of the fractional changes in the body before and after urinary excretion.

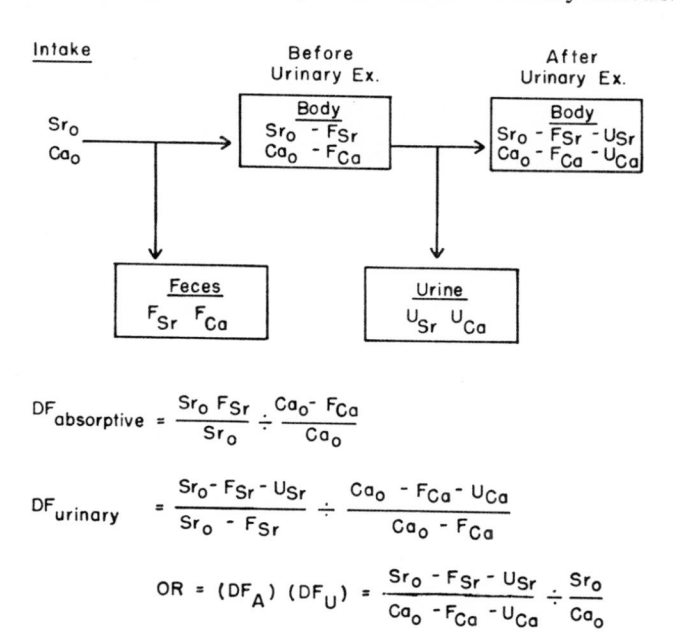

$$DF_{absorptive} = \frac{Sr_0 \; F_{Sr}}{Sr_0} \div \frac{Ca_0 - F_{Ca}}{Ca_0}$$

$$DF_{urinary} = \frac{Sr_0 - F_{Sr} - U_{Sr}}{Sr_0 - F_{Sr}} \div \frac{Ca_0 - F_{Ca} - U_{Ca}}{Ca_0 - F_{Ca}}$$

$$OR = (DF_A)(DF_U) = \frac{Sr_0 - F_{Sr} - U_{Sr}}{Ca_0 - F_{Ca} - U_{Ca}} \div \frac{Sr_0}{Ca_0}$$

FIG. 5. Calculation of discrimination factors when discriminating processes occur in sequence.

If the discriminatory steps are competitive, as illustrated in Fig. 6, the arithmetic is somewhat more complicated. These equations have not been presented before and were derived on the assumption that the competing processes occur simultaneously according to first-order kinetics. The example refers to secretion into milk and urine, but could apply equally to other processes, such as endogenous secretion into the gastrointestinal tract.

It is emphasized that the discrimination factor as defined takes into account both the comparative rates of movement into or out of a compartment and the amounts that are involved.

Discrimination and rate constants

Further understanding comes about from consideration of comparative rates at which Sr and Ca move across membranes. Observed ratios and discrimination factors as defined deal with amounts of Sr and Ca that have

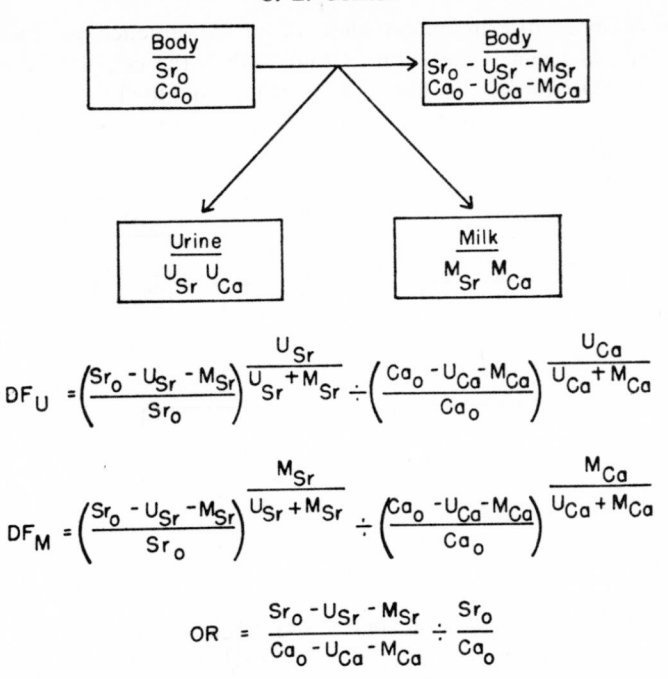

$$DF_U = \left(\frac{Sr_0 - U_{Sr} - M_{Sr}}{Sr_0}\right)^{\dfrac{U_{Sr}}{U_{Sr} + M_{Sr}}} \div \left(\frac{Ca_0 - U_{Ca} - M_{Ca}}{Ca_0}\right)^{\dfrac{U_{Ca}}{U_{Ca} + M_{Ca}}}$$

$$DF_M = \left(\frac{Sr_0 - U_{Sr} - M_{Sr}}{Sr_0}\right)^{\dfrac{M_{Sr}}{U_{Sr} + M_{Sr}}} \div \left(\frac{Ca_0 - U_{Ca} - M_{Ca}}{Ca_0}\right)^{\dfrac{M_{Ca}}{U_{Ca} + M_{Ca}}}$$

$$OR = \frac{Sr_0 - U_{Sr} - M_{Sr}}{Ca_0 - U_{Ca} - M_{Ca}} \div \frac{Sr_0}{Ca_0}$$

FIG. 6. Calculation of discriminating factors when discriminating processes occur simultaneously.

moved across physiological barriers. The principle to be outlined, first developed by Walser and Robinson (1963) for renal discrimination, applies to passage across any membrane, provided the assumptions are fulfilled; however, the discussion is presented in terms of the absorption of the two elements from a site in the gastrointestinal tract (Fig. 7). If Ca is absorbed at a faster rate, then with time there will be an enrichment of Sr in the mixture available for absorption; thus even though the fractional rates might be constant, the relative amounts of the two elements absorbed will vary with time. As shown, two first-order equations give expressions for F_{Sr} and F_{Ca} (the amounts unabsorbed) as a function of time. If it is assumed that the rate constants are related to each other in the same way at each site of movement, then the two equations can be divided to give the relationship:

$$F_{Sr}/Sr_0 = (F_{Ca}/Ca_0)^K \tag{1}$$

in which the factor of time (t) disappears. It can then be shown that:

$$DF_{absorptive} = \frac{1 - (F_{Ca}/Ca_0)^K}{1 - F_{Ca}/Ca_0} \tag{2}$$

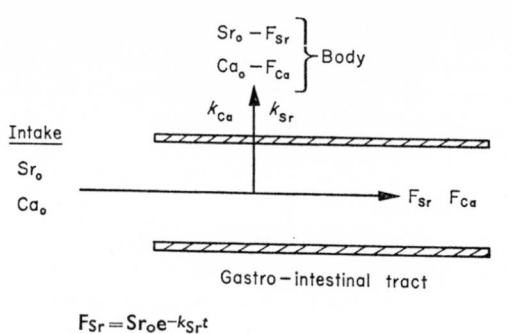

$$F_{Sr} = Sr_0 e^{-k_{Sr}t}$$

$$F_{Ca} = Ca_0 e^{-k_{Ca}t}$$

$$F_{Sr}/Sr_0 = (F_{Ca}/Ca_0)^K \qquad K = k_{Sr}/k_{Ca}$$

$$DF_{absorptive} = \frac{Sr_0 - F_{Sr}}{Sr_0} \div \frac{Ca_0 - F_{Ca}}{Ca_0}$$

$$= \left[1 - (F_{Ca}/Ca_0)^K \right] \div (1 - F_{Ca}/Ca_0)$$

Fig. 7. Expression of discrimination in terms of rate constants.

Equation (1) has been shown to hold for gastrointestinal absorption in rats by Marcus and Wasserman (1965) and for renal tubular absorption in dogs and man by Walser and Robinson (1963) with values of K in these instances of about 0·7.

Values for K, which represent the comparative fractional rate constants for Sr and Ca, can now be estimated from values of $DF_{absorptive}$ and percentage of Ca absorption that are in the literature. From equation (2) it can be inferred that any treatment that alters absorptive discrimination can do so in two ways: by changing the absorption of both alkaline earths proportionally and by changing the comparative rate constants. These two effects can now be distinguished. For example, lactose fed to the rat tends to raise the DF_A value and does so by increasing Ca absorption with no significant change in the rate constants (Marcus and Wasserman, 1965). In contrast, as discussed later, changes with age appeared to result from changes in K, the comparative rate constant.

A few illustrations from among many are presented to demonstrate the applications and limitations of the concepts presented.

CHANGES IN DISCRIMINATION

In early studies, given values of OR were consistently found to fall within ranges of a factor of 2 or less. For instance, the $OR_{body/diet}$ generally was about 0·25 and the $OR_{milk/diet}$ about 0·1. However, further study showed that these values, especially the $OR_{body/diet}$, could be affected by physiological and nutritional factors, and there was interest in changes either that might occur under conditions commonly encountered or that could be induced. Age was one of the interesting physiological factors.

Effect of age on discrimination

Figure 8 presents data for values of $OR_{body/diet}$ in very young children. There seems little question that in the first month or so of life, the $OR_{body/diet}$ is about 0·9 to 1 (Lough *et al.*, 1963); these data are based on direct balance studies and are reasonably reliable. Thereafter the values appear to fall rapidly to the adult value of about 0·25; these values are based on iterative

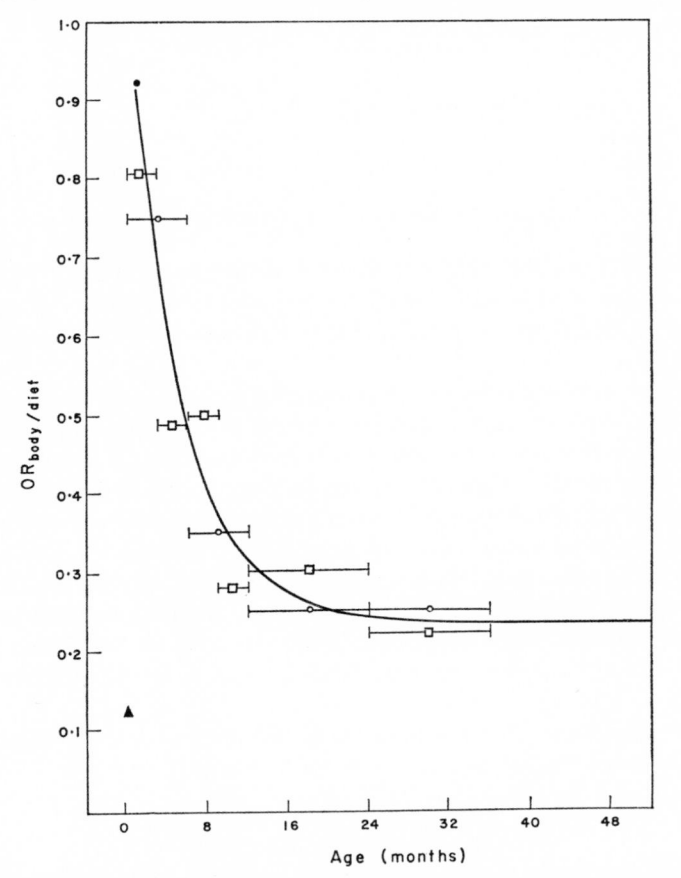

FIG. 8. $OR_{body/diet}$ as related to age of young children. ○, Beninson *et al.* (1964), ^{90}Sr; □, Beninson *et al.* (1964), Sr; ● Lough *et al.* (1963), ^{90}Sr; ▲, Newborn, $OR_{body/mother's\ diet}$.

procedures with survey values of ^{90}Sr, stable Sr and Ca (Beninson *et al.*, 1964). They may have considerable uncertainties, but the trend is clear; also the same pattern has been observed with rats and miniature pigs.

Some balance data on three breast-fed infants (Harrison *et al.*, 1965) permit an estimation of various discrimination factors and these are compared with those of adults (Spencer *et al.*, 1960) in Table I. Liberties have been taken in

TABLE I

Age and Discrimination

	Infants (3) Harrison *et al.* (1965)	Adults (16) Spencer *et al.* (1960)
$OR_{body/diet}$	0·84	0·29
DF_A	1·06	0·42
DF_U	0·79	0·70
%Ca absorp.	58	46
F_{Ca}/Ca_0	0·42	0·54
$K_{absorp.} = k_{Sr}/k_{Ca}$	1·1	0·35

using the data in this way that are discussed later, but the essentials appear to be consistent. It is apparent that the high OR value in the infants resulted

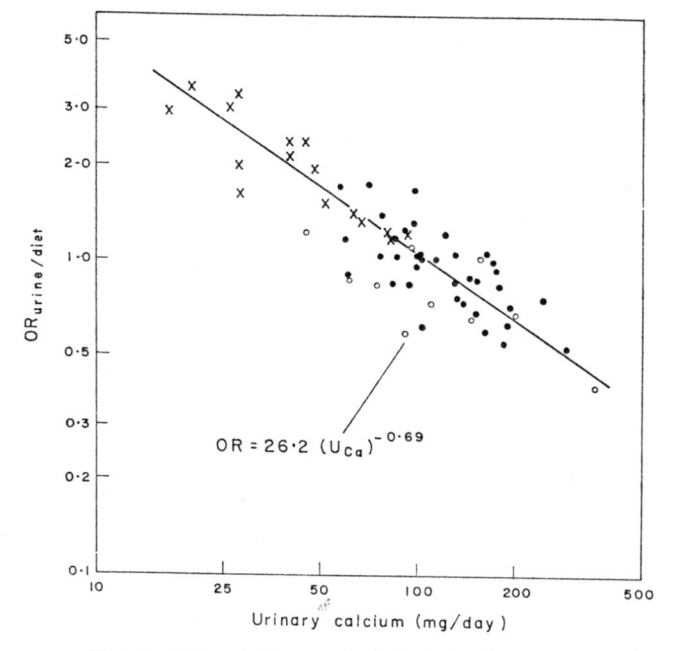

$$OR = 26\cdot2 \, (U_{Ca})^{-0·69}$$

FIG. 9. $OR_{urine/diet}$ as a function of urinary Ca.

primarily from lack of absorptive discrimination ($DF_A = 1\cdot06$); further, this came about not from the overall increase of alkaline-earth absorption, but rather because the rate constant of absorption of Sr was about the same as for Ca ($K_{absorp.} = k_{Sr}/k_{Ca} = 1\cdot1$). Thus it seems that early in life there is a fundamental alteration in the processes that govern the rates of transport of Sr and Ca across the gut. There are not yet sufficient data to ascertain how much of the alteration results from changes in diet and how much from alteration in physiological transport.

$OR_{urine/diet}$ as related to urinary calcium

Recently a controlled-feeding study was done to explore the factors involved and the reliability of using the $^{90}Sr/Ca$ of urine as a measure of dietary intake (Thompson and Comar, 1967). The investigation was carried out with eight 4- to 5-year-old children and a group of thirteen adults. It is known that the $OR_{urine/diet}$ varies with urinary Ca, and the data are presented in Fig. 9 as

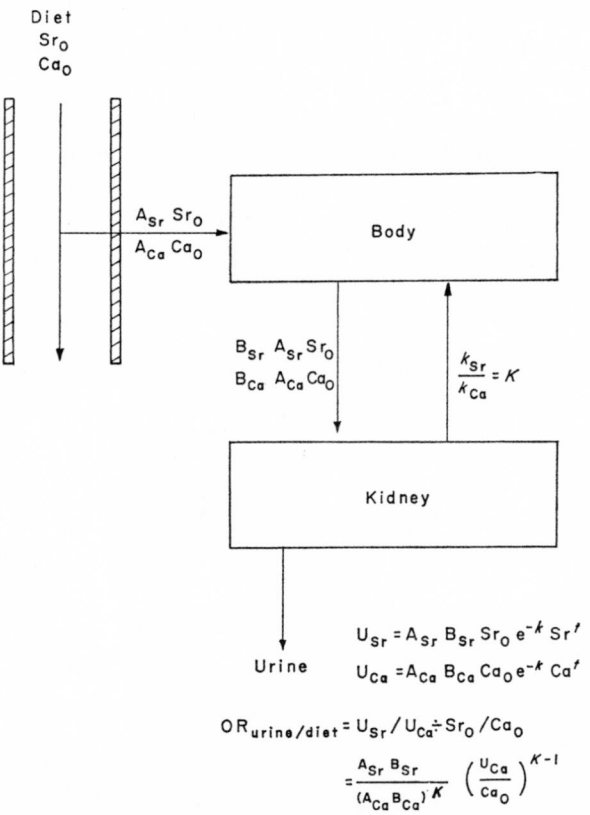

FIG. 10. Theoretical model for relationship between $OR_{urine/diet}$ and urinary Ca.

a log–log graph; for comparison some values of Samachson and Spencer-Laszlo (1962) are included. From the standpoint of feasibility of monitoring, the results support the extensive trials of Czosnowska (1963, 1964) in Poland and in addition cast some light on problems of diurnal and individual variation. However, it is the intention here not to discuss monitoring, but rather a theoretical model (Fig. 10) that appears appropriate. If Sr_0 and Ca_0 are the amounts of the two elements ingested per day, then $A_{Sr}Sr_0$ and $A_{Ca}Ca_0$ are the amounts absorbed into the body per day, and $A_{Sr}B_{Sr}Sr_0$ and $A_{Ca}B_{Ca}Ca_0$ are the amounts available for tubular resorption (A and B represent fractional absorption and fractional filtration into the glomerulus). If tubular resorption is first order, then:

$$U_{Sr} = A_{Sr}B_{Sr}Sr_0 e^{-k_{Sr}t} \tag{3}$$

$$U_{Ca} = A_{Ca}B_{Ca}Ca_0 e^{-k_{Sr}t} \tag{4}$$

Dividing equation (3) by equation (4) yields:

$$U_{Sr}/Sr_0 = A_{Sr}B_{Sr}\left(\frac{U_{Ca}}{A_{Ca}B_{Ca}Ca_0}\right)^K \tag{5}$$

by definition:

$$OR_{urine/diet} = \frac{U_{Sr}/U_{Ca}}{Sr_0/Ca_0} = \frac{U_{Sr}}{Sr_0} \div \frac{U_{Ca}}{Ca_0} \tag{6}$$

Substituting from equation (5):

$$OR_{urine/diet} = \frac{A_{Sr}B_{Sr}}{(A_{Ca}B_{Ca})^K}\left\{\frac{U_{Ca}}{Ca_0}\right\}^{K-1} \tag{7}$$

For purposes of evaluation of the constants of equation (7), the data from the adults and children have been represented in Fig. 11 as log–log graphs of $OR_{urine/diet}$ against U_{Ca}/Ca_0. The least-mean-squares fit gives the following equations:

Adults: $\qquad OR_{urine/diet} = 0.32(U_{Ca}/Ca_0)^{-0.50} \tag{8}$

Children: $\qquad OR_{urine/diet} = 0.54(U_{Ca}/Ca_0)^{-0.46} \tag{9}$

It is first noted that the form of the theoretical equation is indeed a power function, which means that one can justifiably expect a straight line from the log–log graph of $OR_{urine/diet}$ *versus* some function of U_{Ca}. The determined values of K, as calculated from values of $K-1$ in equations (8) and (9) are 0.50 and 0.54, which compare with a value of 0.7 as estimated for dogs and human subjects by Walser and Robinson (1963). The physical meaning of K is that the rate constant of tubular reabsorption of Sr is 0.5 to 0.7 that of Ca, and it is to be noted that the same value of K was obtained for both children and adults.

A further test can be applied to determine if the equation is consistent. The

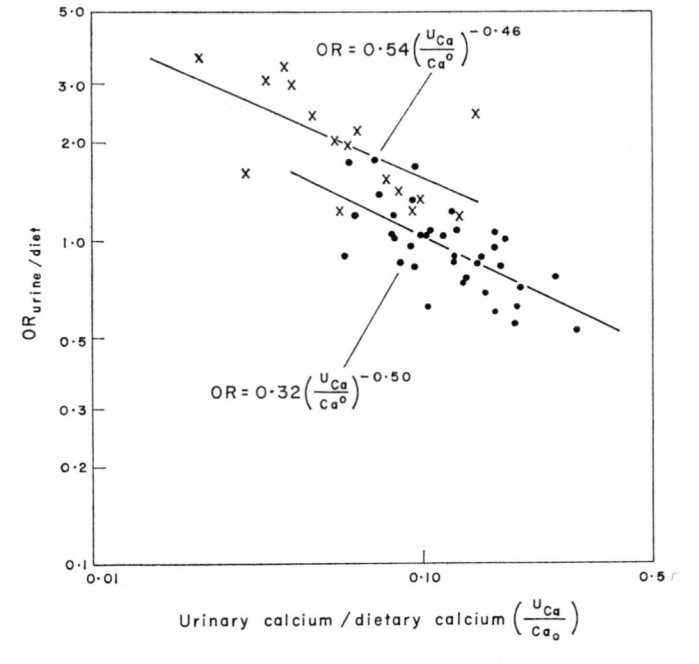

FIG. 11. $OR_{urine/diet}$ as a function of fraction of urinary excretion.

term $\dfrac{A_{Sr}B_{Sr}}{(A_{Ca}B_{Ca})^K}$ was found to be 0·32 and 0·54. If typical values are used for A_{Sr} (0·2), A_{Ca} (0·4) and K (0·5), and it is further assumed that $B_{Sr} = B_{Ca}$, then the value of B_{Sr} and B_{Ca} is calculated to be 1 to 3. This is roughly the same as estimated for the fraction of an intravenous dose of radioactive calcium that is filtered through the glomerulus in 24 h as determined by using a disappearance specific activity curve and an average glomerular filtration rate.

APPLICATION TO NON-STEADY-STATE SYSTEMS

The procedures used in the urine study illustrate a most important question that has a bearing on practically all applications of Sr/Ca ratios. What is the extent of the error that arises from use of $^{90}Sr/Ca$ (a single tracer) instead of the theoretically correct method that would utilize either Sr/Ca or a double tracer under steady-state conditions? The answer, of course, depends to a large extent upon timing.

The Sr and Ca in the body, in simple terms can be thought to be present in two compartments: a small "exchangeable" pool and a large "slowly

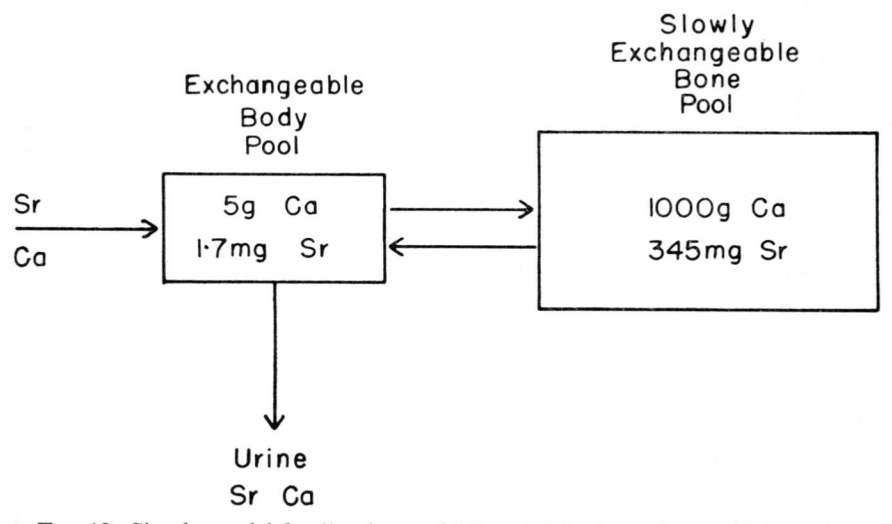

FIG. 12. Simple model for "exchangeable" and "slowly exchangeable" pools of Sr and Ca in the body.

exchangeable" pool (Fig. 12). In the adult on a uniform diet of stable Sr and Ca, the latter are in steady state throughout the system. If ^{90}Sr is introduced it will reach steady state in relatively short times as far as input, exchangeable body pool and urine is concerned. From observations of urinary excretion of ingested stable Sr by man (Harrison *et al.*, 1955) it is estimated that measurements at 10 days after an increased level would not be in error by more than about 10% because of lack of steady state. The specific question may now be considered as to whether it was legitimate to derive an OR$_{urine/diet}$ value from measurements of ^{90}Sr/Ca ratios. In the controlled study, the ^{90}Sr/Ca of the dietary intake as measured was the same as that of the normal diet, which had shown little change over the preceding months. Therefore the OR values are considered a good approximation of the steady-state value.

The more uncertain point is the error that is introduced when such values are applied during a period when ^{90}Sr/Ca levels in the diet are changing. For reasons presented, there is no question but that the use of the OR values and measurements of ^{90}Sr/Ca in urine would underestimate the calculated dietary value during the few days following a sudden increase in dietary ^{90}Sr/Ca. However, the time lag in approach to steady state in terms of past patterns of changes in dietary levels would not, in my opinion, seriously reduce the value of the observations for purposes of routine monitoring.

Another important matter is the legitimacy of application to children, who are not in steady state because they are growing and the sizes of their metabolic pools are increasing. The classical kinetic approach is to limit the period of observation so that the pool size can be considered as constant during the experiment. This is possible with double tracer studies or with balance studies with normal dietary constituents in which measurements are made of intake and output. It may be very useful to consider whether the observation of interest (e.g., Sr/Ca in new bone or in urine) is indeed affected by changes in pool size.

SOME CURRENT RESEARCH

Considerable attention is being given in our laboratory, particularly by Dr R. H. Wasserman and various graduate students, to the detailed mechanism of transport of alkaline earths across the gastrointestinal tract (Taylor and Wasserman, 1965: Marcus, 1964). The approach has been to analyze mucosal preparations for radio-tracers at early times after the start of transport, to study and isolate a factor in the intestinal mucosa that has been found to bind alkaline earths and which is induced by vitamin D, to correlate the rate of absorption with the level of this binding factor, and to relate these matters to the action of vitamin D.

The general picture emerges as follows: Sr and Ca enter the mucosal cells at about the same rate. Ca is moved out at a faster rate, as shown by the comparatively high levels of Sr in mucosal cells at early time intervals. One may speculate that the Ca is transported faster because of the presence of the binding factor which facilitates its movement; this factor tends to bind or complex Ca to a greater extent than it does Sr. In a paper now in press evidence is given that this binding factor is a protein or is associated with a protein.

Another matter of interest has been the possibility of incorporation of alkaline earths into chromosomal material. In a Ph.D. thesis, Dr Rita Herzog has presented definite evidence for the incorporation of calcium in a firmly bound form in the chromosomes of dividing human white blood cells (Herzog, 1966). The comparative incorporation of strontium into the chromosomes and the relationship to DNA synthesis are under investigation.

REFERENCES

Beninson, D., Migliori, H., and Ramos, E. (1964). *In* "Fallout Program Quarterly Summary Report". (HASL–149) U.S.A.E.C. New York Operations Office.

Comar, C. L., and Wasserman, R. H. (1964). *In* "Mineral Metabolism", vol. 2A (Comar, C. L., and Bronner, F., eds.). Academic Press, New York.

Czosnowska, W. (1963). *Nukleonika*, **8,** 779–782.

Czosnowska, W. (1964). *Nukleonika*, **9,** 757–764.

Harrison, G. E., Raymond, W. H. A., and Tretheway, H. C. (1955). *Clin. Sci.* **14,** 681–695.

Harrison, G. E., Sutton, A., Shepherd, H., and Widdowson, E. M. (1965). *Br. J. Nutr.*, **19,** 111–117.

Health and Safety Laboratory (HASL-165) (1966). "Fallout Program Quarterly Summary Report." U.S.A.E.C. New York Operations Office.
Herzog, R. K. (1966). Ph.D. Thesis, Cornell University.
Lough, S. A., Rivera, J., and Comar, C. L. (1963). *Proc. Soc. expl biol. Med.* **112,** 631–636.
Marcus, C. S. (1964). Ph.D. Thesis, Cornell University.
Marcus, C. S., and Wasserman, R. H. (1965). *The Am. J. Physiol.* **209,** 973–977.
Samachson, J., and Spencer-Laszlo, H. (1962). *Nature, Lond.* **195,** 1113–1115.
Spencer, H., Li, M., Samachson, J., and Laszlo, D. (1960). *Metabolism,* **9,** 916–925.
Taylor, A. N., and Wasserman, R. H. (1965). *Nature, Lond.* **205,** 248–250.
Thompson, J. C. Jr., and Comar, C. L. (1967). *Hlth Phys.* **13,** 5–13.
Walser, M., and Robinson, B. H. B. (1963). *In* "The Transfer of Calcium and Strontium Across Biological Membranes", (Wasserman, R. H., ed.). Academic Press, New York.

Time Course of The Transfer of Strontium-90 Through Food Chains to Man

R. S. BRUCE, B. O. BARTLETT AND R. SCOTT RUSSELL

Agricultural Research Council Radiobiological Laboratory,
Letcombe Regis, Wantage, Berkshire

SUMMARY

The contamination of diet with ^{90}Sr is affected by the rate of fallout in the previous year as well as that in the current year. This is due primarily to delays in the transfer of fallout into milk, but the delay between the production and consumption of some foods, notably cereals, must also be taken into account. When the rate of fallout is high, the ratio of ^{90}Sr to Ca in milk is similar to that in the mixed diet. Although the relative importance of milk as a source of ^{90}Sr decreases somewhat when uptake is mainly from the cumulative deposit in the soil, the integrated levels of contamination in milk and diet do not differ greatly. Therefore the ratio of ^{90}Sr to Ca in milk is a good index both of the current exposure to this nuclide when the rate of fallout is high and also of the total dose commitment.

INTRODUCTION

It is now generally recognized that when the rate of fallout is high, dietary contamination with ^{90}Sr depends more on the rate of deposition than on the cumulative deposit of ^{90}Sr in the soil. However, the rate of fallout a year previously may influence dietary contamination as much as, or more than the rate in the more recent past, and a decrease in the rate of fallout, therefore, may not be reflected in dietary levels until after a considerable period. The importance of this lag effect may change depending on the relationship between the annual fallout and the cumulative deposit. Dietary habits are a further source of variation.

These matters are reviewed here in so far as they may be relevant to examination of the relationships between the pattern of dietary contamination and the resultant level of ^{90}Sr in bone.

RELATIONSHIP BETWEEN THE CONTAMINATION
OF MILK AND THE PATTERN OF FALLOUT

Because milk is a major source of ^{90}Sr in diet, especially for young children, it has received most attention in dietary surveys, and the results provide the best basis for studying quantitatively the relationship between fallout data and dietary contamination. The importance of recent fallout in determining the contamination of milk was well established in 1958 when a comparison of the ratios of ^{89}Sr to ^{90}Sr in rain and milk suggested that, on average, there was an interval of about 2 months between the deposition of ^{90}Sr and its appearance in milk (Russell, 1960). Subsequent more detailed analysis of the relationship between monthly fallout and the levels in milk during the summer (when cattle were grazing) showed that the levels of ^{90}Sr in milk were more closely correlated with the deposit 1 to 2 months previously than with that in the more recent past (Mercer *et al.*, 1963).

Until lately, however, data were insufficient to take account of such factors in deriving quantitative relationships between the pattern of fallout and the contamination of milk; it was possible to predict future situations only on a simplified model that envisaged contamination as depending on two components, namely, that due to the current rate of deposition and that due to the cumulative deposit. Equations of the following form have been widely used (United Nations, 1962):

$$C = p_r F_r + p_d F_d \tag{1}$$

where C is the annual mean ratio of ^{90}Sr to Ca in milk (pCi/g), F_r is the deposit of ^{90}Sr in the current year (mCi/km^2), F_d is the cumulative deposit of ^{90}Sr at the middle of the year (mCi/km^2) and p_r and p_d are the "rate" and "soil" proportionality factors.

The average values for p_r and p_d calculated from dietary surveys and measurements of fallout in the United Kingdom between 1958 and 1961 were 0·76 and 0·19, respectively (Bartlett *et al.*, 1963). As would be expected, values for other countries differ according to the climatic and agricultural conditions, though usually not to a great extent (United Nations, 1962).

None the less, it is evident that a two-term equation of this type greatly oversimplified the actual situation. The assumption that the "soil" factor is constant ignores losses of ^{90}Sr due to removal in crops and movement deeper into the soil away from the rooting zone of pasture plants. Moreover, the equation takes no account of the fact that, particularly in the early part of the year, an appreciable fraction of the diet of cattle is derived from stored foods, such as hay or silage that were grown in the previous year. Also, fallout that occurs late in one year may contribute to the direct contamination of pastures grazed in the following spring. The effect of this lag on the time course of contamination of milk has recently been demonstrated (Bartlett and Russell,

1966). Calculations based on equation (1) overestimate the ratio of ^{90}Sr to Ca in milk when the rate in the previous year was relatively low, and underestimate the ratio when the rate of fallout was high. Indeed when the rate of fallout rises steeply, the highest levels of contamination in milk may not be attained until the year after the rate of fallout reaches its peak (Agricultural Research Council, 1965).

The following equation, which contains a third term that takes account of the rate of fallout in the second half of the previous year, has been found to give much closer agreement between the calculated and observed values in individual years (Bartlett and Russell, 1966):

$$C = p_r F_r + p_l F_l + p_d F_d \qquad (2)$$

where F_l is the deposit of ^{90}Sr in the second half of the previous year (mCi/km^2) and p_l is the "lag-rate" factor, the other symbols being defined as for equation (1). When the lag-rate factor was related to the deposit in the whole of the previous year, or in the summer months only, there was not such good agreement. The values for p_r, p_l and p_d in equation (2) derived by least-squares analysis are, respectively, 0·70, 1·13 and 0·11 for the United Kingdom. The levels of ^{90}Sr in milk calculated on this basis (Table I) do not differ by more than about 10% in any year compared with deviations of up to 20% with equation (1).

TABLE I

Comparison of calculated and observed levels of strontium-90 in milk in the United Kingdom

Equation (2) was used to obtain the calculated values

Year	Annual deposit of strontium-90 mCi/km^2	Strontium-90 in milk: Observed pCi^{90}Sr/g Ca	Strontium-90 in milk: Calculated values as % of observed	% of strontium-90 in milk attributable to cumulative deposit in soil
1958	5·3	7·2	97	22
1959	8·1	9·8	107	23
1960	1·9	6·4	94	44
1961	2·1	5·9	90	53
1962	10·7	11·7	108	28
1963	19·1	25·6	96	21
1964	14·9	28·0	101	25

The introduction of this lag-rate factor leads to a lower value for the soil factor than was formerly supposed, and a smaller fraction of the ^{90}Sr in milk is thus attributed to absorption from the cumulative deposit. This now appears to have been responsible for from 20–50% of the contamination in milk in past years, the proportion being highest in years when there was least fallout (Table I).

The foregoing refers to the country-wide situation in the United Kingdom, but considerable local variations may occur. These have been examined for several hill areas in the wetter parts of the country where high rainfall and special agricultural factors together cause the contamination of milk to exceed the average to a considerable extent. Estimates of the proportionality factors for these individual sites are subject to much greater uncertainty than those for the country as a whole; the values obtained ranged between the following limits:

Rate factor: 0·7–1·7
Lag-rate factor: 0·3–2·3
Soil factor: 0·1–0·5

It is evident, therefore, that local conditions may cause considerable divergence from the general trends throughout a region. In the United Kingdom, however, this has a negligible effect on the country-wide average as very little milk is produced in such areas.

OTHER FOODS

Data on the contamination of other foods are much less extensive than those for milk, but the approximate values of the rate and soil proportionality factors, as defined for equation (1), for cereals and vegetables in the United Kingdom have been estimated (Bartlett *et al.*, 1963) as:

Food	Rate factor p_r	Soil factor p_d
Cereals (milled)*	4–8	0·5
Leaf vegetables	0·9	0·3
Root vegetables	0	0·9

* Not including *Creta praeparata*

These values are generally similar to those estimated for the world-wide situation (United Nations, 1962), but between years wide variation is to be expected due to differences in the seasonal distribution of fallout.

When the rate of fallout is high, the content of ^{90}Sr in cereals is determined primarily by "floral" contamination that occurs after the ears have emerged. Consequently its extent depends not on the annual rate of fallout but on the deposition in the 4–6 weeks immediately before harvest. In contrast, the contamination of potatoes and other "root" crops is determined almost entirely by the total deposit in the soil and changes in the rate of fallout are unimportant. This is because ^{90}Sr taken up by the aerial parts of a plant tends to remain at the site of absorption and there is negligible downward translocation. Direct contamination, however, is more important for leaf vegetables when the rate of fallout is high, though the rate factor is smaller than that for cereals.

The rapidity with which changes in the rates of deposition affect the contamination of the foodstuffs which the population consume depends not only on these food-chain factors, but also on the delay between production and consumption of foods. Fresh milk and green vegetables are usually consumed within a short time of their production, but cereals do not enter diet to a large extent until at least the year after harvest; for imported grain the delay may be even longer. The effect of this delay was apparent in 1962 when, in contrast to other foods, the level of contamination in cereal products was lower than in 1961 because the grain from which they were derived had been harvested before the rate of fallout increased late in 1961.

THE TOTAL DIET

From the point of view of radiological assessment, measurements of radioactivity in diet serve two major purposes:

(i) To estimate the current exposure of the population when the rate of fallout is high or increasing rapidly.

(ii) To estimate the integrated total dose or "dose commitment" which members of the population will receive over an extended period as a result of environmental contamination.

Because the average contamination of milk can be estimated with much greater reliability than that of other foods, it is relevant to consider how closely changes in the contamination of the mixed diet can be inferred from those in milk to meet these two requirements.

The ratios of ^{90}Sr to Ca in milk which would be expected over different periods of time after fallout has been deposited have been compared with those in the mixed diet (Table II). In this calculation it has been assumed that the fallout was deposited at a constant rate in a single year, and the proportionality factors referred to earlier in this paper have been used. Over long periods the extent to which ^{90}Sr will be removed from the soil by crops or lost by leaching is a major source of uncertainty, but, following earlier assessments, it has been assumed that such losses lie between 2 and 5% annually (United Nations, 1962; Dolphin et al., 1960).

Table II shows that in the first 5 years, when the ^{90}Sr that enters diet will be about half the eventual total, the average ratio of ^{90}Sr to Ca in milk slightly exceeds that in the mixed diet. This relationship has been apparent from the results of surveys of dietary contamination hitherto; in the United Kingdom between 1958 and 1964 the ratio of ^{90}Sr to Ca in milk was on average 1·1 times that in diet, values for individual years ranging from 0·95 to 1·2.

Table II also shows that although the ratio of ^{90}Sr to Ca in milk will decrease somewhat with time relative to that in other foods, the ratio in milk will nevertheless still provide a reasonable guide to that in the average diet

TABLE II

Average ratios of strontium-90 to calcium in milk and the mixed diet in different periods after the deposition of strontium-90

A deposition of 100 mCi ^{90}Sr/km^2 in year 1 is assumed

Period years	Mean ratio of strontium-90 to calcium in the mixed diet, pCi/g	^{90}Sr/Ca in milk ^{90}Sr/Ca in diet
1–5	32·3	1·05
6–15	7·8	0·81
16–50	2·5	0·81
1–50	6·6	0·93

Notes:

(*a*) Calculated values of the ratios of strontium-90 to calcium in milk, leaf and root vegetables, and cereals have been compounded to give values for the total mixed diet (including mineral calcium added to flour) in the United Kingdom in the manner described in ARCRL (1965).

(*b*) The figures shown are the mean results for alternative calculations in which it was assumed that, in addition to radioactive decay, absorption from the soil decreases by 2% or 5% annually owing to leaching and removal in crops. The two calculations gave closely similar results for the relationship between milk and diet for each period. However, when a 2% loss was assumed, the ratio in diet for the period 1–50 years is about 15% higher than the mean and for a 5% loss the value is correspondingly lower; the divergence between the two calculations is greatest in the later years.

for many years. Moreover, integrated over 50 years the ratio of ^{90}Sr to Ca in milk will differ by not more than 10% from that in the mixed diet; in this time about 90% of the value at infinity would be reached.

Thus, for purposes of radiological assessment it is evident that measurements of ^{90}Sr in milk provide a good guide both to the changing ratios of ^{90}Sr to Ca in the mixed diet in the initial period when the rate of fallout is high, and also to the total dose commitment. Other aspects of the evaluation of radiation doses which may be received from this nuclide, and of the attendant risks, are undoubtedly open to considerably greater uncertainties.

The situation would, however, be somewhat different if it were desired to compare the ^{90}Sr content of diet in individual years when the rate of fallout is low with the quantities of ^{90}Sr deposited in bone. Under these circumstances dietary contamination would be due almost entirely to absorption from soil and it is evident from the relative magnitudes of the soil factors reviewed earlier in this paper that the ratios of ^{90}Sr to Ca in different foods would then differ markedly; for example, the ratio in milk would be only about one-eighth of that in root vegetables. Variations in the composition of diet could thus have a considerable influence on the ratio of ^{90}Sr to Ca in the mixed diet and, as a result, values for the intake of individuals would be likely to vary from the average more than when the rate of fallout is high. In an overall assessment this is, however, a minor consideration as the levels of dietary contamination would then be only a small fraction of those in the early period.

REFERENCES

Agricultural Research Council Radiobiological Laboratory. (1965). Report ARCRL–14, 3–33.

Bartlett, B. O., and Russell, R. S. (1966). *Nature, Lond.*, **209,** 1062–1065.

Bartlett, B. O., Burton, J. D., Ellis, F. B., and Mercer, E. R. (1963). Agricultural Research Council Radiobiological Laboratory Report ARCRL–10, 82–85.

Dolphin, G. W., Loutit, J. F., Marley, W. G., Mayneord, W. V., and Russell, R. S. (1960). *In* "Hazards to Man of Nuclear and Allied Radiations" (Medical Research Council), pp. 109–119. H.M.S.O., London.

Mercer, E. R., Burton, J. D., and Bartlett, B. O. (1963). *Nature, Lond.*, **198,** 662–665.

Russell, R. S. (1960). *In* "Radioisotopes in the Biosphere" (Caldecott, R. S., and Snyder, L. A., eds.), pp. 269–292. University of Minnesota, Minneapolis.

United Nations. (1962). Report of the Scientific Committee on the Effects of Atomic Radiation, General Assembly, Official Records: Seventeenth Session, Supplement No. 16 (A/5216), Annexe F, Part II, pp. 287–346. United Nations, New York.

Strontium-90 from Fallout in Human Bone

J. F. LOUTIT

Medical Research Council Radiobiological Research Unit,
Harwell, Didcot, Berks.

SUMMARY
Analyses of human bones have been made by the U.K. Atomic Energy Authority continuously since 1956. From a review of their data and of observations by the Agricultural Research Council on the contamination of diet by ^{90}Sr, the rate at which bone loses its incorporated strontium has been calculated. The rates vary according to age of subject.

INTRODUCTION

Human bones, predominantly femora, have been analysed for ^{90}Sr each year since 1956 by the United Kingdom Atomic Energy Authority. The results are currently published semi-annually in the Monitoring Report Series of the Medical Research Council (1960–65). Figure 1 illustrates the data obtained for 1964, the last year for which full results are available. Figure 2 shows the combined data for previous years, the earlier results being taken from the A.E.R.E. Reports of Bryant and his colleagues (1957–60).

There is a consistent pattern. The maximum concentrations of ^{90}Sr/g Ca are observed around the age of 1 year. Values decline to plateaux in adolescence that, so far, have been greater than the concentrations found in adults. This pattern must be explicable in terms of the intake of ^{90}Sr, its accretion into bone of those that are growing and its turnover in bone at all ages.

Bryant and Loutit (1961, 1964) have already reviewed the data available up to the end of 1961. It was concluded from observations on the specific activity (pCi ^{90}Sr/mg Sr) that different bones in adults exhibited a variable "rate of replacement" of their strontium content from about 2% to 8% each year. This was in conformity with an estimated mean of about 3% for the whole skeleton derived by Kulp and Schulert (1962) from analyses of cadavers in which total ascertainment was carried out. It was stressed that the derived figures indicate the amount of body's Sr replaced from current diet (or viewed another way, excreted) each year and not total turnover, since re-utilization of resorbed bone salt that is part of total turnover is not measured by any of the procedures. Bryant and Loutit (1961) were unable to produce satisfactory figures for a similar "rate of replacement" in children, though the data indicated that, particularly in early life, it must be so high as to verge towards complete annual replacement.

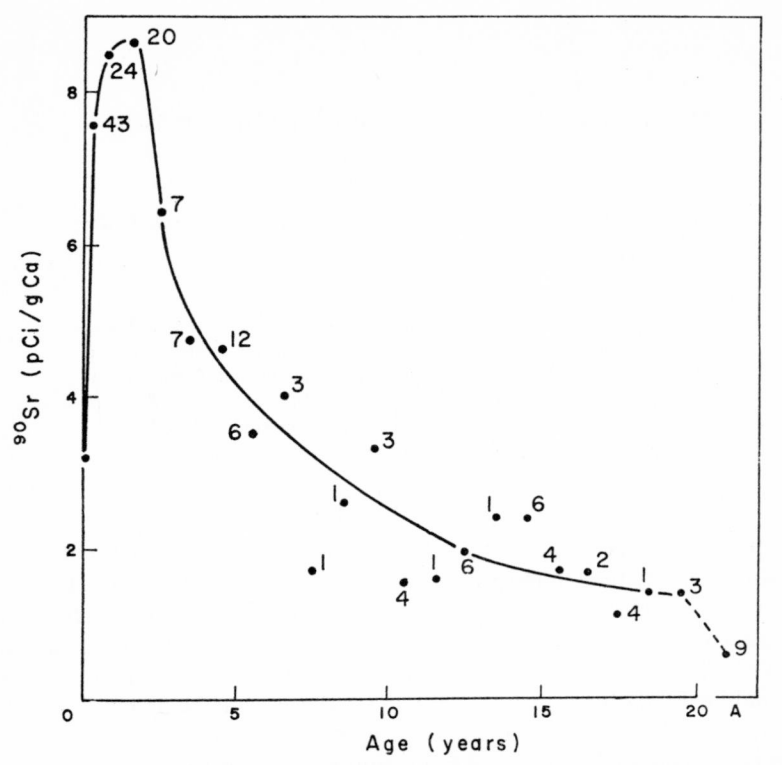

Fig. 1. Mean concentrations of ⁹⁰Sr (pCi/gCa) in bone according to age: U.K.A.E.A. results 1964. A = Adults.

TABLE I

Values used for contamination of national diet in calculations of rate of strontium replacement

Year	National diet pCi ⁹⁰Sr/g Ca
1957	5·5
1958	5·9
1959	9·0
1960	6·4
1961	6·2
1962	9·9
1963	22·8
1964	25·9

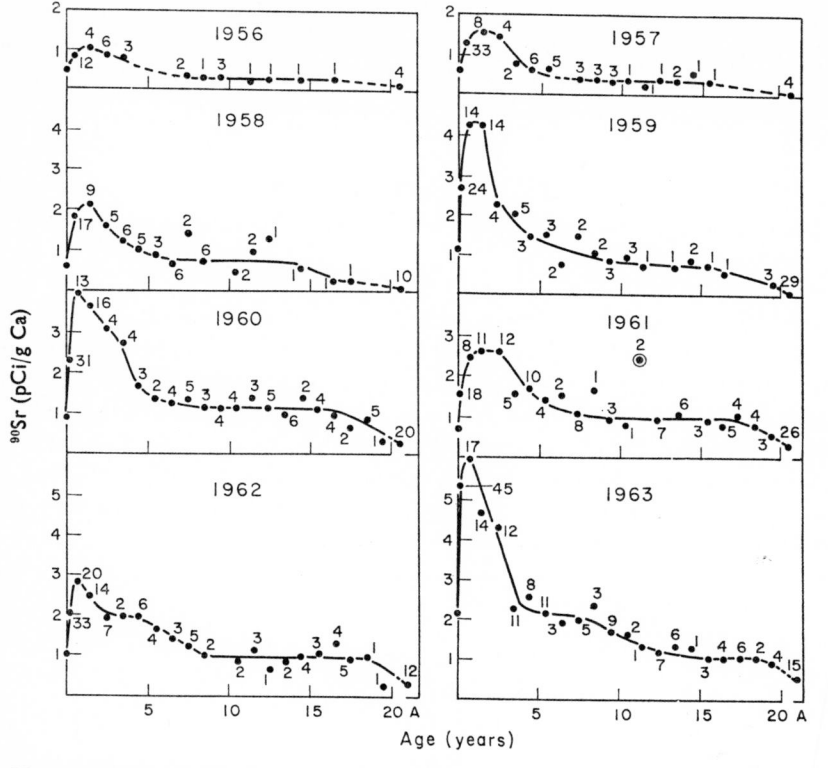

FIG. 2. Mean concentrations of ⁹⁰Sr (pCi/g Ca) in bone according to age: U.K.A.E.A. results 1956–1963. A = Adults.

INTAKE OF STRONTIUM-90

In the latest review, where additional bone results for the years 1962–64 are available, Fletcher *et al.* (1966) have, as formerly, utilized for intake of ⁹⁰Sr the figures obtained for diet by survey since 1958 by the Agricultural Research Council and for 1957 by Bryant *et al.* (1958 a, b) (Table I). They assume, not unreasonably, an average of 1 g Ca/day at all ages and derive a yearly intake of ⁹⁰Sr as 365 times the observed ⁹⁰Sr/Ca ratio of diet. However, experience shows that there is a lag of 6–12 months before bone reflects a rise or fall in the ⁹⁰Sr/Ca value of the national diet. Therefore for relating diet to bone a mean of the current and previous years' values was taken.

ACCRETION

For the purposes of calculation, the values of skeletal calcium according to age accepted for American children (Mitchell *et al.*, 1945) were used for their counterparts in Britain. As it was shown (Fletcher *et al.*, 1966) that in children and adolescents the Sr concentration was nearly uniform between and within the bones analysed, the ^{90}Sr skeletal burden for each cohort at each age was calculable from skeletal Ca $\times$ ^{90}Sr/Ca ratio of femora.

RATE OF REPLACEMENT

The rate of replacement of ^{90}Sr is then calculable from the equation:

$$[^{90}\text{Sr}]_n = (1 - p')\,[^{90}\text{Sr}]_{n-1} + P(1 - p'')D$$

where $[^{90}\text{Sr}]_{n-1}$ and $[^{90}\text{Sr}]_n$ denote body content of strontium-90 at the end of $(n - 1)$th and nth year of age, p' is the fraction of the existing body burden of ^{90}Sr at the beginning of a year excreted during the year, P is the fraction of ingested ^{90}Sr absorbed by the gut from the diet, p'' is the fraction of the absorbed dietary ^{90}Sr of the year excreted during the year and D is the ^{90}Sr ingested in a year derived from an assumed intake of 365 g Ca $\times$ ^{90}Sr/Ca ratio of diet. Values of p' and $P(1 - p'')$ were estimated by Papworth by the method of least squares and plotted in Fig. 3.

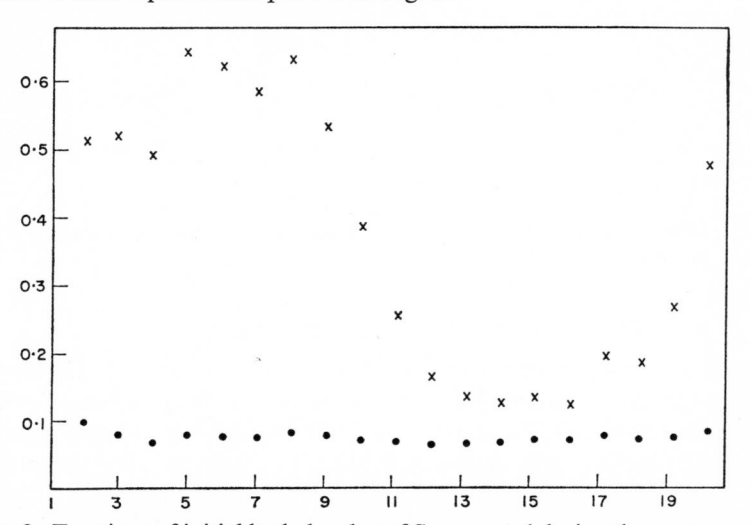

FIG. 3. Fractions of initial body burden of Sr, excreted during the year, according to age—p' ($\times$), and annual dietary intake retained—$P(1 - p'')$ ($\cdot$).

The values, obtained in this way, relating to the experience of 8 years, suggest a remarkably uniform performance. Each year from the ages of 1 to 9 about half the existing body burden of ^{90}Sr is apparently lost. After that the fraction lost falls to 0·1 to 0·2 per year, apart from the final value. Throughout infancy and adolescence the calculations indicate a remarkably constant retention at the end of each year of about 7% of that year's dietary intake.

REFERENCES

Agricultural Research Council Radiobiological Laboratory (1959–1965). Reports 1, 3, 5, 8, 10 and 14. H.M.S.O., London.

Arden, J. W., Bryant, F. J., Henderson, E. H., Lloyd, G. D., and Morton, A. G. (1960). AERE–R 3246. H.M.S.O., London.

Bryant, F. J., and Loutit, J. F. (1961). AERE–R 3718. H.M.S.O., London.

Bryant, F. J., and Loutit, J. F. (1964). *Proc. R. Soc.*, **159B**, 449–465.

Bryant, F. J., Chamberlain, A. C., Morgan, A., and Spicer, G. S. (1957). AERE HP/R 2353. H.M.S.O., London.

Bryant, F. J., Chamberlain, A. C., Spicer, G. S., and Webb, M. S. W. (1958a), *Br. med. J.* i, 1371–1375.

Bryant, F. J., Henderson, E. H., Spicer, G. S., and Webb, M. S. W. (1958b). AERE C/R 2583. H.M.S.O., London.

Bryant, F. J., Henderson, E. H., Spicer, G. S., Webb, M. S. W., and Webber, T. J. (1959a). AERE C/R 2816. H.M.S.O., London.

Bryant, F. J., Cotterill, J. C., Henderson, E. H., Spicer, G. S., and Webber, T. J. (1959b). AERE–R 2988. H.M.S.O., London.

Fletcher, W., Loutit, J. F., and Papworth, D. G. (1966). *Br. med. J.* ii, 1225–1230.

Kulp, J. L., and Schulert, A. R. (1962). *Science*, **136**, 619–632.

Medical Research Council Monitoring Report Series Nos. 1–11. H.M.S.O., London.

Mitchell, H. H., Hamilton, T. S., Steggerda, F. R., and Bean, H. W. (1945). *J. biol. Chem.*, **158**, 625–637.

Predicting Strontium-90 Concentrations in Human Bone

JOSEPH RIVERA

Health and Safety Laboratory,
U. S. Atomic Energy Commission, New York

SUMMARY

An empirical relation between ^{90}Sr/Ca ratios in the human diet and resulting ^{90}Sr/Ca ratios in human bone is derived from observations of ^{90}Sr in vertebrae from accident victims of all ages, collected in New York City, Chicago and San Francisco since March, 1961, and from estimates of dietary ^{90}Sr made at these same cities on a trimesterly basis since March, 1960. The physiological significance of the age-dependent parameters involved in the empirical model are discussed, taking into account stable Sr concentrations found in diet and bone.

INTRODUCTION

Many countries have instituted extensive programs for the collection and analysis of human diet and bone samples for ^{90}Sr. Unfortunately these diet and bone survey programs sometimes fall into the category termed by Andrew McLean as "the dog and lamp-post type survey", that is they serve more to relieve than to illuminate.

Although it is probably true that the public achieves a measure of relief in knowing that the government is keeping a close watch on ^{90}Sr in the diet and in bone, it is the responsibility of the scientists in charge of these monitoring programs to see that some illumination results from the considerable effort being expended.

Scott Russell and his associates of the Agricultural Research Council Radiobiological Laboratory have shown how survey data can provide more than relief. Acting on a suggestion of Eizo Tajima at the United Nations Scientific Committee on the Effects of Atomic Radiation, they have shown that the results of measurements of the concentrations of ^{90}Sr in milk and in the total diet can be predicted from measured or predicted estimates of the annual fallout rate of ^{90}Sr and the cumulative deposit of ^{90}Sr in the soil. We too have used this trick and are astonished at how close the predictions come to the observations.

The discovery of general quantitative relations between survey fallout data and survey diet data is illuminating, since the instances where they do not work have revealed interesting and unexpected phenomena, such as stem-base absorption, unusual chemical composition of soil, peculiar rainfall distribution, and the like.

Several general quantitative relations between ^{90}Sr/Ca ratios in human bone and in human diet have been formulated. The differences between predictions based on these theories and observations may also provide helpful insights into our understanding of human-bone metabolism.

BONE PREDICTION MODELS

Model A

General quantitative relations between the ^{90}Sr content of human diets and the resulting ^{90}Sr concentrations of human bone were deduced by Kulp and Schulert and their associates at Columbia University (1962) from various survey data; bone samples obtained in various places throughout the world, a few diet and milk samples and some rather bold assumptions.

The substance of their interpretation of the relation between diet and bone ^{90}Sr/Ca ratios is given by an equation of the form:

$$\mathrm{Ca}_n X_n = \mathrm{Ca}_{n-1}(0 \cdot 975) X_{n-1} - \mathrm{Ca}_{n-1} X_{n-1}(0 \cdot 975) f + \mathrm{Ca}_{n-1} K Z_{n-1} f$$
$$+ (\mathrm{Ca}_n - \mathrm{Ca}_{n-1}) K Z_{n-1} \qquad (1)$$

where Ca_n is the Ca content of the vertebrae at year n; X_n is the ^{90}Sr/Ca ratio in vertebrae at year n; f is the fraction of vertebrae Ca replaced during the interval from year $n-1$ to year n; K is the bone–diet observed ratio during the interval $n-1$ to n; and Z_{n-1} is the ^{90}Sr/Ca ratio of the diet during the interval $n-1$ to n. The factor $0 \cdot 975$ is introduced to account for the radioactive decay of ^{90}Sr in 1 year.

The values of Ca_n and Ca_{n-1} were assumed to be proportional to the Ca content of the whole skeleton. The variation of the Ca content of the skeleton with age was assumed to be that given by Mitchell et al. (1945). In their approach, Kulp and Schulert (1962) assumed that f and K did not vary with age and were equal to $0 \cdot 035$ and $0 \cdot 25$, respectively.

Bryant and Loutit (1961) criticized most of the assumptions used by Kulp and Schulert (1962) in the construction of this model. They pointed out that a model based on the agreement between observed and predicted bone ^{90}Sr/Ca ratios in 1958 for such an inhomogeneous group as "Western culture" may have been fortuitous, that the discrimination factor between Sr and Ca in going from diet to bone is probably not the same for infants and adults, that the diet values used by Kulp and Schulert may not have been appropriate, since accurate surveys had not been made, and finally that the assumption of equal bone-mineral replacement rates for infants and adults was probably false.

The validity of their criticism of the model proposed by Kulp and Schulert (1962) was demonstrated by them when predicted and observed concentrations of ^{90}Sr in the bones of British children during 1959 were compared. Predicted concentrations using the Kulp and Schulert model for ages from 1 to 2 were about 2·5 pCi/g Ca, and the observed values were about 4·5 pCi/g Ca.

Model B

Bryant and Loutit (1961) attempted to improve the model used by Kulp and Schulert (1962) by estimating the variation of f and K with age. Thus by taking into account changes in specific activities of bones, resulting from changes in specific activities of diets with time, they were able to show that the rate of mineral replacement (f) during the first year of life was probably much greater than that during later years. They also concluded from stable Sr and Ca measurements in the diets and bones of infants and adults that Sr–Ca discrimination (K) during infancy is probably less than that during later life. Attempts to estimate f and K for children beyond infancy by Bryant and Loutit (1961) were generally inconclusive because of the small number of bone samples available up to 1958.

Some recent studies by Beninson et al. (1966) in Argentina, with more and better data on infant diet and bone ^{90}Sr and stable Sr concentrations have shown that reasonable estimates of f and K during the first year of life are probably about 0·50 and 0·35, respectively.

These experiments, which were initiated because observed data did not agree with crude predictions, are an example of how survey data can be used to shed light on metabolic processes. It may be argued, on general physiological grounds, that a more reasonable initial assumption by Kulp and Schulert might have allowed for increased calcium replacement rates for infant bone and diminished bone–diet Sr–Ca discrimination; however the fact is that survey data were used to give some quantitative indications of the differences in bone metabolism by infants and adults.

Model C

Our measurements of ^{90}Sr in human bone began in March, 1961. Since the principal objective of these measurements was to develop a model to predict ^{90}Sr concentrations in human bone and not primarily to document bone ^{90}Sr levels in general, we limited the sampling locations to New York City, Chicago and San Francisco, where we had established, in 1960, quarterly food sampling programs to estimate ^{90}Sr intake. Only specimens from accident victims or from individuals whose metabolism could reasonably be assumed to have been normal before death were accepted for analyses. In every case vertebrae were submitted to us by the co-operating pathologists, since for adults, from the work of Kulp and Schulert, this bone was expected to have the highest ^{90}Sr content.

Average ^{90}Sr concentrations in human vertebrae obtained in New York City, Chicago and San Francisco from March 1961 through 1964 are listed in Table I. Highest ^{90}Sr levels were found in New York, lowest levels were found in specimens from San Francisco and intermediate levels were found in

TABLE I

Average strontium-90 concentrations in human vertebrae in picocuries of strontium-90 per g of calcium

Age group	1961		1962		1963	
	mean	number of samples	mean	number of samples	mean	number of samples
New York City						
0–4	3·07	17	3·31	34	5·60	20
4–20	1·24	21	1·63	38	2·06	28
>20	0·89	23	1·03	14	1·55	22
Chicago						
0–4	2·26	1	1·50	2	3·24	7
4–20	0·93	7	1·33	16	2·05	3
>20	0·60	29	0·79	57	1·10	13
San Francisco						
0–4	0·86	5	1·12	35	2·05	45
4–20	0·87	7	1·06	24	1·39	33
>20	0·52	39	0·72	10	0·98	22

	1964		1965*	
New York City				
0–4	6·15	21	6·73	3
4–20	3·15	35	3·16	21
>20	2·02	25	1·77	5
Chicago				
0–4	1·95	1	..	..
4–20	3·26	5	..	..
>20	1·44	43	..	..
San Francisco				
0–4	2·66	42	2·60	27
4–20	1·65	27	1·85	14
>20	1·23	11	1·34	13

* Preliminary Results

specimens from Chicago. A sharp increase in average ^{90}Sr concentrations in vertebrae obtained in 1963 and 1964 as compared to those obtained in 1961 and 1962 was also observed particularly for specimens from individuals less than 4 years old at death.

The generally good correlation of bone and diet levels is apparent from a comparison of the data in Table I with the diet ^{90}Sr estimates shown in Fig. 1. From this Figure one can see that highest ^{90}Sr intakes have been in New York City, lowest intakes have been in San Francisco, and intermediate intakes have been observed in Chicago.

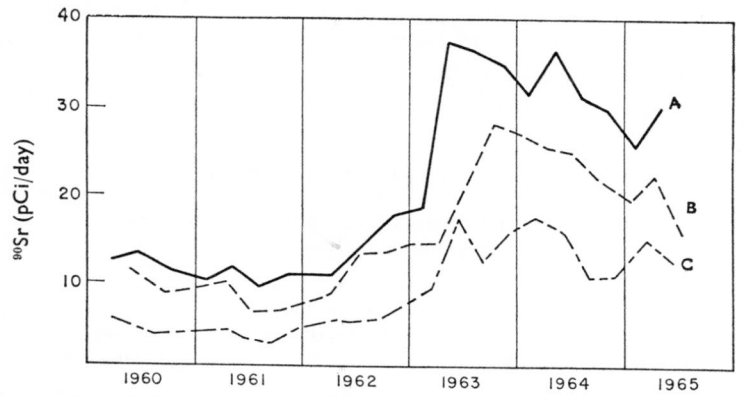

FIG. 1. Estimated daily intake of ^{90}Sr based on quarterly measurements of ^{90}Sr concentrations in 19 composite diet components and on U.S. Department of Agriculture consumption statistics. A, New York City; B, Chicago; C, San Francisco.

One can also see that intakes at all three cities sharply increased early in 1963 and attained new plateaux in 1964, at levels roughly three times as high as those during 1960, 1961 and 1962. Similar data have been obtained on the composition of infant diets and on the ^{90}Sr and Ca concentrations of the components of the diet during the first year of life. Despite the differences in composition between adult and infant diets it has been observed that the ^{90}Sr/Ca ratios of the two diets were practically the same whenever they were compared. The use of these diet and bone data in constructing a model for Sr metabolism seemed justified in view of the obviously good correlations that existed.

From the data obtained by us from 1961 and 1964, we concluded that the model used by Kulp and Schulert (model A) was adequate for predicting bone ^{90}Sr/Ca ratios in adults, but that perhaps it could be improved if estimates for f and K as functions of age could be made. Our approach to the problem of making these estimates for children older than 1 year has been described

c

in detail elsewhere (Rivera and Harley, 1965). Essentially, what we have done is to take the few data that are available as source terms for the formulation of an equation relating $^{90}Sr/Ca$ ratios in bone, age in years and date of death.

From this equation interpolated values for X_n and X_{n-1} (as defined in equation (1)), appropriate diet values for the intervals corresponding to the period between X_n and X_{n-1} (from Fig. 1), and values of Ca_n and Ca_{n-1} (from Mitchell *et al.*, 1945) appropriate to the age being considered, are substituted into equation (1). From data obtained through 1964, we obtained for each age 22 pairs of points X_n and X_{n-1} in New York City, and 22 pairs of points from San Francisco. Each of these 44 pairs of points was substituted into equation (1), and the 44 equations were solved for f and K by multiple-regression analysis. The results obtained are shown in Table II.

TABLE II

Strontium metabolic parameters calculated by multiple regression analysis

Age (years)	Annual turn-over rate $f \pm$ S.E.	Bone-diet observed ratio $K \pm$ S.E.	Calcium* in skeleton (g)
0–1	0·50†	0·35†	100
1–2	0·46 ±0·17	0·24 ±0·08	147
2–3	0·26 ±0·13	0·25 ±0·09	179
3–4	0·21 ±0·11	0·24 ±0·09	201
4–5	0·20 ±0·09	0·23 ±0·08	219
5–6	0·22 ±0·09	0·21 ±0·07	239
6–7	0·25 ±0·08	0·20 ±0·06	264
7–8	0·28 ±0·08	0·19 ±0·05	297
8–9	0·30 ±0·07	0·18 ±0·05	341
9–10	0·30 ±0·08	0·18 ±0·04	396
10–11	0·27 ±0·08	0·18 ±0·05	463
11–12	0·24 ±0·09	0·18 ±0·06	539
12–13	0·20 ±0·10	0·19 ±0·07	624
13–14	0·16 ±0·10	0·20 ±0·08	715
14–15	0·13 ±0·10	0·21 ±0·09	806
15–16	0·12 ±0·09	0·21 ±0·08	894
16–17	0·14 ±0·09	0·18 ±0·08	973
17–18	0·18 ±0·09	0·16 ±0·07	1035
18–19	0·21 ±0·11	0·14 ±0·08	1073
19–20	0·24 ±0·15	0·13 ±0·11	1078

* From Mitchell, *et. al.* (1945).
† Estimated from Beninson *et al.* (1966).

This model attempts to meet some of the objections raised by Bryant and Loutit (1961) to the approach taken by Kulp and Schulert (1962), since we allow for the possibility of varying f and K with age, our samples are taken only in places where we have accurately documented diet ^{90}Sr/Ca estimates and we attempt to use only specimens from accident or other victims of sudden death.

COMPARATIVE EVALUATION OF THE MODELS

Three models have been proposed to account for the variation of ^{90}Sr/Ca ratios in bone with variations of ^{90}Sr/Ca ratio of the diet. The simplest of these is that proposed by Kulp and Schulert (model A) in which it is assumed that no change in the replacement rate of skeletal Ca (f) or the bone–diet Sr–Ca discrimination factor (K) takes place with age.

The second model presented (model B) assumes that f is 0·50 and K is 0·35 during the first year of life and in subsequent years the values of these parameters remain constant at $f = 0·035$ and $K = 0·25$. The third model (model C) assumes that f and K vary with age from 1 to 20 in the manner indicated in Table II.

To test the accuracy of each model, predictions were made with each model of the ^{90}Sr/Ca ratios for each age during 1961, 1962, 1963, 1964 and 1965, in New York City and in San Francisco. The predictions are based on measured

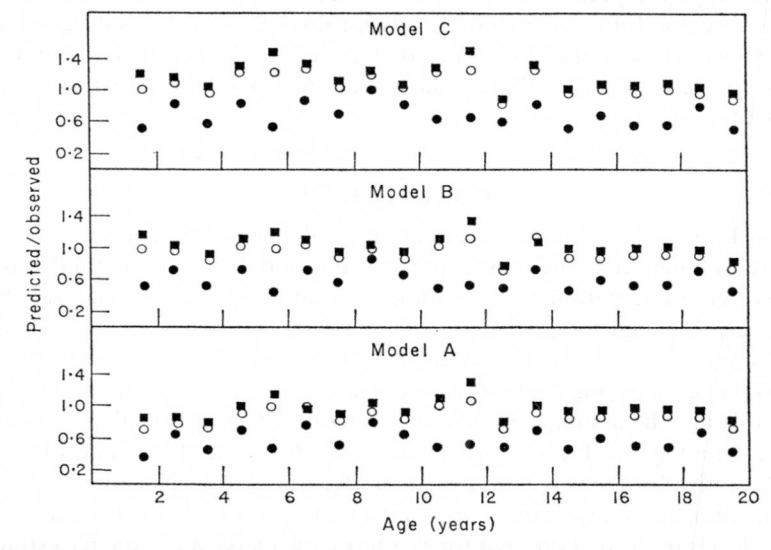

FIG. 2. Predicted/observed slopes for ^{90}Sr/Ca ratios in human vertebrae collected from 1961 to 1965 by using three models of Sr metabolism. ■, New York city; ○, San Francisco; ●, New York City plus San Francisco.

diet ^{90}Sr/Ca ratios in New York City and San Francisco for the period 1960 to 1965 and estimated diet ratios for the period 1956–1960. The contamination of diets by ^{90}Sr before 1956 was assumed to have been negligible. Graphs of the predicted values were then plotted against average observed values for each age, and a straight line through the origin was fitted to the points. The degree of agreement between predicted and observed ^{90}Sr/Ca ratios was judged by the nearness of the calculated slope to unity. The predicted/observed slopes calculated by using models A, B and C for data from New York City alone, San Francisco alone, and combined data from both cities, are shown in Fig. 2.

When data for all ages were considered together the overall predicted/observed slopes obtained were:

Model	New York	San Francisco	Combined
A	0·93±0·03	0·51±0·02	0·86±0·02
B	1·02±0·03	0·57±0·03	0·95±0·03
C	1·15±0·03	0·63±0·03	1·06±0·03

Thus we see that if only data from New York City were considered, model B would have yielded a very accurate prediction of average ^{90}Sr/Ca ratios in children's bones for the period 1961 to 1965, model A would have slightly underestimated ^{90}Sr/Ca ratios and model C would have overestimated ^{90}Sr/Ca ratios. When data from San Francisco alone were considered, the three models consistently gave predictions lower than were observed. Why this happened is not at present fully understood. Notwithstanding this deficiency, the best model seemed to be model C. When data from New York City and San Francisco were combined all three models were reasonably good with model B being the best.

CONCLUSIONS

Based on the very limited information available on ^{90}Sr concentrations in children's bones, it would appear that no substantial changes in f and K from values that these parameters obtain during adulthood ($f=0·035$, $K=0·25$) need be postulated to account for the ^{90}Sr bone concentrations observed in New York City and San Francisco from 1961 through 1965, except for the values that exist during the first year of life.

At this age the assumption of values for f and K of 0·50 and 0·35, respectively, definitely produces better agreement between predicted and observed ^{90}Sr/Ca ratios for the first 3 or 4 years of life. Another conclusion that can be drawn from this study is that with methods now available for the prediction of ^{90}Sr levels in the diet (at least for Western-type diets), a reasonable estimate of the accuracy of predicted average bone ^{90}Sr levels in children based on these predicted ^{90}Sr diet levels is probably about $\pm 25\%$.

It should be kept in mind that the parameters f and K entering into the empirical model for predicting ^{90}Sr levels in human bone may not be exactly the same as the true annual Ca replacement rate or bone–diet observed ratio that might be obtained from a carefully controlled metabolic experiment. These parameters merely serve to relate one set of measurements (average diet estimates) with possible inherent systematic errors to another set of measurements (average vertebral $^{90}Sr/Ca$ ratios). In this process, questions of the accuracy of skeletal growth estimates, effects of re-utilization of exchanged or resorbed bone, etc., are not considered. For purely dosimetric calculations for large populations, however, these questions are irrelevant since the dose rate to bone is calculated from the average ^{90}Sr concentration in bone, and this average can apparently be predicted with reasonable certainty for any given age.

The problem of calculating doses to children's bones resulting from an acute contamination of the diet that might come about from an accidental release of ^{90}Sr, remains to be solved. Since it is unlikely that very much experimental data will ever be available on the Sr metabolism of children, we will probably have to rely on inferences from refined survey type measurements to provide the illumination that this dim area of our understanding requires.

REFERENCES

Beninson, D., Ramos, E., and Touzet, R. (1966). U.S. Atomic Energy Commission Health and Safety Laboratory, New York, Report HASL–165.

Bryant, F. J., and Loutit, J. F. (1961). AERE–R 3718. H.M.S.O., London.

Kulp, J. L., and Schulert, A. R. (1962). Lamont Geological Observatory, Columbia University, Palisades, New York.

Mitchell, H. H., Hamilton, T. S., Steggerda, F. R., and Bean, H. W. (1945). *J. biol. Chem.*, **158**, 625–631.

Rivera, J., and Harley, J. H. (1965). U.S. Atomic Energy Commission Health and Safety Laboratory, New York, Report HASL–163.

Studies in Strontium Metabolism*

I: Correlation of Strontium–90 Levels in Foetal Bone and Maternal Diet

J. M. A. LENIHAN

Western Regional Hospital Board,
9–13 West Graham Street, Glasgow, C.4

SUMMARY

Study of some 500 bone samples, from subjects in the Glasgow region in the age range 0–14 days, suggests that the concentration of ^{90}Sr in the bones of the newborn can be forecast from a knowledge of the concentration in milk forming part of the national diet at various periods up to 18 months before birth. Numerical coefficients that are proposed give close agreement between predicted and observed levels during the years 1961–65.

It is generally observed that the concentration of ^{90}Sr in the bones of the newborn human subjects (expressed in pCi/g Ca) is about 10% of the corresponding figure for the average national diet at the time of birth. Levels of ^{90}Sr in typical diet and in newborn bone have been determined regularly in Britain since 1958. The figures given in Table I, based on measurements of ^{90}Sr in bone samples from England, Scotland and Wales, show that the estimate of a 10% ratio is reasonably correct. This agreement does, however, conceal regional variations in ^{90}Sr levels (Fletcher *et al.*, 1966) as well as considerable fluctuations in dietary intake of ^{90}Sr during any given year.

The Glasgow sample now comprises more than 1200 bone specimens (whole femora) that have been analysed since 1959. This material represents a comprehensive selection from post-mortem examinations made by hospital

* The work described in this and the two succeeding papers forms part of a programme conducted by the University of Glasgow (Department of Child Health), the Royal Hospital for Sick Children (Pathology Department) and the Western Regional Hospital Board (Regional Physics Department), with the support of the Medical Research Council. The workers currently participating in the programme are G. C. Arneil, Wilma M. M. Brown, J. M. A. Lenihan, C. Lam, A. M. Macdonald, Janet M. Warren and Margaret L. Winning.

pathologists in the Glasgow region—a geographically and socially coherent area with a population of about 3,000,000, largely concentrated in the industrial zone of south-west Scotland. More than 90 % of the specimens came from subjects in the 0–5 year age range. Analytical results, and details of the experimental methods, have been published at 6-monthly intervals since 1960, in the Medical Research Council Monitoring Report Series. This paper, and another contributed by J. M. Warren to the present symposium, are the first of a series of communications in which the Glasgow data will be subjected to scrutiny and interpretation.

TABLE I

Strontium-90 in typical diet and in human bone (0–14 day age group)

Year	^{90}Sr (pCi/g Ca) Diet*	Bone†
1958	5·9	0·65
1959	9·0	1·15
1960	6·4	0·86
1961	6·2	0·72
1962	9·9	1·00
1963	22·8	2·14
1964	25·9	3·00

* United Kingdom average.

† U.K.A.E.A. results; samples almost all from England and Wales, with a small number from Scotland.

The following factors are important in choosing the parameters to be compared when seeking a correlation between ^{90}Sr levels in bone and in diet:

1. Definition of foetal bone. It is customary to use the age range 0–14 days for this purpose. Fletcher *et al.* (1966) found significant differences in ^{90}Sr levels between samples obtained in 1964 (though not in earlier years) from subjects aged 0–1 day and from subjects aged 2–14 days. In the Glasgow series this difference was discernible in 1964 but not statistically significant. The age range 0–14 days has therefore been used.

2. Definition of dietary level of ^{90}Sr. An estimate of the average ^{90}Sr content of British diet (covering England, Wales, Scotland and Northern Ireland) has been made each year since 1958. A separate estimate has not been made for Scotland, but published figures for ^{90}Sr levels in milk produced in different parts of the United Kingdom indicate that the dietary intake of ^{90}Sr in Scotland is likely to be higher than the mean value for the whole kingdom. In these circumstances, it will be best to use the ^{90}Sr levels in Scottish milk as an index of dietary intake.

3. Intervals during which correlation is sought. The 0–14 day age group has produced roughly 100 bone specimens in each year since 1961; the smaller supply of material analysed in earlier years does not justify statistical treatment of the kind now being attempted. In view of the sometimes rapid changes in deposition of fall-out during a year, reflected in ^{90}Sr levels in milk, other foodstuffs and bone, it is desirable to compare the milk and bone levels at intervals of less than 12 months. A quarterly comparison would be reasonable, but, as will be seen by reference to Table II, a 6-monthly period is equally suitable in the present circumstances. For reasons not yet wholly clear, there has been little change in bone levels of ^{90}Sr between the first and second quarters of each year and between the third and fourth quarters — but sometimes substantial change between the second and third quarters.

TABLE II

Strontium-90 in human bone, 0–14 day age group, Glasgow region

	^{90}Sr in bone (pCi/g Ca)					
	1961		1962		1963	
Date of death	No. of samples	^{90}Sr mean	No. of samples	^{90}Sr mean	No. of samples	^{90}Sr mean
January–March	41	0·8	42	0·8	12	1·5
April–June	33	0·8	25	1·0	21	1·7
July–September	41	1·0	21	1·4	26	3·4
October–December	26	1·0	14	1·5	20	3·3
	1964		1965			
January–March	19	3·5	30	3·2		
April–June	21	3·5	34	3·2		
July–September	26	4·1				
October–December	19	4·0				

The two sets of data required for the comparison of ^{90}Sr in milk and in bone are summarized in Table III. Milk levels are taken from the regularly published reports of the Agricultural Research Council Radiobiological Laboratory.* Levels of ^{90}Sr in bone are taken from the Medical Research Council's Monitoring Report Series.

It is apparent that the levels of ^{90}Sr in newborn bone and in contemporary milk are not closely correlated. It is known that many foodstuffs derived from

* Levels of ^{90}Sr in Scottish milk were not published on a country-wide basis for 1960 and 1961 in a form allowing immediate recognition of values for each 6-monthly interval, as is possible for later years. This information is, however, readily inferred from the published mean levels for various 12-monthly periods.

milk or cereals do not reach the table for months, or even years, after the raw materials were harvested. The age distribution of the constituent items in a typical diet would be difficult to measure directly, but the following empirical relation is suggested between ^{90}Sr levels in milk and in newborn bone.

TABLE III

Strontium-90 in milk and bone

Year	^{90}Sr (pCi/g Ca) Milk*	bone† (0–14 day age group)
1960 January–June	9·8	
July–December	6·8	
1961 January–June	6·9	0·79
July–December	8·0	1·04
1962 January–June	10·4	0·85
July–December	23·2	1·41
1963 January–June	19·1	1·65
July–December	47·6	3·37
1964 January–June	38·8	3·42
July–December	34·6	4·00
1965 January–June	29·0	3·24
July–December	22·3	

* Average for Scotland.
† Average for Glasgow region.

TABLE IV

Strontium-90 in bone: comparison of predicted and observed levels (Glasgow region)

Year	^{90}Sr (pCi/g Ca) observed	calculated
1961 January–June	0·79	0·80
July–December	1·04	0·73
1962 January–June	0·85	0·85
July–December	1·41	1·47
1963 January–June	1·65	1·64
July–December	3·37	3·21
1964 January–June	3·42	3·27
July–December	4·00	4·06
1965 January–June	3·24	3·40

If M_1, M_2, M_3 ... are the levels of ^{90}Sr (pCi/g Ca) in milk during successive 6-month periods and B_1, B_2, B_3 ... are the corresponding figures for newborn bone during the same periods, then:

$$B_n = 0.02(2M_{n-2} + M_{n-1} + 2M_n)$$

Table IV gives a comparison between observed values of B_n and values predicted on this basis, during the years 1961–65.

REFERENCE

Fletcher, W., Loutit, J. F., and Papworth, D. G. (1966). *Br. med. J.* ii, 1225–1230.

Studies in Strontium Metabolism

2: Strontium–90 in Urine: Correlation with Diet and Age

JANET M. WARREN

Western Regional Hospital Board,
9–13 West Graham Street, Glasgow, C.4

SUMMARY

Levels of ^{90}Sr have been estimated in diet and in urine for two age groups of children. In children aged 3 weeks–3 months with an all milk diet, the ratio [90]Sr (urine)/[90]Sr (diet) had a mean value of $1·16\pm0·21$. For subjects between 2 and 13 years old on a mixed diet, the corresponding figure was $0·51\pm0·15$.

Urinary assay is the most direct method available for estimation of current contamination with ^{90}Sr in the living subject (Comar and Georgi, 1961; Schulert, 1961). More than 100 analyses of this kind have been made in Glasgow since 1963. In the work now to be described, measurements were made on two groups of subjects:

(*a*) Seven infants under the age of 3 months. All were bottle fed on National Dried Milk. Duplicate feeds were made for analysis and urine was collected over a 48-h period.

(*b*) Twelve children between 2 and 13 years old. All received a normal mixed diet with a calcium content of approximately 1 g/day. Urine was collected after an equilibration period of 1 week.

The methods of analysis were based on those described by Bryant *et al.* (1959), Spicer (1961) and by Parker *et al.* (1965). Mixed diet samples were ashed at 600° C. Urine samples required some preliminary treatment to concentrate the Sr and Ca into small bulk. This was most conveniently carried out by the addition of Ca, Sr and Ba carriers and co-precipitation as oxalates.

It was found that relatively large amounts of Ca (300 mg/litre of sample) and Ba (50 mg) added at the initial stage increased the final Sr yield considerably. After removal of the supernate from the co-precipitated strontium and calcium oxalates, the sample was treated in the usual way for the removal of

Ca and Ba and any associated radionuclides before the ^{90}Sr content was assayed by low-level counting. Ca was estimated in all samples by using a flame-photometric method (McIntyre, 1961).

Results are summarized in Tables I and II.

TABLE I

Strontium-90/calcium ratios in the diet and urine of children on a mixed diet

Subject	Age	Date of sampling	Diet (pCi ^{90}Sr/g Ca)	Urine (pCi ^{90}Sr/g Ca)	$\dfrac{^{90}\text{Sr (pCi/g Ca) in urine}}{^{90}\text{Sr (pCi/g Ca) in diet}}$
E.M.	12y	10/63	36·8	15·7	0·43
B.M.	11y	10/63	44·8	31·2	0·70
B.C.	9y	11/63	32·1	15·8	0·49
J.M.	13y	12/63	37·0	13·1	0·35
K.M.	6½y	1/64	38·1	21·4	0·56
C.M.	10y	3/64	40·0	15·0	0·38
L.C.	9½y	3/64	32·8	14·6	0·45
A.D.	10½y	4/64	33·3	13·7	0·41
S.C.	4¾y	4/64	44·0	22·0	0·50
P.G.	4½y	7/64	59·2	76·1	1·29*
B.Mc.	9y	8/64	29·7	25·6	0·86
M.G.	2½y	2/65	33·1	80·4	2·43*
		Mean	36·9	18·8	0·51 (s.d. 0·15)

* Excluded from averages. P.G. was acutely ill with leukaemia. M.G. was suffering from rickets, resistant to vitamin D.

TABLE II

Strontium-90/calcium ratios in the diet and urine of infants with an all-milk diet

Subject	Age	Date of sampling	Diet (pCi ^{90}Sr/g Ca)	Urine (pCi ^{90}Sr/g Ca)	$\dfrac{^{90}\text{Sr (pCi/g Ca) in urine}}{^{90}\text{Sr (pCi/g Ca) in diet}}$
S.M.	3w	2/64	34·5	32·9	0·95
R.P.	5w	2/64	56·5	57·8	1·02
A.G.	7w	3/64	40·6	42·3	1·04
J.G.	3m	1/65	36·0	51·2	1·42
D.M.	5w	1/65	34·6	44·0	1·27
S.T.	7w	3/65	36·1	53·2	1·47
W.A.	3m	3/65	39·3	37·0	0·94
		Mean	39·6	45·5	1·16 (s.d. 0·21)

REFERENCES

Bryant, F. J., Morgan, A., and Spicer, G. S. (1959). AERE–R 3030. H.M.S.O., London.

Comar, C. L., and Georgi, J. (1961). *Nature, Lond.*, **191,** 390–391.

McIntyre, I. (1961). *Adv. clin. Chem.*, **4,** 1–28.

Parker, A., Henderson, E. H., and Spicer, G. S. (1965). AERE–AM 101. H.M.S.O., London.

Schulert, A. R. (1961). *Nature, Lond.*, **189,** 933–934.

Spicer, G. S. (1961). AERE–AM 80. H.M.S.O., London.

Studies in Strontium Metabolism

3: Preferential Calcium–Strontium Renal Tubular Absorption Measured by Activation Analysis

GAVIN C. ARNEIL

Department of Child Health, Royal Hospital for Sick Children, Oakbank, Glasgow, C.4

SUMMARY

Samples of blood and urine were obtained from nine children. The Ca content of each was measured by a flame photometry and the Sr content by activation analysis. The ratio of Sr/Ca in urine and plasma was measured in each child. The ratio of Sr/Ca in the urine was higher than in plasma in eight of the nine children. This probably indicates selective preference of Ca to Sr in renal tubular absorption, possibly amounting to 3·4%, although quantitative measurement is difficult.

INTRODUCTION

The offspring of *homo sapiens* is exposed to hazard from radioactive strontium not only during intra-uterine life, but also in early extra-uterine life and childhood, characterized by rapid accumulation and turnover of Ca and associated cations. Fortuitously, a series of biological preferential absorptive and secretory mechanisms helps reduce the hazard created by his reckless forebears who wantonly produced and scattered ^{90}Sr on the terrestial surface.

Probably the first significant preference affecting the human infant is selective radicular (as opposed to foliar) absorption of Ca. Next comes the preferential absorption of Ca to Sr from the alimentary tract of the cow and preferential secretion of Ca from bovine plasma to milk. All these steps reduce the Sr (and ^{90}Sr) content of such milk whether ingested liquid, dried or evaporated by the infant or his omnivorous and lactating mother. When once the lactiverous infant or omnivorous older child has acquired contaminating ^{90}Sr in the Ca pool, any mechanism that might reduce the ratio of Sr (and ^{90}Sr) to Ca is of great interest.

In the child aged 5–10 years, it is likely that 5–10 g of Ca daily are filtered by the glomeruli and that 99% is re-absorbed. This is the ultrafilterable (ionic and chelated) Ca as opposed to the protein-bound Ca. It is clear that preferential re-absorption of calcium rather than Sr could materially affect the elimination of intracorporeal Sr and ^{90}Sr.

It was therefore decided to make estimations of the Sr and Ca content of blood, plasma and urine at the same time from the children.

EXPERIMENTAL METHODS

The method of analysis is essentially that described by Harrison and Raymond (1955). Samples were collected in plastic syringes and containers (with precautions against contamination) transferred to silicon tubes, weighed and air-dried before ashing at 550–600° C. The tubes were tightly sealed with polyethylene stoppers and subjected to a thermal neutron flux of 10^{12} neutrons/sq. cm/sec. for 3 h, in the Scottish Universities Research Reactor, East Kilbride.

The chemical separation consisted of two to three precipitations of strontium and barium nitrates, in the presence of manganese and copper carriers, followed by an iron scavenge, two to four carbonate precipitations and a further two nitrate precipitations.

The final precipitate was made up to 10 ml of solution and transferred to a polyethylene container for counting. After chemical separation ^{87m}Sr, with a half-life of 2·8 h, was determined by using a Laben multi-channel pulse-height analyser and compared with a standard of "Spec-pure" strontium carbonate activated under the same conditions as the sample. Chemical recovery of Sr was estimated by adding ^{85}Sr in known amount to the activated ash. Strontium can be assayed in 5 g of plasma at a sensitivity of 1×10^{-8} and a total content 0·1–0·4 μg.

Calcium was determined by one method described by McIntyre (1961) by using a Zeiss PMQ II spectrophotometer with a flame attachment. These estimations were carried out on nine children with an age range of from 4 to $11\frac{8}{12}$ years. They were in good health and free of renal or metabolic disease.

RESULTS

The results for Sr and Ca content of plasma and urine are shown in Table I.

TABLE I

Strontium and calcium content in urine and plasma of children

Child	Strontium (ng/g)		Calcium (mg/g)		Strontium $\times 10^3$/calcium	
	urine	plasma	urine	plasma	urine	plasma
D.T.	219	46·5	85·5	119·5	2·6	0·39
J.Q.	170	52·9	115·4	107	1·5	0·49
S.C.	190	35·7	313·0	101	0·61	0·35
E.U.	41·3	90·7	43·3	118	0·95	0·8
S.B.	86	52·7	80·0	108	1·1	0·5
C.McE.	55	27·1	62·0	112	0·9	0·24
A.G.	78	105	98·0	114	0·8	0·9
M.A.	126	29·9	54·0	112	2·3	0·27
L.M.	231	15·4	143·0	103	1·6	0·15

The range of Sr in plasma was 15·4–105 ng/g. In urine the range was from 41·3–231 ng/g. The comparable figures for Ca were 101–119·5 μg/g (plasma) and 43·4–313 μg/g (urine). When $\dfrac{Sr \times 10^3}{Ca}$ ratios were calculated the range for plasma varied from 0·15–0·9. The values for urine were much higher in eight of the nine children, the ratio varying from 0·6–2·6. These results strongly suggest preferential Ca absorption by the renal tubules, if a similar plasma distribution and glomerular filtration of Ca and Sr is assumed. Although the qualitative effect of filtration and re-absorption is obvious, it is difficult to make quantitative comparison. Normally less than 1% of the water and calcium filtered will appear in the urine; a difference of 1% in Sr absorption could double the Sr/Ca ratio in urine as compared with plasma.

If similar distribution of Ca and Sr in plasma and similar glomerular filtrations are assumed then:

(a) Filtered Sr $= \dfrac{\text{filtered Ca}}{\text{plasma Ca}} \times \text{plasma Sr}$

(b) % Sr re-absorption $= \dfrac{\text{filtered Sr} - \text{urinary Sr}}{\text{filtered Sr}} \times 100$

It is most improbable that Ca absorption in healthy children exceeded 99·5% or failed to reach 98% (25–200 mg per day with 5–10 g of filtrate). Table II has been calculated for 99·5%, 99% and 98% Ca absorption. It is considered that 99% is most likely to be near the correct value.

TABLE II

Percentage tubular strontium re-absorption calculated from assumed values for calcium re-absorption

Subject	Assumed calcium re-absorption (%)			Relative Sr/Ca clearance of plasma
	99·5	99	98	
D.T.	96·7	93·4	86·8	6·6
J.Q.	98·5	97·0	94·0	3·0
S.C.	99·1	98·2	96·4	1·8
E.U.	99·3	98·7	97·4	1·3
S.B.	99·0	97·9	95·8	2·1
C.McE.	98·2	96·3	92·6	3·7
A.G.	99·6	99·1	98·2	0·99
M.A.	95·7	91·3	82·6	8·7
L.M.	94·3	88·6	77·2	11·4

The values for relative Sr/Ca clearance from plasma are included, and compare with 3·5 quoted by Barnes et al. (1961).

DISCUSSION

The results indicate a renal tubular preference for Ca, as opposed to Sr, re-absorption in eight of the nine children with an estimated average of 3·4% and certainly not less than 1·7%. This small preference is perhaps in itself not of great value, but might be if pharmacological methods could be found to increase the preference. Ichikawa *et al.* (1964) showed that in newt nephron the re-absorption rate for ^{45}Ca in the proximal tubules exceeds that for ^{85}Sr. Acetazolamide and mersalyl depressed absorption of Ca but not of Sr.

It is hoped that as the technique becomes more sensitive it will be possible to test plasma from both venous and arterial sides of the placental circulation and also the maternal plasma. This will help to extend the findings of Rivera (1963).

REFERENCES

Barnes, D. W. H., Bishop, M., Harrison, G. E., and Sutton, A. (1961). *Int. J. Radiat. Biol.*, **3**, 637–646.
Harrison, G. E., and Raymond, W. H. A. (1955). *J. nucl. Energy*, **1**, 290–298.
Ichikawa, R., Enomoto, Y., and Sakai, F. (1964). *Science*, **144**, 53–54.
McIntyre, I. (1961). *Adv. clin. Chem.*, **4**, 1–28.
Rivera, J. (1963), *Nature, Lond.*, **200**, 269–270.

Strontium Metabolism in Man

V. A. KNIZHNIKOV AND A. N. MAREI
Institute of Biophysics, Ministry of Public Health of the U.S.S.R., Moscow

SUMMARY

This is a survey of recent work on Sr metabolism in the Soviet Union leading to the following conclusions. 1. The daily intake of ^{90}Sr with diet during the years 1963–64 was considerably greater than in the United States. 2. The level of ^{90}Sr in human bone at various age groups in the Soviet Union during these years was about the same as in the United States. It is suggested that the fallout strontium deposited in insoluble form on growing grain crops is less effectively absorbed in the human subject than the soluble Sr compounds present in milk and other dairy products. Bread and grain represent the major source of dietary ^{90}Sr in the Soviet Union, whereas milk and dairy products are the main source in the United States. 3. Addition of Ca to the diet of experimental human subjects reduced the retention and accumulation of ^{90}Sr in bone. 4. Levels of ^{90}Sr in human bone were lower in towns having drinking water with a relatively high fluorine content than in control towns with normal fluorine content.

INTRODUCTION

Earlier reports (Knizhnikov *et al.*, 1965; Marei *et al.*, 1964, 1965) have discussed the passage of ^{90}Sr into the human body with food and water as well as the accumulation of this nuclide in bone in various regions of the Soviet Union.

The present paper reviews the earlier work, along with some additional results that are significant for the study of the passage of ^{90}Sr along food change and its metabolism in the human body.

Levels of stable and radioactive Sr and Ca in the skeleton were determined by the analysis of individual bones as well as of extracted teeth. Teeth from not less than twenty persons were combined into single samples. The relationship of average skeletal levels of ^{90}Sr to the quantities measured in teeth or individual bones was determined in a preliminary survey (Marei *et al.*, 1965). The accuracy attained in measurements on teeth was ± 6–10% for ^{90}Sr, $\pm 12\%$ for stable Sr and $\pm 2\%$ for Ca.

DIETARY INTAKE OF STRONTIUM-90

Tables I and II show the average dietary intake of ^{90}Sr in the Soviet Union.

TABLE I

**Average dietary intake of strontium-90 in the
Soviet Union in 1963 and 1964**

| | ^{90}Sr, 1963 | | ^{90}Sr, 1964 | |
Population	(pCi/day)	(pCi/g Ca)	(pCi/day)	(pCi/g Ca)
Urban	38	67	58	94
Rural	48	62	72	88

TABLE II

**The content of strontium-90 in the diet of groups of population of
various ages in 1964 (average values in the country)**

	^{90}Sr (pCi/day)	Number of samples
Creches	35±23	237
Kindergartens	40±40	360
Boarding schools	57±46	229
Catering houses for workers	60±51	186

The average daily intake of ^{90}Sr in the U.S.S.R. from food was 38–48 pCi in 1963 and 58–72 pCi in 1964. Preliminary data indicate that the daily intake of ^{90}Sr was considerably lower in 1965. Contamination of the urban population diet does not vary much from one part of the Soviet Union to another. There are however more marked differences in contamination of the rural population diet in different republics. These variations correspond to known differences in dietary composition. In 1964 the average daily intake of ^{90}Sr in the Azerbaijan Soviet Socialist Republic was 50 pCi ^{90}Sr or 116 pCi ^{90}Sr/g Ca; in the Estonian Soviet Socialist Republic the corresponding figures were 114 pCi ^{90}Sr/day and 83 pCi ^{90}Sr/g Ca. The traditional diet of the Azerbaijan rural population includes a relatively small amount of milk and a correspondingly low Ca level.

A study of dietary intake in children (Table II) showed that in 1963 the diet in creches (children from 3 months to 3 years of age), in kindergarten (4–6 year old children) in towns contained about half as much ^{90}Sr as the adult diet. In boarding schools (children aged from 7 to 16 years) the dietary intake of ^{90}Sr approached that of adults. The role of individual components of the diet as sources of ^{90}Sr and of stable Sr is illustrated in Fig. 1. The distribution

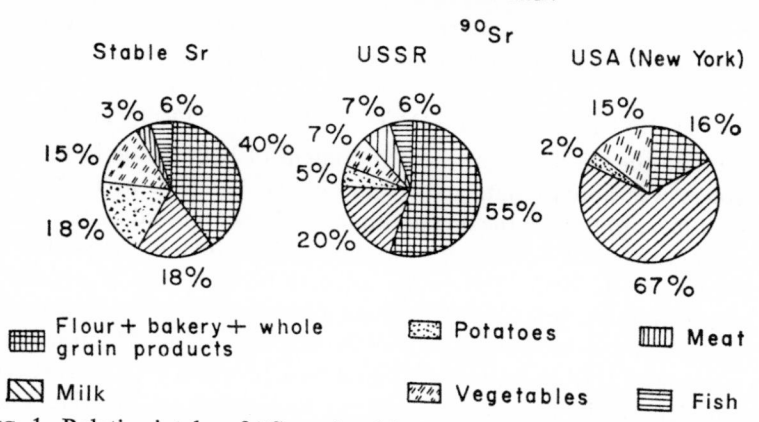

^{90}Sr

Stable Sr   USSR   USA (New York)

Flour + bakery + whole grain products

Milk

Potatoes

Vegetables

Meat

Fish

FIG. 1. Relative intake of ^{90}Sr and stable Sr from dietary constituents, 1963.

shown there is valid for urban and rural populations throughout the Soviet Union. About 75% of the dietary ^{90}Sr is contained in bread and bread products. Milk comes next in importance, and all other foodstuffs supply on average only about 10% of the dietary intake of ^{90}Sr. In communities relying on surface water supplies (as distinct from underground sources) the contribution of drinking water to total intake of ^{90}Sr was less than 5%. There is however some evidence that the intake of stable Sr in drinking water may be many times greater than in the rest of the diet (Knizhnikov and Novikova,

TABLE III

Comparative data on strontium-90 content in the bone tissue in various population age groups in 1963, 1964 and 1965 (pCi/g Ca)

(Mean values in the country)

Age group	1963		1964	
	No. of cases (samples)	^{90}Sr (pCi/g Ca)	No. of cases (samples)	^{90}Sr (pCi/g Ca)
Stillborn	—	—	68 (56)	3·46
0–12 months	6 (6)	5.00	33 (33)	5·93
1–4 years	7 (7)	4·23	9 (9)	6·96
5–19 years	1567 (44)	1·88	5425 (161)	2·23
20 years and over	4142 (347)	1·05	18694 (508)	1·38
1965				
Stillborn	105 (105)	3·48		
0–12 months	39 (39)	4·95		
1–4 years	17 (17)	5·64		
5–19 years	5969 (231)	3·49		
20 years and over	14391 (499)	1·42		

1964). Table III shows estimates of the ^{90}Sr content of bone in various age groups of the population during the years 1963, 1964 and 1965.

INFLUENCE OF DIET ON UPTAKE OF STRONTIUM-90

It is interesting to note that the ^{90}Sr content of bone tissue in the various age groups was approximately the same in the U.S.S.R. as in New York (HASL, 1965a, b) though the dietary intake of ^{90}Sr expressed as pCi ^{90}Sr/g Ca was much greater in 1963 than 1964 in the U.S.S.R. than in the United States (Knizhnikov et al., 1965). The composition of the diet may be significant in this regard. In the U.S.S.R., approximately 55% of the dietary intake of ^{90}Sr is provided by flour, grain, etc., and 20% by milk. In the United States, flour and grain provide 16% of the dietary ^{90}Sr, and milk provides 67%. It is suggested that ^{90}Sr deposited in insoluble form on the surface of growing grain is absorbed by the body less effectively than the soluble Sr compounds present in dairy products and in drinking water. Tables IV and V summarize estimates of ^{90}Sr/Ca ratios in the bone of stillborn children in Moscow and in the maternal diet.

TABLE IV

Observed strontium-90/calcium ratio in the bone tissue of stillborn children and in the maternal diet* (in Moscow)

	Daily dietary intake			pCi ^{90}Sr/g Ca in bone of stillborn children[†]	Observed ratio: stillborn bone / maternal diet
	Ca (g)	^{90}Sr (pCi)	^{90}Sr (pCi/g Ca)		
1964	0·740	57	77	4·18 (34)	0·054
1965	0·740	50	67	3·50 (99)	0·052
1964–1965	0·740	54	73	3·84 (133)	0·053

* Maternal diet is accepted as being similar to the average diet of the adult population in regard to Ca and Sr intake with drinking water.

† In parenthesis is the number of stillborn children investigated, including neonates who died up to the age of 2 weeks.

TABLE V

Observed ratio between stable strontium (mg) and calcium (g) in the bone tissue of stillborn children and in the maternal diet (in Moscow)

Maternal dietary intake Strontium (mg/day)	Sr (mg) / Ca (g)	Bone tissue of stillborn: Sr (mg) / Ca (g)	Observed ratio: skeleton of stillborn / maternal diet	Observed ratio: maternal skeleton / maternal diet
2·0	2·7	0·208 (20)	0·08	0·20

TABLE VI

Discrimination of strontium-90 in relation to calcium in children aged 0–12 months (in Moscow)

Year	^{90}Sr and Ca content in children aged 0–12 months			^{90}Sr and Ca content in bone tissue at the time of birth			Amount excreted from the time of birth (6 months) in the skeleton		Amount deposited in children aged 0–12 months		^{90}Sr/Ca ratio		General discrimination coefficient bone/diet
	^{90}Sr (pCi/g Ca)	Ca in the whole skeleton (g)	^{90}Sr in the whole skeleton (pCi)	^{90}Sr (pCi/g Ca)	Ca in the whole skeleton (g)	^{90}Sr in the whole skeleton (pCi)	Ca (g)	^{90}Sr (pCi)	Ca (g)	^{90}Sr (pCi)	in the added bone	in the diet of children aged 0–12 months	
1964	5·97 (32)	60	358	3·70	28	104	7	26	39	280	7·2	11·8	0·61
1965	4·9 (39)	60	297	3·84	28	108	7	27	39	216	5·5	9·7	0·56
										1964–65 average:	6·4	10·8	0·6

Notes:

1. It is assumed that children's diet consisted of 75% human milk (with content of ^{90}Sr/g Ca = 0·1 of maternal diet (Mitchell *et al.*, 1945) and 25% cow's milk.

2. Bone tissue ^{90}Sr content at the time of birth is assumed to be equal, in children who died in 1964, to the average of these values for the stillborn in 1963–64; in children who died in 1965, the value assumed is the average for the stillborn in 1964–65.

3. ^{90}Sr content in stillborn children in 1963 is taken to be equal to 0·056 of the ^{90}Sr/Ca ratio in the maternal diet.

4. Ca content was obtained from the literature as an average for the age range 0–12 months (Mitchell et. al., 1945).

Table VI shows calculations made to estimate the general discrimination coefficient for ^{90}Sr in relation to Ca in children aged 0 to 12 months. This coefficient is:

$$\frac{^{90}Sr/g \text{ Ca in bone added to the skeleton during the first year of life}}{^{90}Sr/g \text{ Ca in average diet during the same period}}$$

Figure 2 shows estimates of the general discrimination coefficient made for the age range 1 to 4 years. It was considered that 50% of Sr and Ca were exchanged during the first year of life and 40% during the second year (HASL, 1965a, b). The general discrimination coefficient of stable Sr was also determined for adults.

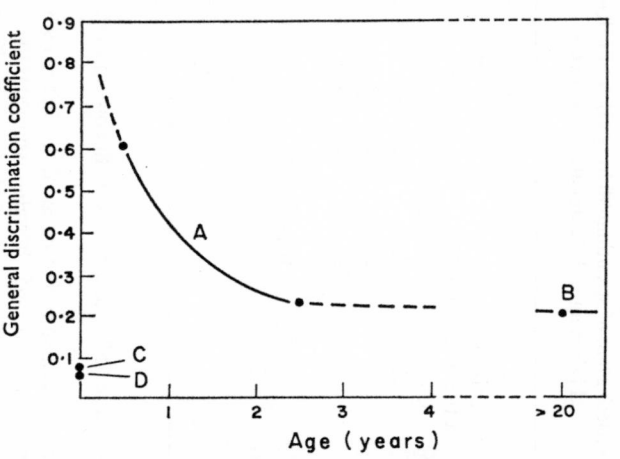

FIG. 2. Indices of relative Sr and Ca accumulation in man (general discrimination coefficient) at various ages. *A*, ^{90}Sr in children; *B*, stable Sr in adults; *C*, stable Sr in foetus; *D*, ^{90}Sr in foetus.

From Tables IV and V, as well as from Fig. 2, it will be seen that the observed ratio, foetal bone tissue to maternal diet, estimated in our investigations is much lower for ^{90}Sr (0·053) than the corresponding figure obtained in the United States (0·08–0·1). For stable Sr, the observed ratio was about 0·08, and not much different from the values reported in the literature for other countries. (Bryant and Loutit, 1964; HASL, 1965b). The lower observed ratio for ^{90}Sr confirms the suggestions already made that the absorption of ^{90}Sr from grain is less effective than the absorption from milk and from drinking water.

INFLUENCE OF CALCIUM INTAKE

There have been several reports in the literature suggesting that increased dietary intake of Ca reduces the absorption and retention of ^{90}Sr by animals (Knizhnikov and Bugryshev, 1963). Conflicting evidence has however been reported from investigations on human subjects (Carr *et al.*, 1962; Cohn *et al.*, 1962; HASL, 1965a, b; Samachson and Spencer–Laszlo, 1963).

An experiment to study this matter was made in a town in Central Asia. This town has two sources of drinking water, significantly different in Ca content. Part of the population uses surface water supplies with a low Ca content. Another part of the population obtains drinking water from artesian wells with a higher Ca content. In other respects, especially in regard to dietary intake, the two groups are very similar. Ca and ^{90}Sr were estimated in bone samples (sternum) obtained post mortem from adults who lived in the town for not less than 5 years. Results of analysis of drinking water are summarized in Table VII. Estimates of the dietary intake of Ca and ^{90}Sr are given in Table VIII. Levels of ^{90}Sr in bone tissue in the two groups of the population are

TABLE VII

Calcium, fluorine and strontium-90 content in sources of drinking water

Water source	Ca (mg/l)	F (mg/l)	^{90}Sr (pCi/l)
Surface water	35·5±4·0	0·17±0·002	3·66±0·7
Artesian wells	106·3±10·6	0·14±0·02	—

Notes:
1. The mean annual ^{90}Sr content in the well water did not exceed $4·0 \times 10^{-13}$ Ci/l.
2. Ca estimations are based on examinations of 240 samples; ^{20}F, and ^{90}Sr, 15 samples.

TABLE VIII

Dietary intake of calcium and strontium-90 in adults

Population groups	Ca intake (mg/day) with food	with water	Sr intake, ^{90}Sr (pCi/day) with food	with water	Total
Receiving water from open (surface) sources	753	78	17·5	1·0	18·5
Receiving water from underground sources	753	233	17·5	—	17·5

Note: ^{90}Sr data were obtained in 1962.

presented in Table IX. From these data it may be concluded that the level of ^{90}Sr in bone tissue was lower (on average by about 16%) in subjects who received additional Ca with the drinking water.

TABLE IX

Strontium-90 in samples of bone tissue obtained from adult population supplied with water from different sources

Water derived from	Daily Ca content of the diet (food + water) (mg)	^{90}Sr content of bone tissue number of cadavers examined	pCi/g ash	pCi/g Ca
Surface sources	831	51	0·37	1·06±0·036
Underground sources	986	108	0·31	0·97±0·017

The difference in ^{90}Sr content in bone for the two population groups was statistically significant. It should be emphasized that the dietary intake of Ca was adequate for physiological requirements in both groups.

Further study of the protective role of Ca in relation to bodily absorption and retention of ^{90}Sr was made in a group of twenty-three men, aged between 25 and 35 years, who were placed on a strict regime for 1 month and were given a controlled diet containing 660 mg Ca/day. From the 9th day to the end of the observation period, twelve members of the group received an additional 680 mg Ca/day in the form of lactate foodstuff. At the same time all members of the group were given 5·5 mg stable Sr, in milk, each day. Excreta were collected periodically and analysed for stable Sr, ^{90}Sr and Ca. These elements

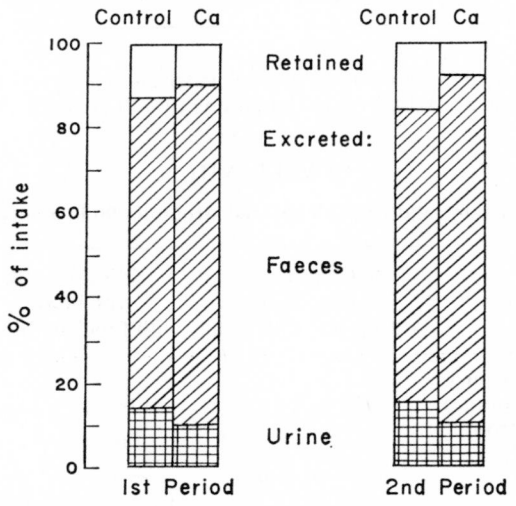

Fig. 3. Comparison of Sr assimilation and excretion by adult men with augmented Ca diet and control diet.

were also estimated in the daily diet (Fig. 3). During the state of equilibrium, before the enrichment of the diet with stable Sr, about 20% of the daily intake of Sr was excreted with the urine. Of this 20%, about 5% represented freshly ingested Sr and the remainder came from material stored in the body from earlier intake. These data correspond to findings reported in the literature (Carr *et al.*, 1962; Samachson and Spencer-Laszlo, 1963).

INFLUENCE OF FLUORINE ON STRONTIUM UPTAKE

Investigations have also been made (Knizhnikov, 1961) to study the suggestion that prolonged use of drinking water with comparatively high fluorine content diminishes the deposition of ^{90}Sr in the skeleton of experimental animals. Pairs of towns in various districts of the country were selected for this purpose. The localities in each pair were chosen so that, apart from fluorine content of drinking water, all factors capable of influencing Sr metabolism were similar (Knizhnikov and Novikova, 1964). Particular care was taken to match significant factors related to drinking water supplied including mineral content, hardness, Ca, Mg and stable Sr content. A total of 12,974 adult teeth and 2,117 children's teeth were examined for stable Sr. Bone tissue obtained post mortem from sixty-two adults and five children

TABLE X

Strontium-90 content of bone tissue of adult population of paired towns differing by fluorine concentration in drinking water (pCi ^{90}Sr/g Ca)

Geographical region	Town	Dietary intake of ^{90}Sr in 1964 (pCi/day)	^{90}Sr content of the skeleton 1963	1964	Statistical significance (P) 1963	1964
Southern Kazakhstan	Control	62	1·18±0·09	1·37±0·067		
	"Fluorine"	68	0·82±0·039	1·10±0·033	<0·01	<0·01
Central Kazakhstan	Control	71	1·21±0·12	1·27±0·067		
	"Fluorine" I	68	1·09±0·06	1·13±0·167	<0·5	<0·5
	"Fluorine" II	71	0·97±0·12	—	<0·05	—
Northern Kazakhstan	Control	82	1·24±0·09	1·47±0·03		
	"Fluorine"	85	1·0 ±0·09	1·33±0·05	<0·1	<0·05
Central Ukraine	Control	58	0·94±0·027	1·23±0·04		
	"Fluorine"	58	0·79±0·039	0·77±0·04	<0·05	<0·01

Note: Coefficient of "teeth/skeleton" normalization is accepted for the material of 1963 as 0·33, and for 1964 as 0·30 (Marei *et al.*, 1965).

TABLE XI

Accumulation multiple of stable strontium and its discrimination in respect to calcium in adult population as a function of fluorine content in drinking water

Geographical region	Town	Fluorine content of water (mg/l)	Dietary intake (mg/day) Sr	Ca	Sr in teeth (mg/g ash)	Sr in the whole skeleton	Accumulation multiple Ca	Sr	Observed ratio
Southern Kazakhstan	Control	0·1	2·26	637	0·228±0·014	643	1640	285	0·175
	"Fluorine"	0·8	2·34	629	0·150±0·001	420	1650	180	0·109
Central Kazakhstan	Control	0·5	3·1	719	0·203±0·032	572	1450	185	0·128
	"Fluorine" I	1·0	3·1	659	0·17 ±0·015	498	1580	161	0·102
	"Fluorine" II	1·5	3·1	629	0·155±0·015	434	1650	140	0·085
Northern Kazakhstan	Control	0·6	2·39	620	0·163±0·011	460	1680	193	0·115
	"Fluorine"	4·0	2·59	615	0·125±0·009	350	1700	135	0·081
Central Ukraine	Control	0·7	3·27	746	0·21±0·013	600	1400	183	0·132
	"Fluorine"	1·2	2·82	607	0·175±0·005	490	1710	174	0·102

Notes:

1. $\text{Accumulation multiple} = \dfrac{\text{Ca(Sr) in the skeleton}}{\text{Ca(Sr) in the daily diet (water + food)}}$.

2. $\text{Observed ratio} = \dfrac{\text{Sr/Ca in the skeleton}}{\text{Sr/Ca in the diet}}$ or $\dfrac{\text{Sr accumulation multiple}}{\text{Ca accumulation multiple}}$.

3. Dietary intake of Sr varied in different towns from 1·6 to 1·8 mg/day in food and 0·6 to 1·5 mg/day in water.

were analysed to determine the correlations between the stable Sr levels in the teeth and in other parts of the skeleton. [90]Sr was estimated in 22,704 adult teeth, combined in 408 samples and in 4,385 children's teeth combined in ninety-nine samples. Determination of stable Sr in food products and in pathological material was carried out by Mr A. A. Kuznetsov.

All of these teeth were collected in the years 1963 and 1964. Ca levels were also determined in these samples. The total Ca content of the skeleton at different ages was estimated from data given in the literature (Mitchell *et al.*, 1945).

The fluorine content of the main dietary constituents varied between one town and another, but not to any significant extent; the limits were 0·48 and 0·52 mg fluorine/day. In these circumstances a total daily intake of fluorine depended almost entirely on concentration of this element in drinking water. The results summarized in Table X indicate that levels of ^{90}Sr in bone were lower in the cities using water with a relatively high fluorine content than in the control towns; in most instances this difference was statistically significant. A similar trend was apparent from measurements of stable Sr content in bone and in diet (Table XI and Fig. 4), both for adults and for children.

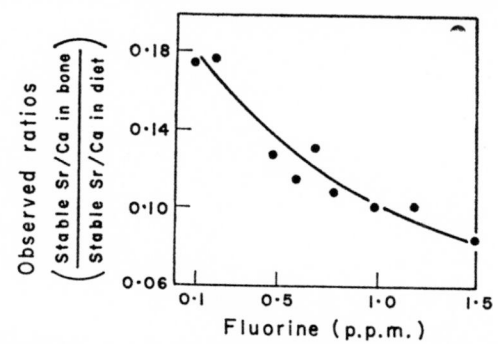

FIG. 4. Observed stable Sr–Ca ratios in bone tissue and diet of adult population in relation to fluorine content in drinking water.

There was, however, no significant difference in the Ca content of teeth obtained from people in the towns using drinking water with relatively high fluorine content and in the control towns.

REFERENCES

Bryant, F. J., and Loutit, J. F. (1964). *Proc. R. Soc.* **B159**, 449–465.
Carr, T. E., Harrison, G. E., Loutit, J. F., and Sutton, A. (1962). *Br. med. J.* ii, 773–775.
Cohn, S. H., Spencer, H., Samachson, J., Feldstein, A., and Gusmano, E. A. (1962). *Proc. Soc. exp. Biol. Med.* **110**, 526–528.
HASL–155 (1965a). U.S.A.E.C.

HASL–163 (1965b). U.S.A.E.C.

Knizhnikov, V. A. (1961). *Med. Radiol.* **2**, 58–62.

Knizhnikov, V. A., and Bugryshev, P. F. (1963). *Problemy Nutrit.* **6**, 56–62.

Knizhnikov, V. A., and Novikova, N. Ya. (1964). Hyg. Sanit. **8**, 93–95.

Knizhnikov, V. A., Stepanov, Yu. S., Petukhova, E. V., and Barkhudarov, R. M. (1965). *Atomizdat.*

Marei, A. N., Knizhnikov, V. A., and Yartsev, E. I. (1964). *Atomizdat.*

Marei, A. N., Knizhnikov, V. A., and Yartsev, E. I. (1965). *Atomizdat.*

Mitchell, H. H., Hamilton, T. S., Steggerda, F. R., and Bean, H. W. (1945). *J. biol. Chem.* **158**, 625–637.

Samachson, J., and Spencer-Laszlo, H. (1963). *Nature, Lond.* **200**, 593–594.

Discussion

QUESTION: What was the fluorine content of the water given to the experimental animals? Was there any evidence of renal damage?

ANSWER: The fluorine concentration in the water was 5–15 p.p.m. At the highest fluorine levels there were some changes in the teeth, but no renal damage. There were also some blood and other changes, which were not thought relevant to the main issue.

Chronic Accumulation of Strontium-90 in the Body as Predicted from the Retention of a Single Tracer Dose

JULIAN LINIECKI

Institute of Occupational Medicine, Lodz, Poland

SUMMARY

By using a power function for Sr retention $R_t = At^{-b} = 0.6t^{-0.24}$ with data on the absorption efficiency of this element from the gastrointestinal tract, ^{90}Sr concentration in bones of adults (as expected from contamination of the diet from 1954 to 1963) is calculated for New York City, Poland and Great Britain. Good agreement with direct measurements is apparent.

Errors due to extrapolation of the power function beyond the time-range of direct observation are discussed. It is concluded that uncertainty in the extrapolation is of little importance for the prediction of ^{90}Sr accumulation in the body under conditions of chronic intake.

INTRODUCTION

In a previous publication (Liniecki and Karniewicz, 1963) the accumulation of ^{90}Sr from fallout in adult humans was compared with predictions based upon different kinetic models. It was shown that some models (ICRP, 1960; Dolphin and Eve, 1963) predict accumulation of ^{90}Sr in the body at levels higher than those actually measured. At the same time it was found that estimates of expected ^{90}Sr activity based upon extrapolated power function of Sr retention were very close to the values observed in the period from 1954 to 1961. Good agreement was also given by the operational model of Kulp and Schulert (1962), who assumed replacement rate of calcium in the skeleton at 2–3% per year and $OR_{bone-diet} = 0.25$.

In this paper comparison between prediction (based on metabolic studies with isotopic tracers) and observed values of the ^{90}Sr/Ca ratio in human bone is continued for the period from 1954 to 1963.

METHOD OF CALCULATION OF EXPECTED BODY BURDEN

In calculation of expected accumulated activity of ^{90}Sr in the body, the following assumptions were made:

1. Calculation was limited to adults, for whom it may be assumed that the time dependence of retention of a single dose of Sr does not vary

D

substantially with age. This, together with the assumption of additivity of the retention of all doses of a tracer, permits one to calculate accumulation of ^{90}Sr in the body by integrating a function of the retention of a single tracer dose.

2. Difficulty in description of Sr retention is especially pronounced when long-term predictions have to be obtained by extrapolation of data from relatively short direct observations. Irrespective of lack of physiological interpretation of the so-called "power function of retention" of alkaline earth elements, it seems that there is no alternative as yet available describing Sr retention over an extended period.

The power function for Sr retention in adults, deduced from data reported in the literature (Cohn et al., 1962; Bishop et al., 1960; IRCP, 1960) is:

$$R_t = At^{-b} = 0.6t^{-0.24}$$

It may be noted that MacDonald et al. (1964) have reported values of the constant b for several adults (with normal Ca metabolism) varying from 0.27 to 0.29. At chronic or intermittent exposure to ^{90}Sr, the body burden at any time is predicted by step-wise calculation of activity accumulated in intervals for which the average rate of intake is known.

Retention up to the end of 1963 from intake in consecutive years from 1954 to 1963 has been calculated separately to yield values for total body content Q. Figures used for comparison with observed values, which represent approximate levels in the middle of calendar years, were obtained by linear interpolation.

Correction for extra-skeletal strontium-90 in the body

The total ^{90}Sr in the body is composed of that present in skeleton (Q_s) and that in soft tissues and body fluids. Retention when described by means of power function does not separate these components. The extra-skeletal amount of ^{90}Sr, under conditions of chronic intake, was calculated indirectly by assuming that ^{90}Sr/Ca ratios in soft tissues and in dietary supply were in equilibrium. For the extra-skeletal Ca, a round figure of 10 g was assumed; accuracy of this assumption is not of critical importance for calculation of Q_s.

Dietary intake of strontium-90

Average yearly values reported for ^{90}Sr/Ca ratios in the diet were multiplied by mean daily intake of Ca to give the daily intake of ^{90}Sr.

The data used were as follows:

(a) New York City: values published by Kulp and Schulert (1962) for the period 1954–59 and by Rivera (1961–64) for the years 1960–63. Daily consumption of Ca was taken as 1 g.

(b) *Great Britain*: values were taken from Bryant et al. (1958) and Agricultural Research Council Radiobiological Laboratory Reports (1959–64). Values for the years 1954–57 are based upon extrapolations made by Bryant and Loutit (1961). Daily intake of Ca was taken as 1·085 g.

(c) *Poland*: values reported by Czosnowska (1964a, b) were used. In the years 1961 and 1962 average intake of ^{90}Sr amounted to 12·8 and 18·3 pCi/day. For lack of better data in the years for which direct measurements were not available, average intake of ^{90}Sr was estimated by assuming a constant relationship between contamination of an average diet in Poland and in the United Kingdom, as it was found in 1961 and 1962.

Data on average daily intake of ^{90}Sr in New York City, Poland and in Great Britain, used for the present calculation, are presented in Table I.

TABLE I

Average yearly values of strontium-90/calcium ratio in the diet and daily intake of strontium-90 during the period 1953–1963

Year	New York		Poland		United Kingdom	
	pCi/g Ca	pCi/day	pCi/g Ca	pCi/day	pCi/g Ca	pCi/day
1953	0·8	0·8	..	..	..	..
1954	1·2	1·2	..	3·8*	2·0	2·1
1955	2·0	2·0	..	7·7*	4·0	4·3
1956	3·4	3·4	..	9·8*	5·0	5·4
1957	4·9	4·9	..	10·6*	5·5	5·9
1958	5·8	5·8	..	11·8*	6·1	6·5
1959	12·9	12·9	..	18·1*	9·3	10·0
1960	12·0	12·0	..	12·5*	6·4	6·9
1961	9·7	9·7	17·4	12·8	6·2	6·7
1962	12·8	12·8	25·7	18·3	9·9	10·7
1963	30·0	30·0	..	44·5*	22·8	24·6

* Values obtained by extrapolation.

Fraction of strontium-90 absorbed from gastrointestinal tract into the blood (f_1)

Calculations reported here are based on $f_1 = 0·2$, the figure representing a consensus of values reported in the literature (Harrison, 1963: Spencer et al., 1956, 1958; Samachson and Spencer, 1962; Samachson, 1963; Spencer-Laszlo et al., 1964; Irving, 1957). Alternative calculations have also been made by using the value $f_1 = 0·3$ recommended by the International Commission on Radiological Protection (1960).

Skeletal calcium

The average amount of Ca in the skeleton (q) is taken as 1030 g.

Derivation of average skeletal strontium-90/calcium ratio from data on single bone

It is well known that ^{90}Sr distribution among different parts of the skeleton in adults, whose exposure to this isotope started after cessation of growth, is not uniform. It was necessary therefore to obtain empirical coefficients, relating average skeletal concentration to that in different bones, if values found in the latter have to be used for comparison with calculated values of Qs/q.

In the years 1958–59 Kulp *et al.* (1960) had determined ratios between the concentration of ^{90}Sr in vertebrae, ribs and femora and that in the whole skeleton. The ratios found were 2·1, 1·0 and 0·45, respectively. Bryant and Loutit (1961) confirmed these observations.

Rivera (1964) gave similar values for persons deceased in 1960; ratios for long bones (diaphyses) skeleton and vertebrae skeleton were 1·83 and 0·7, respectively. Marei *et al.* (1964) obtained corresponding values of 1·72 and 0·41, and for femoral epiphyses a ratio of 0·93.

It is difficult to judge whether the apparent changes over the last 5 years are significant.

RESULTS AND DISCUSSION

Results of comparison between expected and observed values of $^{90}Sr/Ca$ ratios in the human skeleton have been collected in Table II and are summarized in Fig. 1. Agreement for New York City and Poland is satisfactory, especially when values of Qs/q are calculated under assumption of $f_1 = 0·2$.

For the United Kingdom the agreement is less satisfactory. The following considerations may be relevant:

(i) The number of adult bones analysed was small. For the years 1958 and 1959, when the number of specimens was somewhat larger, agreement between the measurements and the power-function model is better.

(ii) Non-uniformity of sampling (as regards the parts of bones taken for analysis) may be inferred from the original reports (Medical Research Council, 1959–63). Uniformity of sampling is especially important with respect to long bones in which the $^{90}Sr/Ca$ ratio may vary by a factor of two between epiphysis and diaphysis (Marei *et al.*, 1964; Rivera, 1964). That ^{90}Sr accumulation in humans in Great Britain is similar to that seen in other countries is indicated by the fact that Qs/q, calculated from data on vertebrae (available in large number in 1961 only), fits predictions from the power-function model with $f_1 = 0·2$ (Fig. 1).

TABLE II

Average strontium-90/calcium ratio (pCi/gCa) in adult human skeleton, predicted from power-function model ($f_1 = 0.2$ and 0.3), compared with observed values

(Number of samples in parentheses)

Year	New York City Predicted $f_1 = 0.2$	$f_1 = 0.3$	Observed measured	normalized	Poland Predicted $f_1 = 0.2$	$f_1 = 0.3$	Observed measured	normalized	United Kingdom Predicted $f_1 = 0.2$	$f_1 = 0.3$	Observed measured	normalized
1953	<0.01	0.01	0.01 (5)	0.01*	..	..	..	..	..	..	..	..
1954	0.01	0.02	0.02 (40)	0.02*	0.02	0.03	..	..	<0.01	0.01	..	..
1955	0.03	0.04	0.04 (86)	0.04*	0.08	0.12	..	..	0.04	0.06	0.04 (7)	0.09‡
1956	0.05	0.08	0.06 (174)	0.06*	0.16	0.25	..	..	0.09	0.13	0.08 (3)	0.18‡
1957	0.09	0.14	0.10 (97)	0.10*	0.25	0.38	..	..	0.14	0.21	0.11 (4)	0.24‡
1958	0.14	0.20	0.15 (74)	0.15*	0.33	0.50	..	..	0.18	0.27	0.09 (10)	0.20‡
1959	0.22	0.33	0.19 (70)	0.19*	0.45	0.69	0.80 (58)	0.38†	0.25	0.38	0.12 (20)	0.27‡
1960	0.32	0.50	..	..	0.56	0.86	1.00 (49)	0.48†	0.33	0.50	0.23 (13)	0.51‡
1961	0.40	0.61	0.83 (7)	0.40†	0.63	0.96	1.16 (161)	0.55†	0.35	0.53	0.25 (15)	0.55‡
1962	0.47	0.72	1.00 (14)	0.48†	0.72	1.1	1.48 (160)	0.70†	0.40	0.60	0.68 (48) / 0.35 (7)	0.33† / 0.78†
1963	0.64	1.0	1.55 (23)	0.74†	0.98	1.5	1.65 (76)	0.79†	0.54	0.82	0.48 (9)	1.0‡

Normalized from: * whole skeletons; † vertebrae; ‡ femora.

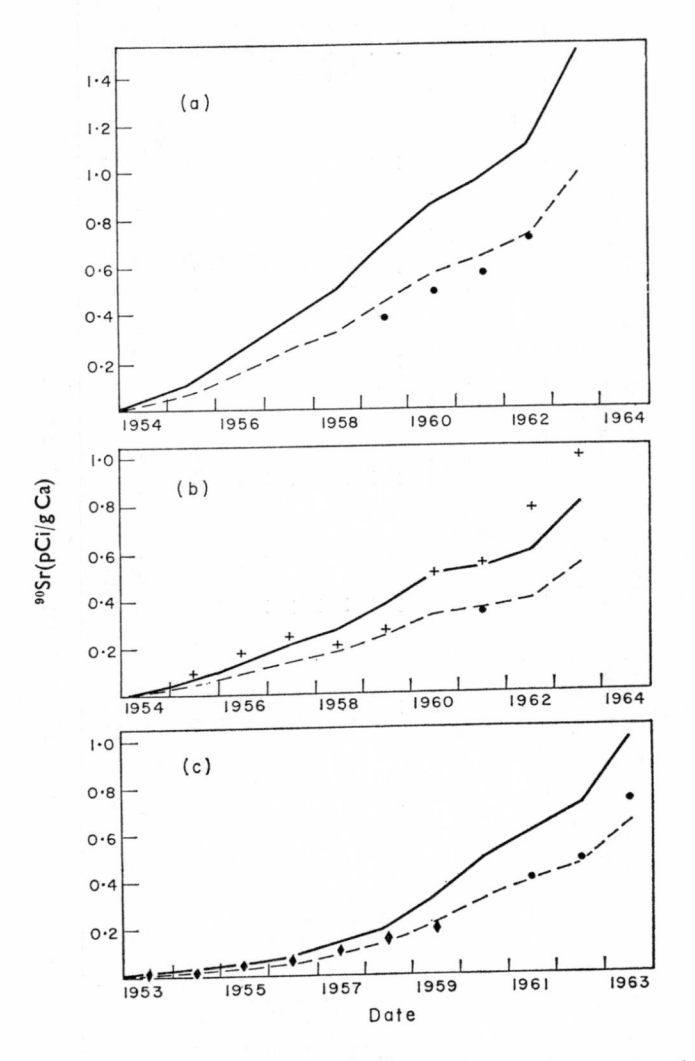

FIG. 1. Comparison of predicted power-function model and observed values of skeletal average $^{90}Sr/Ca$ ratio in adults in the period 1953–63: continuous lines, $f_1 = 0.3$; broken lines; $f_1 = 0.2$. Observed values normalized from: ●, vertebrae; +, femora; ◆, whole skeletons. (a) Poland; (b) United Kingdom; (c) New York City.

It is encouraging to see that, in spite of the approximate nature of the assumptions made, metabolic data derived from human experiments of relatively short duration may be used with reasonable confidence for predictions of ^{90}Sr accumulation in the body over a period as long as 10 years. This supports the basic assumption that Q may be calculated on the basis of the additivity of retentions from separate doses of ^{90}Sr.

Extrapolation errors

Unlimited extrapolation of the power function is incompatible with the metabolic steady state that obtains in adults with respect to stable Sr (Thurber *et al.*, 1958) and radium (Walton *et al.*, 1959). Marshall (1964) postulates that, at a certain time, t_y, after a single dose of alkaline earth tracer, metabolic equilibrium will be reached. From this time on, retention should become monoexponential, because behaviour of the tracer is dominated by a monoexponential decrease of activity in the compartment with the slowest turnover. For Sr a value of t_y in man was estimated by Marshall to be of the order of 3,000 days.

To investigate whether the change of retention pattern could be detected in environmental studies, accumulation of ^{90}Sr in the body under conditions of constant feeding was calculated, assuming values of t_y of 700, 1000, 2000 and 3000 days.

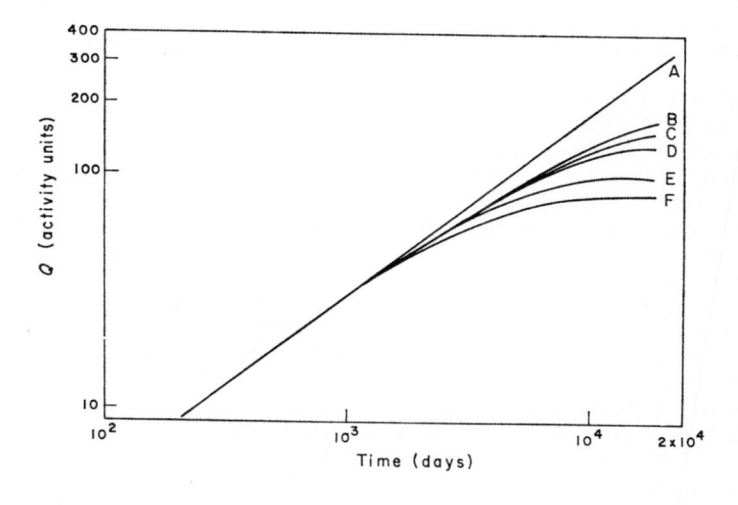

FIG. 2. Chronic accumulation of ^{90}Sr in body of an adult under conditions of constant intake of 1 unit of activity/day. Prediction based on the power-function model with t_y values of: curve *B*, ∞; curve *C*, 3,000; curve *D*, 2,000; curve *E*, 1,000; curve *F*, 700. For comparison, curve *A* shows the predicted accumulation of stable Sr.

This was done by integrating against time a retention function, composed of a power function up to time t_y and changing over to a monoexponential function at t_y. The calculation was performed by accepting Marshall's assumption that the rate constant λ_n of the terminal monoexponential phase of retention is related to constant b of the power function by the relationship:

$$\lambda_n = \frac{b}{t_y}$$

Results of calculation for values of T up to 18,250 days (40 years) are presented in Fig. 2.

It is obvious that possible errors due to extrapolation of the power function cannot be detected in environmental studies over a 10-year period.

REFERENCES

Agricultural Research Council Radiobiological Laboratory. (1959–64). Annual reports.

Bishop, M., Harrison, G. E., and Raymond, W. H. A. (1960). *Int. J. Radiat. Biol.* **2**, 125–141.

Bryant, F. J., and Loutit, J. F. (1961). AERE–R 3718. H.M.S.O., London. 1961.

Bryant, F. J., Chamberlain, A. C., Spicer, G. S., and Webb, M. S. W. (1958). *Br. med. J.* i, 1371–1375.

Cohn, S. H., Spencer, H., Samachson, J., and Robertson, J. S. (1962). *Radiat. Res.* **17**, 173–185.

Czosnowska, W. (1964a). *Nukleonika*, **9**, 471–481.

Czosnowska, W. (1964b). *Nukleonika*, **9**, 745–754.

Dolphin, G. W., and Eve, I. S. (1963). *Physics. Med. Biol.* **8**, 193–203.

Harrison, G. E. (1963). *In* "Diagnosis and Treatment of Radioactive Poisoning", pp. 119–129. I.A.E.A., Vienna.

International Commission on Radiological Protection (1960). Report of ICRP Committee II.

Irving, J. T. (1957). "Calcium Metabolism", Wiley, New York.

Kulp, J. L., and Schulert, A. R. (1962). *Science*, **136**, 619–632.

Kulp, J. L., Schulert, A. R., and Hodges, E. J. (1960). *Science*, **132**, 448–454.

Liniecki, J., and Karniewicz, W. (1963). *Nukleonika*, **8**, 401–410.

MacDonald, N. S., Figueroa, W. G., and Urist, M. R. (1964). U.S.A.E.C., U.C.L.A. Sch. Med. Lab. Nucl. Med. Report 12–538, 1–25.

Marei, A. N., Knizhnikov, V. A., and Yartsev, E. I. (1964). *Atomizdat*.

Marshall, J. H. (1964). *J. theoret. Biol.* **6**, 386–412.

Medical Research Council, (1959–63). Monitoring Report Series No. 1–No. 9, H.M.S.O., London.

Rivera, J. (1961–64). U.S.A.E.C. Health and Safety Laboratory, New York, reports.

Rivera, J. (1964). U.S.A.E.C. Health and Safety Laboratory, New York, report HASL–144, 278–280.

Samachson, J. (1963). *J. appl. Physiol.* **18**, 824–828.

Samachson, J., and Spencer, H. (1962). *Nature, Lond.* **195**, 1113–1115.

Spencer, H., Brothers, M., and Berger, E. (1956). *Proc. Soc. exp. Biol. Med.* **91,** 155–157.

Spencer, H., Samachson, J., Kabakow, B., and Laszlo D. (1958). *Clin. Sci.* **17,** 291–301.

Spencer-Laszlo, H., Samachson, J., Hardy, E. P., and Rivera, J. (1964). *Rad. Res.* **22,** 668–676.

Thurber, D. L., Kulp, J. L., and Hodges E. (1958). *Science,* **128,** 256–257.

Walton, A., Kologrivov, R., and Kulp, J. L. (1959). *Hlth Phys.* **1,** 409–416.

Effect of Age on Radioactive and Stable Strontium Accumulation in Mule Deer Bone

G. C. FARRIS, F. W. WHICKER AND A. H. DAHL

Department of Radiology and Radiation Biology,
Colorado State University,
Fort Collins, Colorado

SUMMARY

Metacarpal and femur bones of Colorado mule deer (*Odocoileus hemionus hemionus*) of various age classes were collected in order to study radioactive and stable Sr distribution within a deer herd. Average ^{90}Sr and stable Sr concentrations in bone of six age classes were significantly different. The highest levels of radioactive Sr were in the younger age classes and stable Sr concentrations were larger in the older age groups. Considerable variation in the radioactive and stable Sr content of metacarpals was observed within age classes. No significant difference in radioactive or stable Sr content was observed between the sexes of deer younger than 17 months. Sex was not compared in older age classes because of limited sample size. Estimates of Sr turnover rates indicated values of about 19% per year for juvenile and 6% per year for adult deer.

INTRODUCTION

Many investigations have been performed on Sr deposition in food and bones of humans and domestic animals. However, there are only a few published references concerning the accumulation of stable and radioactive Sr in bones of native animals. There are even fewer studies reporting Sr distribution by age groups within a population of animals resulting from continuous ingestion.

Intelligent appraisal of the potential effects of ^{90}Sr on populations requires knowledge on differential concentration mechanisms between age groups, particularly when one considers possible effects of age on radiation sensitivity. In addition, this type of information can be useful to those interested in basic bone metabolism as affected by age.

This report presents concentrations of ^{90}Sr and stable Sr in bones from individuals of different age groups and sexes within a wild population of mule deer (*Odocoileus hemionus hemionus*). Estimates of Sr turnover rates in juvenile and adult deer are also given.

MATERIALS

The area from which sampling was carried out is known as the Cache la Poudre drainage and is located in north-central Colorado on the eastern slope of the Rocky Mountains. The study area comprises about 520 sq miles and ranges between 5,200 and 13,000 ft in elevation.

The samples were divided into two classifications, A and B. The A series represented bones that were collected weekly from April, 1961, to April, 1965, from a uniform distribution of locations within the study area by personnel of the Colorado Department of Game, Fish and Parks. The B series of bones (amounting to 63 individuals) were obtained from hunters during the Colorado big-game season in October, 1962. Deer ages were estimated by the tooth replacement and wear method (Robinette *et al.*, 1957).

ANALYTICAL METHODS

Strontium-90 analysis

The procedures used for radiochemical analysis of bone for ^{90}Sr were those described by Farris (1965). In brief, the analyses were carried out on 5-g samples of bone ash by collecting the Sr with a phosphate precipitation, separation from Ca with repeated fuming nitric acid precipitations, and removal of radiochemical impurities by barium chromate and ferric hydroxide scavenges.

The purified Sr solution was allowed to stand until secular equilibrium of ^{90}Y was established. Yttrium was separated from the ^{90}Sr solution as the hydroxide and converted to the oxalate, which was mounted on a plastics disc for counting.

Counting was done with a Sharp low background gas-flow β-counter, with counting arrangements such that the "counting error" of the ^{90}Y counts was not greater than 5% (U.S.A.E.C., 1961).

Stable strontium

Stable Sr determinations followed the atomic-absorption procedure described by Trent and Slavin (1964). For the stable Sr analysis of 0·5-g ash sample was accurately weighed into a beaker and dissolved with 2 ml of conc. hydrochloric acid. The solution was evaporated to dryness and then dissolved in 5 ml of M hydrochloric acid. Insoluble fractions were removed by filtration. The clear solution was transferred to a 50-ml calibrated flask and made up to volume with water. A 3-ml portion of this solution and 2·5 ml of lanthanum solution (10%) were added to a 25-ml calibrated flask, which was filled to the mark with water. The absorption of the solution was then measured with the model 303 Perkin–Elmer Atomic Absorption Spectrophotometer with standard conditions for strontium.

Calcium

Ca analysis were carried out by titrating a 1-g sample of bone ash with the chelating agent disodium dihydrogen ethylenediaminetetra-acetate (EDTA) with calcein as an indicator.

RESULTS AND DISCUSSION

In order to gain an estimate of the individual variation and the effect of age, the B series of samples were divided into six age groups. The following age groups were arbitrarily defined and specimens obtained from at least four individuals for each group: fawns (4–5 months), yearlings (16–17 months), 2 year olds (28–29 months), 3–4 years (40–53 months), 5–6 years (64–77 months) and over 7 years (88–113 months).

Data are reported in terms of pCi ^{90}Sr/g of ash, μg stable Sr/g of ash and pCi ^{90}Sr/mg of stable Sr. Since the Ca content of the bones of all ages was found to be quite uniform (mean Ca 37%/g of ash), analysis of the data in terms of Sr/g Ca would not alter the conclusion.

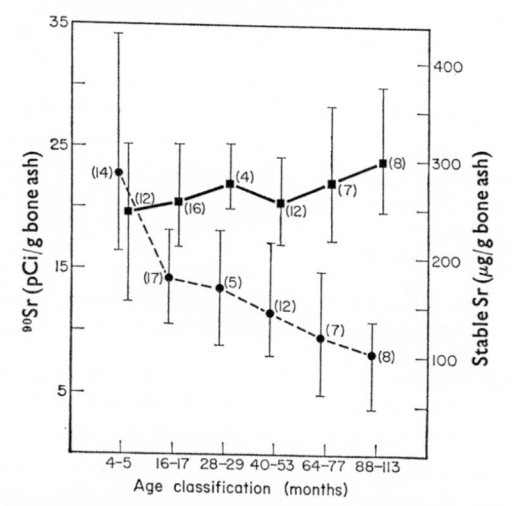

FIG. 1. Variation of radioactive and stable Sr levels of mule deer metacarpals of various age classes taken from the Cache la Poudre drainage during the Colorado big-game season, October, 1962. Points represent group mean; vertical lines represent the range of values obtained; and the numbers in parentheses are the number of individuals. Continuous line, stable Sr; broken line, ^{90}Sr.

Strontium-90

The range and mean values obtained for the various age groups of the B series are presented in Fig. 1. The largest and smallest ranges of values among animals within an age class were found in the 4–5 and 88–113 months age classes, respectively. The variances of the different age classes were tested for homogeneity by using Bartlett's test, and found to be acceptably homogeneous (P>0·05).

The analysis of variance presented in Table I indicated a significant difference among the means of the various age classes with the higher means occurring in the younger age classes. The result that radiostrontium levels were related to age was expected because the bones in the younger animals were in a more active state of net bone formation and the fallout rate was at a higher level than 3 or 4 years earlier, when the adult animals were in the same stage of development.

TABLE I

Analysis of variance of strontium-90 levels of mule deer metacarpals of various age classes (both sexes) taken from the Cache la Poudre drainage during the Colorado big-game season, October, 1962

Means (in terms of pCi ^{90}Sr/g of ash)

	Age class (months)					
	4–5	16–17	28–29	40–53	64–77	88–113
Sample size	14	17	5	12	7	8
Mean	22·7±1·6*	14·1±0·6	13·4±1·8	11·5±0·8	9·5±1·5	8·2±0·8

Analysis of variance:

Source	DF	SS	MS
Ages	5	1,535	307†
Individual/Ages	57	791	13·9

* Mean±1 standard error of the mean.
† Statistically significant at the 1 % level.

Owing to limited sample size in older age groups only deer up to 17 months of age were examined for differences in ^{90}Sr as related to sex. The analysis of variance is presented in Table II and indicated the means of the fawn and yearling age groups to be significantly different. However, sex was found not to be an influencing factor in strontium deposition in the fawn and yearling age classes. Absence of differences between sexes for ^{90}Sr in deer mandibles has been reported by Schultz and Flyger (1965). The nonsignificant age class–sex interaction in our data implies that differences between these age classes is independent of sex.

The ⁹⁰Sr concentrations reported in deer bones in this study were higher than those found in Maryland white-tailed deer (Schulz and Flyger, 1965) and Columbian black-tailed deer from California (Longhurst, 1964) during

TABLE II

Analysis of variance of strontium-90 concentrations in metacarpals of both sexes of 4–5 and 16–17 month old deer

Means (in terms of pCi ⁹⁰Sr/g of ash)

		Age class (months)	
		4–5	16–17
Female			
	Sample size	5	8
	Mean	24·5±3·1*	14·0±1·0
Male			
	Sample size	5	8
	Mean	20·3±1·2	13·9±0·8

Analysis of variance:

Source	DF	SS	MS
Age	1	439·3	439·3†
Sex	1	17·5	17·5
Age × Sex	1	26·6	26·6
Individuals	22	305·6	13·9

* Mean±1 standard error of the mean.
† Statistically significant at the 1 % level.

the same time periods. Our values were considerably higher than the levels reported in sheep by Häggröth and Hoglund (1961) and Bryant *et al.* (1959). The higher levels in Colorado deer may be attributed in part to higher levels of fallout due to elevation and geological location and the type of forage plants which the Colorado deer consume. Metabolic differences between species could also conceivably be involved to some extent.

⁹⁰Sr concentrations in bones of the 4–5 month old deer during 1962 were found to be a factor of 15 higher than reported for children in the United States in 1962 (Rivera and Harley, 1965). Higher levels of Sr activity are found in deer because there are more physical and biological barriers that remove Sr from the diet of man thus resulting in less Sr deposition in the human skeleton.

Stable strontium

Stable Sr concentrations in bone from the deer collected ranged from 158 to 375 μg Sr/g ash, but showed no correlation with the concentration of radiostrontium. It seems likely that this was observed because the level of ^{90}Sr was changing in the environment while stable Sr would have been relatively uniform through time.

Mean concentrations of stable Sr and ranges of the six age groups are presented in Fig. 1. The analysis of variance, Table III, indicated statistically significant differences among the means of the various age classes with the highest levels of stable Sr concentrations being found in the older age groups. This might be the pattern for ^{90}Sr if the dietary level were relatively constant over a period of many years. Considerable variation in stable Sr levels was observed within age classes with 80% of the total sum of squares in Table III due to individual variability.

TABLE III

Analysis of variance of stable strontium levels of mule deer metacarpals of various age classes (both sexes) taken from the Cache la Poudre drainage during the Colorado big-game season, October, 1962

Means (in terms of μg stable Sr/g of ash)

	Age class (months)					
	4–5	16–17	28–29	40–53	64–77	88–113
Sample size	12	16	4	12	7	8
Mean	244±14*	263±8	276±19	266±8	283±17	300±15

Analysis of variance:

Source	DF	SS	MS
Ages	5	17,529	3,506†
Individual/Ages	53	76,975	1,452

* Mean±1 standard error of the mean.
† Statistically significant at the 5% level.

Early workers in the stable Sr field reported the strontium content of bone to be relatively constant in almost all individuals (Hodges *et al.*, 1950; Turekian and Kulp, 1956). However, in our data, with the exception of the 40–53 month age group, the stable Sr levels appeared to build up at a decreasing rate with age. This seems reasonable because data presented by Frost (1964) for humans suggest that the net difference between bone formation and bone resorption decreases with age. Observations similar to ours have been reported by Bohman *et al.* (1962) in cattle and Sowden and Stitch (1957), Bryant and Loutit (1964), and Rivera and Harley (1965) in humans.

As with the ^{90}Sr data, owing to sample size, only deer up to 17 months of age were examined for differences in stable Sr levels as related to sex. The analysis of variance is presented in Table IV. The means of the fawn and yearling age groups were not significantly different and no statistically significant difference was observed between males and females of these younger deer.

TABLE IV

Analysis of variance of stable strontium concentrations in metacarpals of both sexes of 4–5 and 16–17 month old deer

Means (in terms of μg stable Sr/g of ash)

	Age class (months)	
	4–5	16–17
Female		
Sample size	5	8
Mean	229±20*	257±10
Male		
Sample size	5	8
Mean	241±21	269±13

Analysis of variance:

Source	DF	SS	MS
Age	1	4,972	4,972
Sex	1	924	924
Age × Sex	1	1	1
Individuals	22	31,375	1,426

* Mean±1 standard error of the mean.

Specific activity

Mean values and ranges of ^{90}Sr specific activity (pCi ^{90}Sr/mg stable Sr) of the six age groups are presented in Fig. 2 with the ^{90}Sr specific activity following the same trend as the ^{90}Sr data plotted in Fig. 1.

The analysis of variance presented in Table V indicated a significant difference among the means of the various age classes with approximately 50% of the age sum of squares due to differences between the fawn and yearling age classes and the older age groups. The higher levels of ^{90}Sr specific activity in the younger age classes reflects the lower stable Sr concentration and the changing rate of fallout at a time of active bone formation. The individual comparisons reported in Table V support the idea that deer older than 28–29 months undergo very little change in the ^{90}Sr specific activity in bone.

It was interesting that the largest and smallest variances among animals within age classes were found in the fawn and yearling age classes, respectively. The individual variance of the fawns was 13·7 times higher than that of

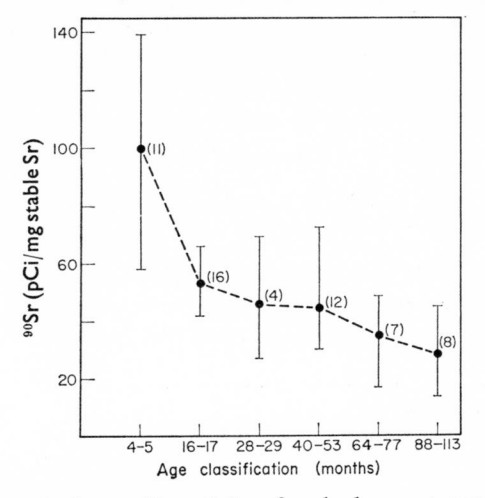

FIG. 2. Variation of ^{90}Sr specific activity of mule deer metacarpals of various age classes taken from the Cache la Poudre drainage during the Colorado big-game season, October, 1962. Points represent group mean; vertical lines represent the range of values obtained; and the numbers in parentheses are the number of individuals.

TABLE V

Analysis of variance of strontium-90 specific activity of mule deer metacarpals of various age classes (both sexes) taken from the Cache la Poudre drainage during the Colorado big game season, October, 1962

Means (in terms of pCi ^{90}Sr/mg of stable Sr)

	Age class (months)					
	4–5	16–17	28–29	40–53	64–77	88–113
Sample size	11	16	4	12	7	8
Mean	98·7±7·0*	53·1±1·8	45·9±8·7	44·4±3·2	34·4±5·2	28·1±3·6

Analysis of variance:

Source	DF	SS	MS
Age	5	31,446	6,289†
Individual/Age	52	11,630	224

Individual comparisons:

Comparison	DF	SS
4–5, 16–17 *vs* Rest	1	16,297†
4–5 *vs* 16–17	1	13,592†
16–17 *vs* 28–29	1	102
16–17 *vs* 40–53, 64–77, 88–113	1	2,322†
64–77 *vs* 88–113	1	154
16–17 *vs* 40–53	1	491
40–53 *vs* 64–77	1	427

Mean ± 1 standard error of the mean.   † Statistically significant at the 1 % level.

the yearlings. One could speculate that the increased variability in fawns could arise from the wide range of forage species available to the doe and fawn, from differential placental transfer of Sr and from differential weaning periods. The yearling deer may show less variability because by migrating and roaming over a wider range of forage plant communities their integrated Sr intake should be more uniform. This observation of increased ^{90}Sr/stable Sr variance in fawns seems to be related more closely to variations in ^{90}Sr than in stable Sr. Data on ^{90}Sr and Ca concentrations in forage indicate large variations between species and locations. Data on variations in stable Sr in forage are incomplete but, nevertheless, one would expect a more homogeneous distribution of the stable isotope.

Strontium turnover rates

From ^{90}Sr specific activity measurements of the diet and femur bone, during the periods June, 1962–May, 1963 and June, 1963–May, 1964, estimates of Sr turnover rates in juvenile and adult deer were made by the following equation (Rivera, 1964):

$$f = \frac{W_n - W_{n-1}}{Z_n - W_{n-1}} \tag{1}$$

where f is the annual turnover rate of bone; W_n is the ^{90}Sr specific activity (pCi ^{90}Sr/mg stable Sr) of bone at the end of year n; W_{n-1} is the average ^{90}Sr specific activity of the bone at the end of year $n-1$ and Z_n is the ^{90}Sr specific activity of the diet during the year n.

Measurements of stable Sr and ^{90}Sr activity of fifteen rumen samples taken from deer during the June, 1963–May, 1964, period were used to estimate the ^{90}Sr specific activity of the diet (244·7 pCi ^{90}Sr/mg stable Sr). The femur bone averages for adult deer for the 1962–63 and 1963–64 periods were 48·1 and 60·4 pCi ^{90}Sr/mg stable Sr, respectively. By use of these specific activity values and equation (1) the annual turnover rate for adult deer was calculated to be approximately 6% (which corresponds to a biological half-life of 11 years), which is in near agreement with the femur-shaft annual turnover rate of 4% as reported by Rivera (1964) for adult humans. A turnover rate for juvenile deer was calculated from femur-bone averages of 105 and 132 pCi ^{90}Sr/mg stable Sr for the 1962–63 and 1963–64 periods and gave a value of 19%/year (which corresponds to a biological half-life of 3·3 years). The difference in turnover rates for juvenile and adult deer reflects the magnitude of overall bone activity, which would include both formation and resorption processes. In summary, it appears that it is the behavioral or metabolic characteristics associated with age, as well as the Sr and Ca level of the diet, which primarily govern the observed level of Sr in the bone.

Research supported by Contract No. AT(11–1)–1156 between the U.S. Atomic Energy Commission and Colorado State University.

REFERENCES

Bohman, V. R., Wade, M. A., and Blincoe, C. (1962). *Science*, **136**, 1120–1121.

Bryant, F. J., and Loutit, J. F. (1964). *Proc. R. Soc.*, **159B**, 449–465.

Bryant, F. J., Dwyer, L. J., Martin, J. H., and Titterton, E. W. (1959). *Nature, Lond.* **184**, 755–760.

Farris, G. C. (1965). "Strontium-90 in antlers and selected bones of Colorado mule deer." M.S. Thesis. Colorado State University, Fort Collins, Colorado.

Frost, H. M. (1964). *In* "Bone Biodynamics" (Frost, H. M., ed.), pp. 315–333. Little, Brown and Co., Boston, Massachusetts.

Häggröth, S., and Hoglund, G. (1961). *Expl. Cell Res.* **24**, 80–87.

Hodges, R. M., MacDonald, N. S., Nusbaum, R., Stearns, R., Ezmirlian, F., Spain, P. C., and McArthur, C. (1950). *J. biol. Chem.* **185**, 519–524.

Longhurst, W. M. (1964) AEC Contract No. UCD–34, P104–1. Univ. California, Davis.

Rivera, J. (1964). *Radiol. Hlth. Data*, **5**, 98–99.

Rivera, J., and Harley, J. H. (1965). HASL–163, U.S. Atomic Energy Commission, Health and Safety Laboratory, New York.

Robinette, W. L., Jones, D. A., Rogers, G., and Gashwiler, J. S. (1957). *J. Wildl. Mgmt*, **21**, 134–153.

Schultz, V., and Flyger, V. (1965). *J. Wildl. Mgmt*, **29**, 39–43.

Sowden, E. M., and Stitch, S. R. (1957). *Biochem. J.*, **67**, 104–109.

Trent, D., and Slavin, W. (1964). Atomic Absorption Newsletter No. 22. Instrument Division, Perkin–Elmer Corporation, Norfolk, Connecticut.

Turekian, K. K., and Kulp, J. L. (1956). *Science*, **124**, 405–407.

U.S.A.E.C. (1961). NYO–4700. U.S.A.E.C. Health and Safety Laboratory, New York.

Discussion

QUESTION: Have you analysed antlers? If your deer are similar to the Scottish variety, the antlers will grow very fast at some seasons of the year and may be eaten after they have been shed, thereby recycling the strontium.

ANSWER: In yearling deer, where both bone and antler tissue are growing rapidly, the levels of ^{90}Sr are about the same. In an older animal, the antler activity may be considerably higher, particularly if the recent rate of fallout deposition has been high. We do not know how much Sr or Ca is taken from bone to form antler tissue, but we hope to make experiments with ^{85}Sr and to see whether the rate of transfer to antler is influenced by the rate of Ca intake at the time.

Skeletal Distribution of Strontium and Calcium and Strontium/Calcium Ratios in Several Species of Fish

I. L. OPHEL AND J. M. JUDD

Environmental Research Branch, Biology and Health
Physics Division, Atomic Energy of Canada Limited,
Chalk River, Ontario, Canada

SUMMARY

The Sr and Ca contents of different bony tissues have been measured in three species of fish. The Sr/Ca ratios in the spinal column of five species have been compared with the corresponding ratios in the water.

INTRODUCTION

Very few studies have been made of the Sr content and the Sr/Ca ratio in tissues of freshwater fishes. Those analyses that have been published have often been made on single specimens of a species and have not been related to the Ca and Sr content of the water. We are unaware of any published studies on the distribution of these elements within the fish skeleton.

MATERIALS AND METHODS

A total of five species of fish from two different lakes have been used. From Perch Lake (Ophel, 1963) situated on Atomic Energy of Canada property near Chalk River, we obtained yellow perch [*Perca flavescens* (Mitchill)]. Species from Lake Huron, one of the Great Lakes, were yellow perch, northern redhorse sucker [*Moxostoma aureolum* (LeSeur)], longnose or sturgeon sucker [*Catostomus catostomus* (Forster)], gizzard shad [*Dorosoma cepedianum* (LeSeur)] and carp [*Cyprinus carpio* Linnaeus]. Of the Lake Huron species, the specimens of perch and redhorse sucker were netted in the Douglas Point area, near Kincardine, Ontario. The other species came from near Sarnia, Ontario.

Specimens of bone were prepared by steaming the gutted fish until the flesh could be pulled easily away from the bones. The bones were blotted with tissues and weighed (fresh weight), air dried for several days and weighed again (dry weight) and then ashed in porcelain crucibles (24 h at 600° C) and weighed (ash weight). Sr and Ca (in solutions of the ash) were determined by using a direct flame emission spectrophotometric method (Judd and Coveart, 1965). All bone concentrations discussed in this paper are based on ash weight. Sr/Ca ratios are expressed throughout as atoms of Sr per thousand atoms of Ca.

CALCIUM AND STRONTIUM IN THE LAKE WATERS

Weekly samples of water from both lakes were obtained in the summer of 1965, during the time the fish were collected. These were combined into monthly composites and analyses were made on these composites. The results are given in Table I.

TABLE I

Strontium and calcium content of lake waters

Source	Date	Sr (µg/ml)	Ca (µg/ml)	Ratio µg Sr/mg Ca	Ratio Sr atoms/ 1000 atoms Ca	Mean Ratio Sr atoms/ 1000 atoms Ca
Lake Huron	June '65	0·12	27·2	4·4	2·01	
	July '65	0·11	26·7	4·1	1·87	1·89
(Douglas	Aug. '65	0·10	25·5	4·0	1·83	
Point)	Sept. '65	0·11	27·0	4·1	1·87	
Perch Lake	June '65	0·030	6·1	4·9	2·24	2·33
	Aug. '65	0·034	6·4	5·3	2·42	

The Ca content of Lake Huron water was higher than that of Perch Lake by a factor of 4·2, and the factor for Sr content was 3·4. The Sr/Ca ratio (atoms Sr/1000 atoms Ca) for Lake Huron water was 1·89; for Perch Lake it was 2·33.

These Sr/Ca ratios are somewhat higher than those found by Templeton and Brown (1964) for United Kingdom waters of similar Ca content.

DISTRIBUTION OF CALCIUM AND STRONTIUM IN THE SKELETON

The distribution of the two elements was examined by dividing the bony tissues of each fish into seven parts. These were designated (1) scales, (2) skull, (3) ribs, (4) anterior vertebrae, (5) posterior vertebrae, (6) gill arches and (7) fins. For this study we used ten perch from Perch Lake; ten perch and five redhorse suckers from Lake Huron, which entailed 175 determinations for Sr and the same number for Ca. The Perch Lake fish ranged from 16·0 to 30·0 cm for length and 42 to 350 g in fresh weight. Lake Huron perch ranged from 17·5 to 26·5 cm and from 62 to 295 g. Corresponding ranges for the redhorse suckers were 18·5 to 31·0 cm and 100 to 630 g.

Table II and III summarize the calcium and strontium results.* Standard deviations are given to indicate the range of the data.

* Complete tabulations of individual results will be available, on request, from the authors.

TABLE II

Calcium content of fish bone

Location and species*	Scales (mg Ca/g ash ±s.d.)	Skull (mg Ca/g ash ±s.d.)	Ribs (mg Ca/g ash ±s.d.)	Anterior Vertebrae (mg Ca/g ash ±s.d.)	Posterior Vertebrae (mg Ca/g ash ±s.d.)	Gill Arches (mg Ca/g ash ±s.d.)	Fins (mg Ca/g ash ±s.d.)
Perch Lake Perch (10)	362 ±18·7	372 ±15·6	370 ±16·1	364 ±16·3	361 ±13·1	367 ±16·1	363 ±13·4
Lake Huron Perch (10)	364 ±16·6	364 ±19·4	368 ±22·6	370 ±20·7	365 ±19·8	374 ±14·8	371 ±19·4
Lake Huron Redhorse Suckers (5)	327 ±31·9	362 ±9·4	352 ±8·3	358 ±10·8	367 ±8·6	365 ±11·2	365 ±9·2

* Figures in parentheses are number of fish.

TABLE III

Strontium content of fish bone

Location and species*	Scales (µg Sr/g ash ±s.d.)	Skull (µg Sr/g ash ±s.d.)	Ribs (µg Sr/g ash ±s.d.)	Anterior Vertebrae (µg Sr/g ash ±s.d.)	Posterior Vertebrae (µg Sr/g ash ±s.d.)	Gill Arches (µg Sr/g ash ±s.d.)	Fins (µg Sr/g ash ±s.d.)
Perch Lake Perch (10)	413 ±20·1	410 ±28·1	399 ±38·8	373 ±24·7	383 ±20·6	378 ±16·7	389 ±24·1
Lake Huron Perch (9)†	295 ±36·7	299 ±39·1	294 ±49·4	277 ±38·5	272 ±39·8	290 ±42·8	312 ±36·7
Lake Huron Redhorse Sucker (5)	389 ±29·6	361 ±41·7	360 ±19·7	321 ±28·7	353 ±35·6	347 ±26·9	368 ±27·6

* Figures in parentheses are number of fish.
† One Lake Huron perch was not included in the group. Its Sr content was considerably higher: scales, 469; skull, 461; ribs, 404; anterior vertebrae, 384; posterior vertebrae, 384; gill arches, 405; fins, 442 µg Sr/g ash.

The Ca content, expressed in mg/g ash, was uniform in different bony tissues of the same fish. No significant differences were found between the Ca content of the perch from the two lakes, although the lakes themselves differed markedly in this respect. The redhorse sucker tissues were comparable with the two perch series, except for the lower Ca content of the scales. This low mean value was mainly due to a single fish.

The strontium results, expressed in μg/g of ash (Table III), showed much greater variation than the Ca. Each species appeared to have a characteristic Sr content, as shown by comparison of the perch and redhorse sucker series from Lake Huron. The differences between the Sr content of perch from Perch Lake and those from Lake Huron were striking. However, as the Ca content of the perch bone from both lakes was essentially the same (Table II), it is to be expected that the Sr content of the Lake Huron perch bone would be lower, reflecting the difference in the Sr/Ca ratios in the two lake waters (Table I).

STRONTIUM-TO-CALCIUM RATIOS IN FISH AND WATER

In Table IV we have calculated the Sr/Ca ratios in the bone of all the species of fish. Analyses from the anterior and posterior vertebrae have been used for comparison because these are the only analyses available at present for three of the species. In addition, the standard deviations of the Ca and Sr means given in Table IV have been rounded off to the nearest whole number.

All of the Sr/Ca ratios of the bones were smaller than those for the corresponding lake waters (Table I). If it is assumed that the lake water was the main source of the Ca and Sr in these fish, then a discrimination against Sr occurred in the passage of the two elements from water to bone.

This may be expressed as an "observed ratio" (Comar *et al.*, 1956) where:

$$\text{observed ratio (OR)} = \frac{\text{Sr/Ca ratio in fish bone}}{\text{Sr/Ca ratio in water}}$$

or, in generalized form:

$$\text{OR} = \frac{\text{Sr/Ca ratio in sample}}{\text{Sr/Ca ratio in precursor}}$$

Observed ratios calculated in this way are listed in Table V.

The greatest discrimination against Sr was found in perch with OR $=0.19$; the least in carp with an OR $=0.66$. Other species had ratios of 0.23 (longnose sucker) and 0.24 (redhorse sucker and gizzard shad).

Freshwater fish can derive Sr and Ca from two sources: food and water. Discrimination between Sr and Ca can occur in several places during the passage of the two elements into the bone, e.g., in gill absorption, intestinal absorption and renal excretion. Experimental studies with Ca and Sr tracers, have been made of some of these processes, but the results are conflicting.

TABLE IV

Strontium/calcium ratios in fish bone

Lake and species*	Sr content (μg/g ash$\pm$s.d.)		Ca content (mg/g ash$\pm$s.d.)		Mean Sr/Ca Ratio (atoms Sr/1000 atoms Ca$\pm$s.d.)
	anterior vertebrae	posterior vertebrae	anterior vertebrae	posterior vertebrae	
Perch Lake Perch (10)	373 ± 25	383 ± 21	364 ± 16	361 ± 13	$0\cdot48\pm0\cdot03$
Lake Huron Perch (10)	$277\dagger\pm39$	$272\dagger\pm40$	370 ± 21	365 ± 20	$0\cdot36\pm0\cdot07$
Lake Huron Longnose Sucker (10)	298 ± 28	316 ± 36	344 ± 10	341 ± 11	$0\cdot41\pm0\cdot04$
Lake Huron Redhorse Sucker (5)	321 ± 29	353 ± 36	358 ± 11	367 ± 9	$0\cdot43\pm0\cdot04$
Lake Huron Gizzard Shad (10)	340 ± 100	344 ± 97	335 ± 9	337 ± 9	$0\cdot46\pm0\cdot13$
Lake Huron Carp (3)	437‡ 1163 1269	385‡ 1243 1295	352‡ 355 355	380‡ 350 355	$1\cdot25\pm0\cdot6$

* Figures in parentheses are number of fish.
† Mean of 9 fish (see Table III footnote).
‡ This fish was a gravid female containing a large quantity of eggs. Other two fish were males.

TABLE V

Strontium/calcium observed ratios in fish bone

Lake and Species	Observed ratio $=\left(\dfrac{\text{Sr/Ca in fish bone}}{\text{Sr/Ca in lake water}}\right)$
Perch Lake	
Perch	0·21
Lake Huron	
Perch	0·19
Longnose sucker	0·23
Redhorse sucker	0·24
Gizzard shad	0·24
Carp	0·66 (0·87*)

* Observed ratio if single low value is excluded (See Table IV).

Ichikawa (1960) found an OR = 0·7 in the transfer of Sr and Ca from food to bone in rainbow trout. In the same species, he found an OR = 0·36 between water and bone for gill uptake from a "natural water" and an OR = 0·2 for the same process in an "artificial water". In gill-uptake studies with three other species of freshwater fish, Rosenthal (1957) was unable to demonstrate a significant discrimination between the two elements.

With the evidence that is available, it is not possible to determine the reasons for the differences in the observed ratios of the Lake Huron species of fish. The fact that such differences do exist is of considerable interest, and may have some application in the fields of palaeoecology and taxonomy.

Of interest is the variability of the Sr/Ca ratio shown by the Lake Huron gizzard shad (Table IV). Individual fish varied between an OR of 0·15 and 0·39. These wide differences in discrimination were not found in the other species of fish examined except, perhaps, in the carp. Only three individuals of this species were available for estimation, these gave ORs of 0·28, 0·86 and 0·88. The reasons for this variability are not known although, in the case of the carp, the low OR was from a female containing many ripe eggs.

If the major part of the skeletal Ca and Sr were derived from the diet then the Sr/Ca ratio in the diet would have a controlling influence on the bone ratio. The greater bone/water discrimination (smaller OR) found in such carnivorous fish as perch (OR = 0·19) may result from a diet in which the OR is already less than unity.

TABLE VI

Strontium/calcium ratios in stomach contents of fish samples collected from Lake Huron

Sample	Sr in sample (μg)	Ca in sample (mg)	μg Sr/mg Ca	Ratio (atoms Sr/1000 atoms Ca)
Perch 12A	2·03	1·10	1·84	0·84
Perch 12B	1·75	1·18	1·48	0·68
Perch 12C	2·03	1·16	1·75	0·80
Carp J 13	1·07	0·21	5·1	2·33
Carp J 11	0·80	0·13	6·1	2·78
Carp J 12	1·26	0·22	5·7	2·60

In the case of fish that feed extensively on aquatic plants, such as carp, the OR is greater (0·66). It has been found (Templeton and Brown, 1964) that aquatic plants do not discriminate against Sr, i.e. the plant-to-water OR is greater than unity.

The higher OR found in carp might therefore be expected. A study of the

Sr/Ca ratios in the food (stomach contents) is necessary before an explanation of the difference in ORs found in the various species of fish examined can be attempted. Some preliminary results of such a study are given in Table VI.

We thank J. Dawson and G. Grant for assistance with the analytical work.

REFERENCES

Comar, C. L., Wasserman, R. H., and Nold, M. M. (1956). *Proc. Soc. exp. Biol. Med.* **92,** 859–863.
Ichikawa, R. (1960). *Rec. oceanogr. Wks, Japan,* **5**(2), 120–131.
Judd, J. M., and Coveart, A. E. (1965). Atomic Energy of Canada Limited, Publication No. AECL–2518.
Ophel, I. L. (1963). *In* "Radioecology", Proceedings of the First National Symposium on Radioecology, September, 1961, pp. 213–216. Reinhold, New York.
Rosenthal, H. L. (1957). *Science,* **126,** 699–700.
Templeton, W. L., and Brown, V. M. (1964). *Int. J. Air Wat. Pollut.* **8,** 49–75 (also *errata,* **8,** 609–610).

Influence of Dietary and Hormonal Factors on Radiostrontium Metabolism in Man

HERTA SPENCER, ISAAC LEWIN
AND JOSEPH SAMACHSON
Metabolic Research Unit,
Veterans Administration Hospital,
Hines, Illinois

INTRODUCTION

Much data is available in experimental animals concerning many aspects of Sr metabolism in comparison to the metabolism of Ca. It would be presumptuous to attempt to review the extensive and excellent work performed by the investigators present at this meeting, and I shall therefore limit myself to the presentation of some of the work performed in our Research Unit in man, and refer to the work of others as it applies to the work to be presented.

This paper will deal with the effect of some dietary factors such as calcium, phosphorus, stable strontium, lactose, lysine and of vitamin D on radiostrontium metabolism in man. The effect of some hormonal factors on radiostrontium metabolism will be described, i.e., the changes in radiostrontium metabolism observed during different phases of thyroid function, the effect of parathyroid extract (PTE), estrogen, androgen, progestin, anabolic agents and corticosteroids. All studies have been performed under constant and strictly controlled dietary conditions on the Metabolic Research Ward. Tracer doses of ^{85}Sr and ^{45}Ca (or of ^{47}Ca) were given either orally, intravenously or by both routes. In order to characterize the metabolic status of the patients under study, balances of calcium, phosphorus and nitrogen were performed throughout the tracer studies. The plasma levels and the urinary and fecal excretions of the radioisotopes were determined. The intestinal absorption of these radioisotopes was calculated from the fecal ^{85}Sr and ^{45}Ca excretions, making allowance for the endogenous fecal excretions, which were determined following the intravenous injection of ^{85}Sr to the same patient. The true absorption of ^{85}Sr was then determined. For example, if 82·5% of the dose was excreted in stool in 12 days after the oral administration of [85]Sr and 12·7% of the dose was excreted in stool after the intravenous injection of [85]Sr, the true absorption of [85]Sr was calculated as follows:

$$x = \text{absorbed } {}^{85}\text{Sr}$$
$$y = \text{unabsorbed } {}^{85}\text{Sr}$$
$$x + y = 100\%$$
$$0{\cdot}127x + y = 82{\cdot}5\%$$
$$\overline{0{\cdot}873x \qquad = 17{\cdot}5\%}$$
$$x \qquad = 20{\cdot}0\%$$

Therefore, the true absorption of [85]Sr was 20% of the orally administered dose, 80% was unabsorbed and 2·5% of the dose passed with the stool as endogenous fecal [85]Sr. The absorption of [85]Sr can be determined reliably from the plasma levels of the radioisotopes determined at 4, 8 and 24 h following the administration of a single oral and intravenous dose (Samachson, 1963) as follows:

$$\% \text{ Absorption} = \frac{\text{Plasma value after oral dose}}{\text{Plasma value after intravenous dose}} \times 100$$

Figures 1 and 2 show the good agreement of the [85]Sr absorption values obtained from the ratios of single and combined plasma values at 4, 8 and 24 h following the oral and intravenous administration of [85]Sr as compared to the [85]Sr absorption calculated from fecal [85]Sr excretions in the same persons.

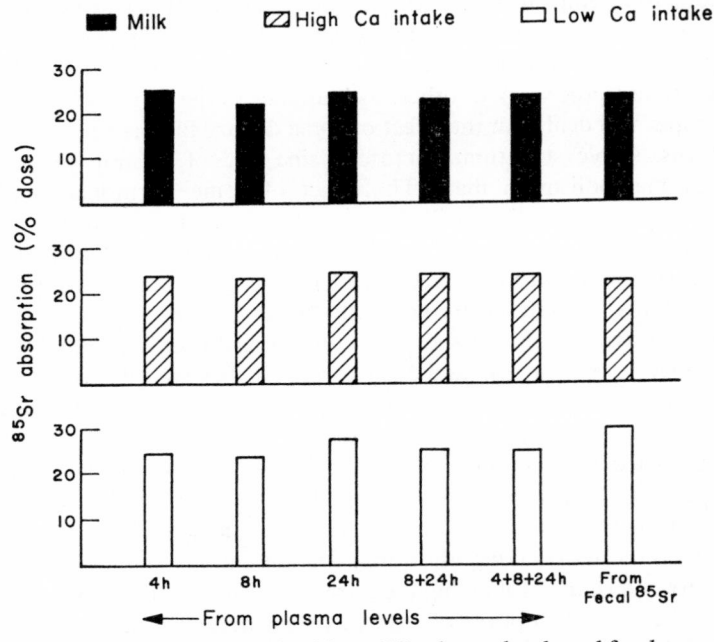

FIG. 1. [85]Sr absorption determined from [85]Sr plasma levels and fecal excretions.

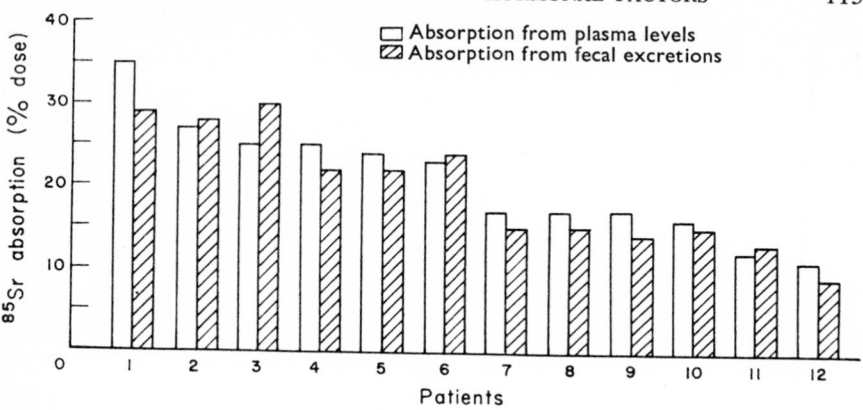

FIG. 2. Absorption of [85]Sr determined from [85]Sr plasma levels and fecal excretions.

EFFECT OF HIGH CALCIUM INTAKE ON RADIOSTRONTIUM ABSORPTION AND ON CALCIUM–STRONTIUM DISCRIMINATION

Reports in the literature indicate that the addition of Ca to the diet decreased the absorption of radiostrontium in rats (MacDonald *et al.*, 1955; Wasserman *et al.*, 1957), that added Ca decreased the absorption of radioactive Ca more than of radioactive Sr (Palmer *et al.*, 1958a) and that a marked reduction in the absorption of radioactive Sr could be achieved by adding phosphate to the high Ca diet (MacDonald *et al.*, 1955; Palmer *et al.*, 1958b; Wasserman and Comar, 1960). *In vivo* perfusion experiments have shown that the addition of Ca decreased the absorption of [45]Ca to a considerably greater extent than the absorption of [85]Sr (Mraz, 1962). In young growing rats a five-fold increase in Ca was required to decrease the deposition of radiostrontium in bone by one half (Hegsted and Bresnahan, 1963). In studies performed in man in this Research Unit, the addition of Ca decreased the absorption of [85]Sr in only five out of thirteen patients who received Ca supplements for time periods ranging from 3 to more than 500 days (Spencer *et al.*, 1961).

The Ca/Sr discrimination at the renal and intestinal level has been shown to exist in animals (MacDonald *et al.*, 1957; Comar *et al.*, 1956) and in man (Comar *et al.*, 1957; Spencer *et al.*, 1960). However, the influence of the dietary Ca intake on Ca/Sr discrimination has not been examined. This aspect was investigated in man by performing [85]Sr and [45]Ca absorption studies during both low and high Ca intake in the same patients. The low Ca intake was approximately 150 mg/day. In the high Ca studies, the Ca intake was increased approximately ten-fold by adding calcium gluconate tablets to the constant low Ca diet, all other constituents of the diet remaining unchanged.

During the low Ca intake, the urinary excretion of [85]Sr was greater than of [45]Ca in each of the patients studied; the average urinary [85]Sr/[45]Ca excretion ratio was 1·8. Examples of these excretions are shown in Fig. 3. Similarly, the fecal [85]Sr excretion was higher than that of [45]Ca during low Ca intake in each case, the excretion of [85]Sr ranging from 76% to 83%, and the excretion for [45]Ca ranging from 48 to 67% in the same patients. In a relatively large series of patients the average fecal [85]Sr/[45]Ca excretion ratio was 2·6 during low Ca intake (Spencer *et al.*, 1960).

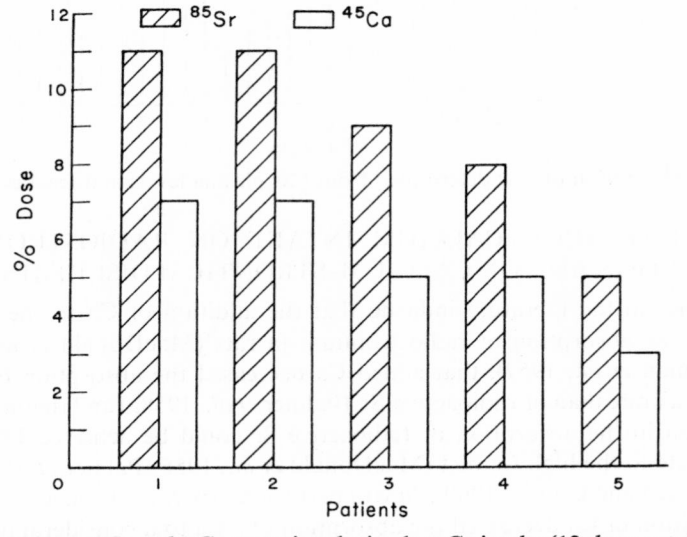

FIG. 3. Urinary [85]Sr and [45]Ca excretion during low Ca intake (12-day cumulative).

The absorption of [45]Ca and of [85]Sr during low and high Ca intake was compared. During low Ca intake, the absorption of both [45]Ca and [85]Sr varied in the different patients depending on the Ca metabolism of the individual and the percentage absorption of [45]Ca ranged from 28 to 55% of the dose. During high Ca intake, the percentage absorption of [45]Ca decreased in each of the patients to about one half the absorption during low Ca intake, the range of absorption being 11 to 25% in the same persons. The absorption of [85]Sr was considerably lower than the absorption of [45]Ca during low Ca intake in each patient and the average [45]Ca/[85]Sr absorption ratio was 2·6. During high Ca intake the absorption of [85]Sr, determined from the fecal [85]Sr excretions decreased in some but not in all patients, resulting in an average [45]Ca/[85]Sr absorption ratio of only 1·4. The similarity of the magnitude of the percentage absorption of [45]Ca and of [85]Sr during high Ca intake in five out of six patients is illustrated in Fig. 4. Figure 5 illustrates the marked decrease in the average percentage of [45]Ca absorption during high Ca intake, while the decrease in the average [85]Sr absorption was only slight.

Since the absorption of ^{45}Ca, determined from the fecal excretions of the radioisotope, decreased more than the absorption of ^{85}Sr during high Ca intake, it was therefore expected that the plasma levels of ^{45}Ca would decrease more than the plasma levels of ^{85}Sr during high Ca intake following the oral

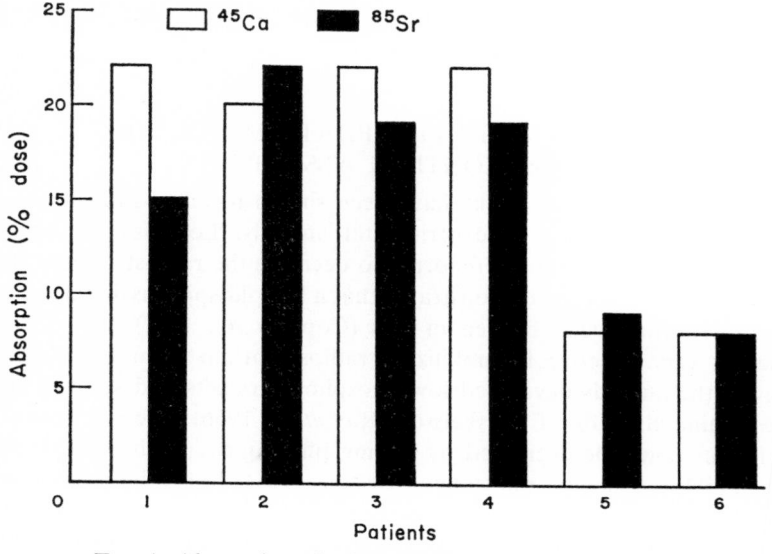

FIG. 4. Absorption of ^{45}Ca and ^{85}Sr during high Ca intake.

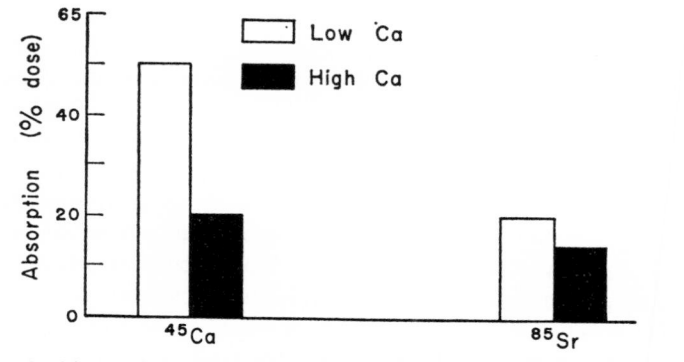

FIG. 5. Absorption of ^{45}Ca and ^{85}Sr during low and high Ca intake.

administration of both tracers. The plasma levels of both ^{45}Ca and ^{85}Sr were lower during high Ca intake than during low Ca intake; the decrease in the plasma level of ^{45}Ca was particularly marked, while the plasma level of ^{85}Sr was only slightly lower than during low Ca intake. This indicated a greater decrease in the absorption of ^{45}Ca than of ^{85}Sr. These changes in the plasma

E   S.M.

levels of ^{45}Ca and ^{85}Sr were in good agreement with the changes in fecal excretion of the two radioisotopes during high Ca intake, the fecal ^{45}Ca excretion increasing to a much greater extent than the fecal ^{85}Sr excretion.

There was little change in the urinary ^{85}Sr and ^{45}Ca excretion during low and high Ca intake and therefore the urinary ^{85}Sr/^{45}Ca excretion ratio was similar during the intake of both the low and high Ca diet. The average urinary ^{85}Sr/^{45}Ca excretion was 1·8.

EFFECT OF HIGH PHOSPHATE INTAKE ON RADIOSTRONTIUM ABSORPTION

Changes in phosphorus intake have been shown to affect radiostrontium absorption and retention in experimental animals. Low as well as high phosphorus intakes have been reported to decrease the radiostrontium body burden in rats. It has been demonstrated that a low phosphorus diet decreased the radiostrontium body burden in rats (Copp et al., 1947) and that this regimen is very effective in mobilizing radiostrontium from the skeleton. However, the animals developed low-phosphorus rickets and suffered from severe demineralization of the skeleton (Ray et al., 1956). The ^{90}Sr uptake in bone in rats could be decreased by adding phosphorus to the diet (Palmer et al., 1958b). Calcium phosphate (Bruce, 1963) and other phosphates (Volf, 1965) have been shown to decrease the absorption of radioactive strontium in rats, whereas an increase in the intake of both phosphorus and of Ca in the diet led to a decrease in the ^{90}Sr body burden in rats (MacDonald et al., 1955) and this combined regimen was more effective than the use of excess Ca (Palmer et al., 1958b; Wasserman and Comar, 1960).

In view of the reported changes in radiostrontium metabolism induced by added phosphate in animals, the effect of phosphorus on the absorption of radiostrontium was studied in man in this Research Unit. In order to assess the influence of the Ca intake and of the Ca/P ratios, the studies were performed during both low and high Ca intake (Spencer et al., 1964; 1965b). In the high phosphorus study, phosphorus was added as glycerophosphate to the constant low Ca diet, raising the phosphorus intake from an average of 600 mg/day to an average of 1600 mg/day, all other constituents of the low Ca diet remaining unchanged. In the high Ca study, calcium gluconate tablets were added to the low Ca diet, raising the Ca intake from an average of 202 mg/day to an average of 1724 mg/day. The Ca/P ratio of the low calcium–low phosphorus diet was approximately 1:4; of the low calcium–high phosphorus intake 1:10; of the high calcium–low phosphorus intake 2:1; and of the high calcium–high phosphorus intake 1:1.

Figure 6 shows the ^{85}Sr plasma levels before and during the addition of phosphate to the low and high Ca diets. During the addition of glycerophosphate to the low Ca diet, the ^{85}Sr plasma levels decreased as compared to the [85]Sr plasma values in the control study, indicating a decrease of [85]Sr absorption during the addition of phosphate. During the high calcium–low phosphorus control study, the [85]Sr plasma level was lower than during the low calcium–low phosphorus control study in the same patient, indicating a decrease of [85]Sr absorption during high Ca intake. When phosphate was added to the high Ca diet, there was a further decrease of the [85]Sr plasma levels. The decrease in the [85]Sr plasma levels on addition of phosphorus to the low and high Ca diet indicates that added phosphorus decreased the

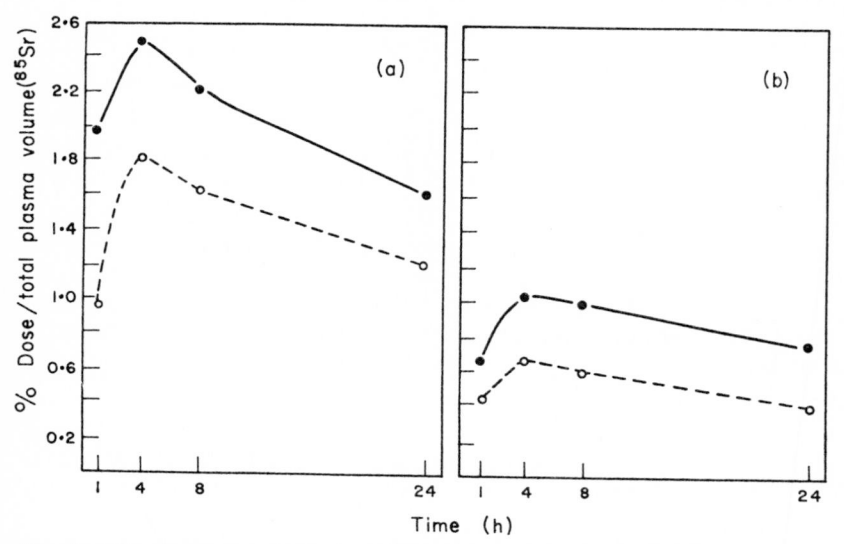

FIG. 6. [85]Sr plasma levels during high and low phosphorus intake: (a) low Ca diet; (b) high Ca diet. Continuous lines, low phosphorus; broken lines, high phosphorus.

absorption of [85]Sr. The decrease in the [85]Sr plasma level was noted in three of four patients given glycerophosphate during both the low and high Ca intake. It might be expected that in those cases in which plasma levels of [85]Sr are lowered, indicating a decrease in absorption, the fecal [85]Sr excretion would increase. However, because of the low values of [85]Sr absorption in the control studies in most of the cases, small changes in absorption could not be determined. The plasma levels of the radioisotopes are much more precise indicators of changes in absorption than the fecal excretions, especially when the absorption is low (Samachson, 1963).

There was also a decrease in urinary excretion of [85]Sr and of Ca during the addition of phosphorus to the low and high Ca diet (Figs. 7 and 8); this decrease may be a consequence of the low absorption of [85]Sr during the intake of added phosphate.

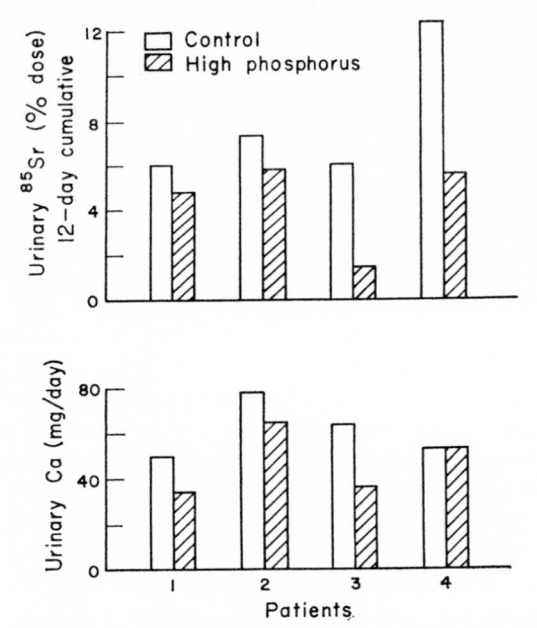

FIG. 7. Effect of high phosphorus intake on urinary [85]Sr and Ca excretion: low Ca intake.

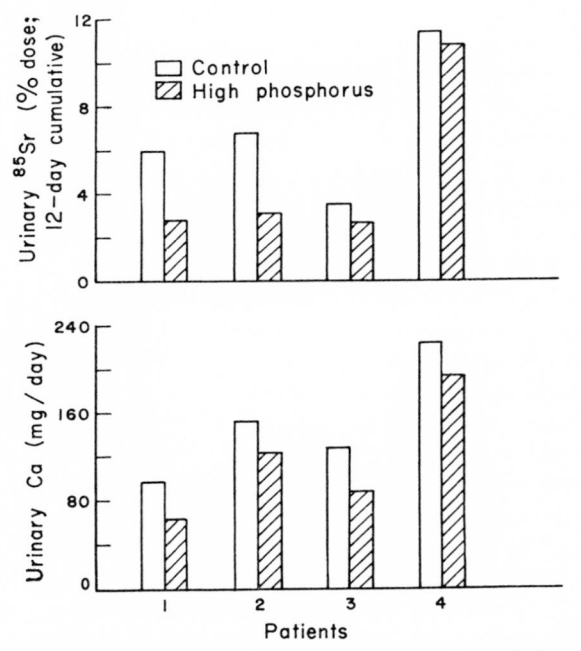

FIG. 8. Effect of high phosphorus intake on urinary [85]Sr and Ca excretion: high Ca intake.

EFFECT OF STABLE STRONTIUM ON
RADIOSTRONTIUM ABSORPTION

Some investigators have reported that the feeding of stable Sr to animals had no effect on radiostrontium absorption and retention, whereas others have reported that the feeding of stable Sr can decrease the absorption of radiostrontium. Stable Sr was reported to have no effect in rabbits (Kidman *et al.*, 1950) and in rats (Harrison *et al.*, 1957; Kawin, 1959). Other investigators have found that the absorption of both ^{45}Ca and ^{85}Sr in rats could be decreased to the same extent by the use of stable Sr (Mraz, 1962). Pre-feeding with stable Sr decreased the deposition of radiostrontium in bone and the total body retention in young rats (Hegsted and Bresnahan, 1963; Teree *et al.*, 1965). It has been postulated that small amounts of stable Sr may actually increase the absorption of radioactive Sr and Ca, whereas large amounts are needed to decrease the uptake of ^{89}Sr and ^{45}Ca in bone (Hegsted and Bresnahan, 1963). In contrast to the controversial results obtained with orally administered stable Sr, intravenously or intraperitoneally injected stable Sr has been reported to decrease the body burden in rats, in mice and in man (Carlqvist and Nelson, 1960; Catsch, 1957; Smith and Bates, 1965; Spencer *et al.*, 1965a).

A limited number of studies have been performed in this Research Unit in man on the effect of orally administered stable Sr on the absorption of radioactive Sr. Tracer doses of ^{85}Sr were given orally before and during the administration of stable Sr as the lactate. In the first study, a rather large amount of stable Sr (3000 mg/day) was administered, in order to compare the effectiveness of this amount with an approximately equivalent amount of Ca (1500 mg/day). The ^{85}Sr absorption study was performed after 7 months of stable Sr supplementation, and the results were compared to those of the control study. The fecal ^{85}Sr excretion was of similar magnitude in the stable Sr study as in the control study, and the plasma levels following the oral administration of the radioisotope were actually higher than in the control study. These data indicate that stable Sr did not decrease the absorption of ^{85}Sr and it may have actually increased. This large amount of stable Sr was, however, not well tolerated and, therefore, smaller doses of stable Sr (approximately 600 mg/day) were used in subsequent studies. This amount of stable Sr was given for many months before the ^{85}Sr absorption studies were repeated in the same patients; for instance, in one patient the ^{85}Sr absorption studies were performed after 240 days and after 390 days of pre-feeding with stable Sr. The fecal ^{85}Sr excretions in the two stable Sr studies were similar to those in the control study, again indicating that no change could be detected in the absorption of ^{85}Sr despite the prolonged pre-feeding with stable Sr. Here again, the ^{85}Sr plasma levels were distinctly higher in the two stable Sr studies than in the control study indicating greater absorption of ^{85}Sr.

The urinary [85]Sr excretion increased following the oral administration of [85]Sr during the ingestion of stable Sr. This increase accompanied an increase in urinary Ca excretion during the administration of stable Sr; the increase in urinary Ca was most likely due to the exchange of Sr with the Ca in bone (Mazzuoli *et al.*, 1961).

The data obtained with orally administered stable Sr indicate that the absorption of [85]Sr was not altered by pre-feeding with stable Sr and that neither the length of time of pre-feeding nor the amount of stable Sr influenced these results. Although the fecal [85]Sr excretions remained unchanged during the intake of stable Sr, the [85]Sr plasma levels were higher than in the control study. Usually, an increase in the plasma levels indicates an increase in absorption. In this case, the higher plasma levels were not associated with a change in fecal [85]Sr excretion and this may indicate that radiostrontium is not as readily accepted by bone when stable Sr is present in the blood stream, or that the absorption of [85]Sr may actually be higher during the intake of stable Sr, the lack of change in fecal [85]Sr excretion being due to greater excretion of endogenous fecal [85]Sr. Further studies are necessary to clarify this problem.

EFFECT OF LACTOSE, LYSINE AND OF VITAMIN D ON RADIOSTRONTIUM ABSORPTION

Effect of lactose and lysine

Certain carbohydrates and amino acids have been shown to increase the absorption of Ca and Sr in animals under certain experimental conditions (Wasserman *et al.*, 1956; Lengemann and Comar, 1961). In order to determine whether these substances have a similar effect in man, the effect of lactose and

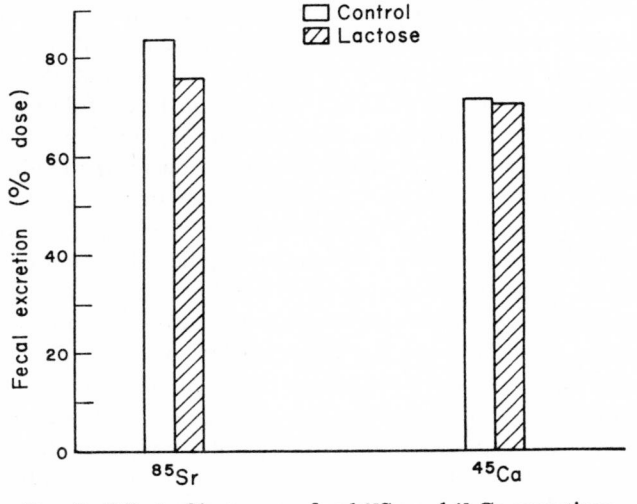

FIG. 9. Effect of lactose on fecal [85]Sr and [45]Ca excretions.

of lysine on the absorption of Ca, ^{85}Sr and ^{45}Ca has been studied (Greenwald *et al.*, 1963; Spencer and Samachson, 1963). The dose of lactose ranged from 20–50 g/day and the dose of lysine from 10–30 g/day; both lactose and lysine were given for several weeks before the radiostrontium absorption studies. Tracer doses of both ^{85}Sr and of ^{45}Ca were given orally in the control studies and during the administration of lactose and of lysine. These studies were performed in patients who had a low absorption of both Ca and Sr in order to facilitate the detection of any improvement in absorption. Figure 9 shows that 50 g of lactose given per day had no appreciable effect on the fecal excretions of ^{85}Sr or ^{45}Ca. Similarly, Fig. 10 shows that 10–20 g of lysine given daily did

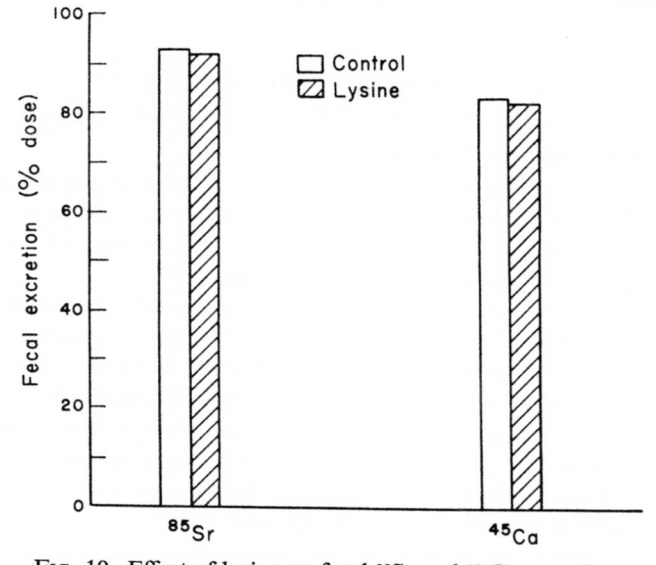

Fig. 10. Effect of lysine on fecal ^{85}Sr and ^{45}Ca excretions.

not affect the fecal excretions of ^{85}Sr or ^{45}Ca. In addition, there was no change in the ^{85}Sr plasma levels during the administration of lactose and lysine. These two parameters indicate that there was no appreciable improvement in the absorption of Sr or of Ca in man.

Effect of vitamin D

The effect of vitamin D on radiostrontium absorption has been studied in young rachitic rats (Greenberg, 1945); a dose of 10,000 I.U. of vitamin D promoted the absorption of Ca, but not of Sr, doubling the uptake of ^{45}Ca by bone while only slightly raising the uptake of radiostrontium. Another report indicates, however, that vitamin D resulted in increased utilization of both ^{89}Sr and ^{45}Ca in rats and chicks (Patrick and Bacon, 1957). A dose of

5,000 I.U./kg of vitamin D administered 15 min after the injection of [85]Sr had no effect on the retention of [85]Sr in rats (Smith and Bates, 1965).

The effect of vitamin D on the absorption of [85]Sr has been studied in a limited number of patients in this Research Unit. A tracer dose of [85]Sr was given orally before and during the administration of vitamin D; 50,000 I.U. of vitamin D was given daily for several weeks before the [85]Sr absorption study was performed. During the administration of vitamin D the [85]Sr plasma levels were the same as or slightly lower than those of the control studies. The fecal [85]Sr excretions did not decrease during the administration of [85]Sr. These results indicate that this amount of vitamin D, given for several weeks, had no effect on the absorption of radiostrontium in adult man.

EFFECT OF SEX HORMONES AND OF CORTICOSTEROIDS ON RADIOSTRONTIUM METABOLISM

Effect of female and male hormones

Little information is available on the effect of male or female sex hormones on radiostrontium metabolism in experimental animals and man. Estrogen (oestrone) administered at the time of the injection of [85]Sr did not alter the radiostrontium retention in rats (Smith and Bates, 1965).

Since estrogens affect Ca metabolism in man, the effect of female and, for comparative purposes, of male sex hormones were investigated on radiostrontium metabolism in man in this Research Unit. When oral tracers of [85]Sr were given during the administration of diethylstilbestrol (15 mg/day), the plasma levels were about the same as or somewhat lower than in the control studies, indicating that the absorption of [85]Sr had not increased during the administration of estrogen. Also, the fecal [85]Sr excretions showed no change during the administration of diethylstilbestrol, in agreement with the [85]Sr plasma levels, again indicating a lack of change in the absorption of [85]Sr. A consistent effect of estrogens is the decrease in urinary Ca excretion and this decrease was accompanied by a decrease in [85]Sr excretion.

The effect of newer synthetic hormones on [85]Sr absorption in man has also been studied. One of these compounds, 3-methoxy-16α-methyl-1,3,5-[10]-estratriene-16β-diol (mytatrienediol), a weakly estrogenic compound, was as effective as diethylstilbestrol in decreasing the urinary excretion of Ca (Spencer et al., 1959) and of [85]Sr, as is shown in Table I. The decrease in urinary Ca and radiostrontium excretion during the administration of estrogenic compounds may have been due to inhibition of bone resorption (Budy et al., 1952) and/or to an increase in tubular reabsorption of Ca and Sr (Spencer et al., 1959).

The effect of two other synthetic hormones on Ca and radiostrontium metabolism was also investigated. The use of a weakly androgenic agent, 17α-ethyl-19-nortestosterone, led to a decrease in urinary Ca excretion,

TABLE I

Effect of mytatrienediol on urinary strontium-85 excretion

Patient	Cumulative ^{85}Sr excretion (% Dose)	
	control	mytatrienediol
1	41·7	22·6
2	54·3	33·7
3	64·7	49·8

The control and experimental study lasted 9 days each in patient 1, and 15 days each in patients 2 and 3.

accompanied by a decrease in radiostrontium excretion (Spencer *et al.*, 1957). Figure 11 shows the decrease in the daily urinary ^{85}Sr excretions and in the 12-day cumulative urinary ^{85}Sr excretions. Similar results were obtained with a progestational compound, 17α-ethynyl-19-nortestosterone, which decreased the urinary excretion of Ca and of ^{85}Sr.

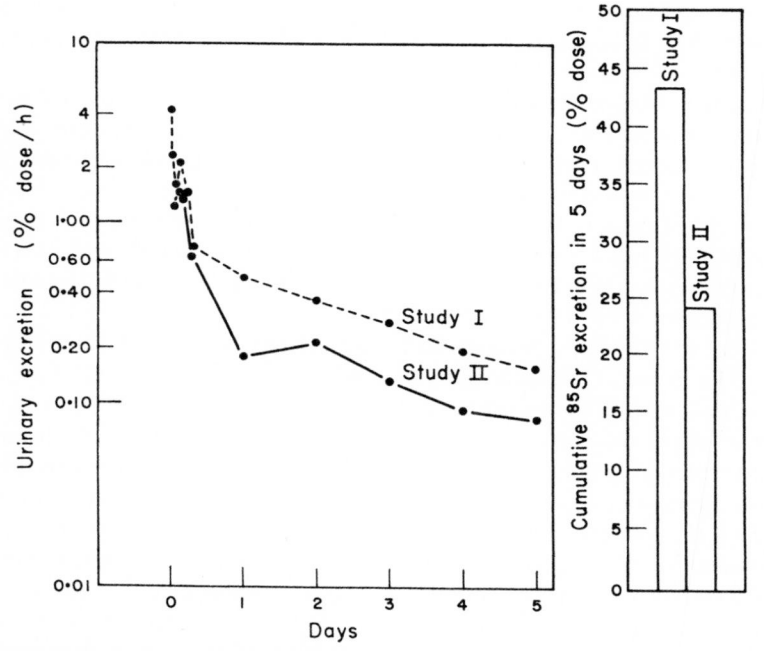

Fig. 11. Effect of 17-ethyl-19-nortesterone on the rate and cumulative urinary ^{85}Sr excretion. Study I = control; study II = experimental.

E2   S.M.

Testosterone propionate improves protein anabolism quite consistently, but it has a variable effect on Ca metabolism. During the administration of testosterone propionate the urinary Ca excretion may or may not decrease. Testosterone propionate and another androgen, fluoxymesterone, affected neither Ca nor radiostrontium metabolism, and both the urinary and fecal ^{85}Sr excretions remained about the same during the administration of these hormones as in the control studies.

Effect of corticosteroids

Little information is available on the effect of corticosteroids on radiostrontium metabolism. Corticoids are known to induce a loss of protein and of Ca. In a study in rats, corticosteroids were found to counteract the effect of PTE on the deposition of ^{45}Ca and ^{89}Sr in the kidney, but had no effect on urinary ^{89}Sr or ^{45}Ca excretion (Bacon et al., 1956). It has been reported that the negative Ca balance in man, induced by corticosteroids, could be changed to a positive balance by the use of testosterone, and that the improvement in Ca balance was accompanied by a decrease in urinary ^{85}Sr excretion, whereas the fecal ^{85}Sr excretion remained unchanged (Skoog et al., 1962). It has also been reported that corticosteroids decrease the tubular re-absorption of Ca resulting in the high urinary excretion of Ca (Laake, 1960). This increase in calciuria may be associated with the increased urinary excretion of radiostrontium.

The effect of several corticoids was studied on radiostrontium metabolism in man. A tracer dose of ^{85}Sr was given before and during the administration of corticosteroids and the ^{85}Sr plasma levels and the urinary and fecal excretions of ^{85}Sr and of Ca were determined.

Methylprednisolone (40 mg/day) increased the urinary Ca excretion from an average of 300 mg/day to more than 400 mg/day. The urinary ^{85}Sr excretions were high in the control study, 25 % of the dose being excreted on the first day of the intravenous injection of ^{85}Sr and 60 % were excreted in 12 days. Eleven days after the start of corticosteroid therapy, a second tracer dose of ^{85}Sr was given. The urinary ^{85}Sr excretion was even higher, 38 % on the first day and 70 % of the dose were excreted in 12 days.

The effect of the corticoid triamcinolone on ^{85}Sr and Ca metabolism was also investigated in man. In two ^{85}Sr studies performed during the administration of triamcinolone, the changes in ^{85}Sr excretion reflected the changes in urinary Ca excretion. In the control study, the urinary Ca excretion was approximately 100 mg/day. Following the administration of triamcinolone for 11 days, the urinary Ca excretion increased to approximately 200 mg/day, and after another 26 days of triamcinolone therapy the calciuria rose to more than 300 mg/day. As the administration of this corticosteroid was continued, the urinary Ca excretion reached levels of 480 mg/day. In the control study, the urinary [85]Sr excretion was 18% of the injected dose on the first day and the 16-day cumulative urinary [85]Sr excretion was 38%. On the first day of the [85]Sr study performed after 11 days of triamcinolone administration, the urinary [85]Sr excretion had increased to about 25%, and on the first day of an [85]Sr study performed after 37 days of triamcinolone therapy this excretion was 36%. Table II shows the progressive increase in the urinary [85]Sr excretion in the two studies performed during the administration of this corticosteroid; most striking was the 16-day cumulative urinary [85]Sr excretion of approximately 74% in the [85]Sr study performed after 37 days of corticoid therapy as compared to an excretion of 38% in the control study.

TABLE II

Effect of triamcinolone on radiostrontium excretion

Study	Urinary [85]Sr (% dose)	
	day 1	day 16*
Control	18	38
Triamcinolone I†	25	58
Triamcinolone II†	36	74

* 16-Day cumulative. A single dose of [85]Sr was given intravenously in each study.
† Study I was performed after 11 days and Study II after 37 days of triamcinolone administration.

The structure of the corticosteroid apparently influences Ca metabolism and presumably, therefore, radiostrontium metabolism as well (Spencer et al., 1962, 1963. In contrast to the effect of triamcinolone, the compound 6-α-fluorotriamcinolone acetonide did not increase the urinary Ca excretion: the calciuria was actually lower during the administration of 6-α-fluorotriamcinolone than in the control study. Concomitantly, the urinary [85]Sr excretion also decreased in the two patients receiving this corticoid while the fecal [85]Sr excretion remained unchanged. Also, there was no change in fecal [85]Sr excretion during the administration of any of the other corticosteroids.

The studies of the effects of corticosteroids have again shown that changes in urinary [85]Sr excretion are associated with changes in urinary Ca excretion and that corticosteroids increase [85]Sr excretion only if the urinary Ca excretion is increased.

EFFECT OF CHANGES IN THYROID AND PARATHYROID FUNCTION ON RADIOSTRONTIUM METABOLISM IN MAN

Certain endocrine dysfunctions, such as hyperthyroidism or Cushing's syndrome, are associated with changes in Ca metabolism. In view of the qualitative relationship between Ca and Sr metabolism in man (Spencer et al., 1960; Samachson and Spencer-Laszlo, 1962), the metabolism of radiostrontium was studied in different phases of thyroid function. In addition, the effect of PTE on radiostrontium metabolism was studied in patients with hypoparathyroidism (Spencer et al., 1966).

Absorption of radioactive strontium during hyperthyroidism and hypothyroidism

The absorption of [85]Sr was determined during spontaneous hyperthyroidism and during the state of hypermetabolism induced by the administration of thyroid extract (3 grains/day). Subsequently, the absorption of [85]Sr was determined during euthyroidism induced by a therapeutic dose of ^{131}I and during myxedema which ensued following the discontinuation of the thyroid medication (Spencer et al., 1966). A tracer dose of [85]Sr was given orally in the hyperthyroid and hypermetabolic state and during euthyroidism and myxedema. The [85]Sr plasma levels were lower than normal in the hyperthyroid and hypermetabolic state and the fecal [85]Sr excretions were higher, indicating that the absorption of [85]Sr was low during these states. During euthyroidism and during myxedema which followed the correction of hyperthyroidism and the discontinuation of thyroid extract, respectively, the [85]Sr plasma levels were markedly increased. The increase in the [85]Sr plasma levels was accompanied by a corresponding decrease in fecal [85]Sr excretion (from 80 to 62% and from 81 to 53% in two patients, respectively), indicating a marked increase in [85]Sr absorption during euthyroidism and myxedema. The fecal Ca excretion also decreased to very low levels following the correction of the hypermetabolic and the hyperthyroid state. A further indication of the very high absorption of Ca during euthyroidism and myxedema was the finding that the very low fecal Ca excretion consisted almost entirely of endogenous fecal Ca.

Effect of PTE on radiostrontium absorption in man

Reports in the literature indicate that PTE did not affect the uptake of radiostrontium in bone or soft tissue in rats 24 h after the administration of a single dose of PTE (Tweedy, 1945). However, when a second dose of PTE was given in this study, the fecal radiostrontium excretion decreased and the urinary excretion increased, indicating that PTE increased the absorption of radiostrontium in intact rats. An increase in urinary ^{89}Sr and [45]Ca excretion during PTE administration has been subsequently reported (Bacon et al., 1956). Parathyroidectomy in rats did not change the intestinal absorption of [45]Ca, [85]Sr or of ^{32}P nor did it alter the transfer of [85]Sr across everted intestinal segments (Wasserman and Comar, 1961). PTE used in doses of

500 U.S.P. units/kg in the late removal of ^{85}Sr was ineffective in rats (Smith and Bates, 1965).

The effect of PTE on radiostrontium and calcium metabolism was investigated in patients with hypoparathyroidism (Spencer *et al.*, 1966). Tracer doses of ^{85}Sr were given orally before and during the administration of 200 I.U. of PTE/day. In the control studies, the ^{85}Sr plasma levels were very low and the fecal ^{85}Sr excretions were very high, approximately 93% of the administered dose, indicating that the intestinal absorption of ^{85}Sr was very low in hypoparathyroidism. During the administration of PTE, the plasma levels of ^{85}Sr were somewhat lower than in the control study and the fecal ^{85}Sr excretion did not change, indicating that PTE had not improved the absorption of ^{85}Sr, in line with the lack of changes in the fecal excretion of ^{85}Sr. The absorption of ^{45}Ca was also very low in the hypoparathyroid state, ranging from 14 to 21%, and there was no change during PTE administration. The fecal Ca excretion also remained unchanged. These studies have shown that parathyroid extract that was shown to be biologically active had little, if any, effect on the absorption of ^{85}Sr and of Ca in patients with hypoparathyroidism. The major change was the increase in urinary Ca excretion during PTE administration, the Ca balances becoming negative primarily due to the increase in calciuria.

CONCLUSIONS

Several aspects of the metabolism of radiostrontium in man have been described. The addition of Ca to the diet had an inconsistent effect on radiostrontium absorption, decreasing the absorption in some patients but not in others. The addition of phosphate to both the low and the high calcium intake resulted in a decrease in the absorption of radiostrontium. Stable Sr did not decrease the absorption of radiostrontium, and further investigations are needed in order to determine whether stable Sr might not increase the absorption of radiostrontium in man. Lactose, lysine and vitamin D had no effect on radiostrontium absorption. The changes in radiostrontium metabolism induced by the administration of estrogen, androgen, progestin and corticosteroids paralleled the changes in Ca metabolism. The most striking change due to endocrine dysfunctions was the increase in radiostrontium absorption following the correction of spontaneous hyperthyroidism and of induced hypermetabolism. In the absence of normally functioning parathyroid glands, i.e., in patients with hypoparathyroidism, the absorption of both radioactive strontium and calcium was very low and was not affected by parathyroid extract.

Supported by research grant RH–00222 from the Division of Radiological Health, U.S. Public Health Service and in part by U.S. Atomic Energy Commission contract AT(11–1)–1231.

REFERENCES

Bacon, J. A., Patrick, H., and Hansard, S. L. (1956). *Proc. Soc. exp. Biol. Med.* **93,** 349–351.

Bruce, R. S. (1963). *Nature, Lond.* **199,** 1107–1108.

Budy, A. M., Urist, M. R., and McLean, F. C. (1952). *Am J. Path.* **28,** 1143–1167.

Carlqvist, B., and Nelson, A. (1960). *Acta radiol.* **54,** 305–315.

Catsch, A. (1957). *Experientia,* **13,** 312–313.

Comar, C. L., Wasserman, R. H., and Nold, M. M. (1956). *Proc. Soc. exp. Biol. Med.* **92,** 859–863.

Comar, C. L., Wasserman, R. H., Ullberg, S., and Andrews, G. A. (1957). *Proc. Soc. exp. Biol. Med.* **95,** 386–391.

Copp, D. H., Axelrod, D. J., and Hamilton, J. G. (1947). *Am. J. Roentg.* **58,** 10–16.

Greenberg, D. M. (1945). *J. biol. Chem.* **157,** 99–104.

Greenwald, E., Samachson, J., and Spencer, H. (1963). *J. Nutr.* **79,** 531–538.

Harrison, G. E., Jones, H. G., and Sutton, A. (1957). *Br. J. Pharmac. Chemother.* **12,** 336–339.

Hegsted, D. M., and Bresnahan, M. (1963). *Proc. Soc. exp. Biol. Med.* **112,** 579–582.

Kawin, B. (1959). *Experientia,* **15,** 313–318.

Kidman, B., Tutt, M., and Vaughan, J. (1950). *J. Path. Bact.* **62,** 209–227.

Laake, H. (1960). *Acta endocr.* **34,** 60–64.

Lengemann, F. W., and Comar, C. L. (1961). *Am. J. Physiol.* **200,** 1051–1054.

MacDonald, N. S., Noyes, P., and Lorick, P. C. (1957). *Am J. Physiol.* **188,** 131–137.

MacDonald, N. S., Spain, P. C., Ezmirlian, F. and Rounds, D. E. (1955). *J. Nutr.* **57,** 555–563.

Mazzuoli, G., Biagi, E., and Coen, G. (1961). *Acta med. scand.* **170,** 21–30.

Mraz, F. R. (1962). *Proc. Soc. exp. Biol. Med.* **110,** 273–275.

Palmer, R. F., Thompson, R. C., and Kornberg, H. A. (1958a). *Science,* **127,** 1505–1506.

Palmer, R. F., Thompson, R. C., and Kornberg, H. A. (1958b). *Science,* **128,** 1505–1506.

Patrick, H., and Bacon, J. A. (1957). *J. biol. Chem.* **228,** 569–572.

Ray, R. D., Stedman, D. E., Wolff, N. K. (1956). *J. Bone Jt. Surg.* **38,** 637–654.

Samachson, J. (1963). *Clin. Sci.* **25,** 17–26.

Samachson, J., and Spencer-Laszlo, H. (1962). *J. appl. Physiol.* **17,** 525–530.

Skoog, W. A., Adams, W. S., and MacDonald, N. S. (1962). *Metabolism,* **4,** 421–433.

Smith, H., and Bates, T. H. (1965). U.K.A.E.A. PG Report 662 (CC), pp. 10–25.

Spencer, H., and Samachson, J. (1963). *J. Nutr.,* **81,** 301–306.

Spencer, H., Lewin, I., and Samachson, J. (1962). *Cancer Res.* **3,** 244.

Spencer, H., Lewin, I., and Samachson, J. (1963). *Rad. Res.* **19,** 211.

Spencer, H., Lewin, I., and Samachson, J. (1965a). *J. nucl. Med.* **6,** 338.

Spencer, H., Lewin, I., and Samachson, J. (1966). *Fed. Proc. Fedn. Am. Socs. exp. Biol.* **25,** 368.

Spencer, H., Li, M., and Samachson, J. (1961). *J. clin. Invest.* **40,** 1339–1345.

Spencer, H., Menczel, J., and Samachson, J. (1964). *Proc. Soc. exp. Biol. Med.* **117,** 59–63.

Spencer, H., Kabakow, B., Samachson, J., and Laszlo, D. (1959). *J. clin. Endocr. Metab.* **19,** 1581–1596.

Spencer, H., Li, M., Samachson, J., and Laszlo, D. (1960). *Metabolism*, **9**, 916–925.

Spencer, H., Menczel, J., Lewin, I., and Samachson, J. (1965b). *J. Nutr.* **86**, 125–132.

Spencer, H., Berger, E., Charles, M. L., Gottesman, E. D., and Laszlo, D. (1957). *J. clin. Endocr. Metab.* **17**, 975–984.

Teree, T., Gusmano, E. A., and Cohn, S. H. (1965). *J. Nutr.* **87**, 399–406.

Tweedy, W. R. (1945). *J. biol. Chem.* **161**, 105–113.

Volf, V. (1965). *Experientia*, **21**, 571–574.

Wasserman, R. H., and Comar, C. L. (1960). *Proc. Soc. exp. Biol. Med.* **103**, 124–129.

Wasserman, R. H., and Comar, C. L. (1961). *Endocrinology*, **69**, 1074–1079.

Wasserman, R. H., Comar, C. L., and Nold, M. M. (1956). *J. Nutr.* **59**, 371–383.

Wasserman, R. H., Comar, C. L., and Papadopoulou, D. (1957). *Science*, **126**, 1180–1182.

Kinetics of Strontium-85 Deposition in the Skeleton During Chronic Exposure

J. RUNDO

Health Physics and Medical Division,
Atomic Energy Research Establishment,
Harwell, Berks

SUMMARY

The skeletal retention of [85]Sr was measured in five healthy men who had been on a constant diet that included milk containing [85]Sr for 21–32 days. The retention curves were analysed into three exponential components that indicated average half-lives and average relative compartment sizes of 2·7 days (0·64), 19·2 days (0·18) and 750 days (0·18). The equations for the skeletal content during the intake period gave values that were in respectable agreement with the observed levels in the two subjects for whom experimental data were available.

INTRODUCTION

An experiment to determine the relative availability of Sr ingested in bread and in milk was carried out by G. E. Harrison and his colleagues at the Medical Research Council's Radiobiological Research Unit, Harwell. Five healthy adult males (aged 30–57) were on a special constant diet consisting principally of bread and milk, for periods of 21–32 days. The bread was made from grain grown on an experimental reserve by the Radiobiological Laboratory of the Agricultural Research Council, and it contained a relatively high amount of naturally incorporated ^{90}Sr, and the milk came from a cow that had received an intravenous dose of [85]Sr. The results of the assessment of the availability to the subjects of the two radioisotopes of Sr in the different foods, have been described (Carr *et al.*, 1962, 1965). Estimates of the body contents of [85]Sr were made by body radioactivity measurement in the Health Physics and Medical Division at A.E.R.E. The results of the measurements of the long-term retention of [85]Sr have been published (Rundo *et al.*, 1964; Rundo and Lillegraven, 1966). These results showed that, after the cessation of the exposure the retention could be described mathematically equally well by the sum of three exponential functions of time, or by the integrated form of a sum of an exponential function and a power function of time.

It will be recalled that after an intravenous injection of ^{85}Sr the retention could be described accurately from 1 day to 1 year by the sum of an exponential function of time and a power function of time (Bishop *et al.*, 1960). Although the power function is useful for long-term prediction of retention, especially in radiological protection, it has no physiological significance, whereas a retention equation consisting of the sum of several exponential functions of time has the advantage that the individual components can be ascribed to certain compartments within the body. In this paper the results are presented of a simple analysis of the three-component exponential function retention equations obtained for the five subjects.

THEORETICAL CONSIDERATIONS

Consider first the case in which a single tracer dose (*m* units) of some bone-seeking element is administered intravenously, and assume that the skeletal retention r_t at any time t (days) may be described by the equation:

$$r_t = m\sum_i a_i e^{-\lambda_i t} \tag{1}$$

where λ_i is a biological elimination constant and $\sum_i a_i = 1$. It must be emphasised that in this simple analysis it is assumed that the i compartments represented by the exponential terms are independent of one another.

Now, $r_0 = m$, which implies that there must have been instantaneous transfer from blood to bone. Since this is impossible, the retention equation is not valid at short times after injection, say $t < 1$. Equation (1) must be modified when we are concerned with ingestion, and the retention is then given by the equation:

$$r_t = mf\sum_i a_i e^{-\lambda_i t} \tag{1a}$$

where f is the absorbed fraction of the ingested dose. Since $r_0 = mf$, the implication is that there was instantaneous transfer from the gut to the bone, since r_t describes the skeletal retention. Again the equation is invalid for small values of t.

A mathematical description of the retention during chronic oral exposure to m units/day is obtained by integrating equation (1a) with respect to time, the accuracy of the resulting equation depending on how closely the pattern of intake approximates to a continuous one. Provided that one does not look for differences in the retention at intervals of time of less than 1 day:

$$R_t = mf\sum_i \frac{a_i}{\lambda_i}(1 - e^{-\lambda_i t}) \tag{2}$$

where R_t is the integrated retention. When exposure is terminated after t_m days, the retention at time t $(t > t_m)$ is given by the equation:

$$R_t = mf \sum_i \left[\frac{a_i}{\lambda_i} (1 - e^{-\lambda_i t_m}) e^{-\lambda_i (t - t_m)} \right]$$

$$= mf \sum_i [A_i e^{-\lambda_i (t - t_m)}] \tag{3}$$

where $A_i = \frac{a_i}{\lambda_i} (1 - e^{-\lambda_i t_m})$. Thus a graph of the retention as a function of time after the end of the exposure can be resolved to give values for λ_i and fA_i, provided that the values of λ_i are sufficiently different from one another. From these values of λ_i and fA_i, fa_i may be deduced, and since $\sum_i a_i = 1$, f and a_i may also be obtained.

EXPERIMENTAL METHODS

Body radioactivity measurements for the estimation of the content of ^{85}Sr were made by scintillation γ-ray spectrometry with four crystals of thallium-activated sodium iodide, 10·8 cm in diameter by 5 cm thick. To reduce the counter background, the measurements were made inside a lead shield with a wall thickness of 10·2 cm. The equipment, which has been described in detail (Rundo, 1958), was calibrated from the results of measurements of the radiation from each subject, 8 days after the end of the exposure period, with a single crystal.

This was placed at a distance of 1 metre from a special bed that formed an arc of a circle with the detector at the centre. Systematic measurements were made of the radiation from a "point" source of ^{85}Sr of known activity, in a tissue-equivalent phantom of various configurations. From the results of these measurements, the body content of each subject was deduced by a method that has been described fully by Lillegraven and Rundo (1965).

Estimates were also made of the skeletal content in the presence of large amounts of unabsorbed ^{85}Sr in the gastrointestinal tract. Initially this was only done during the first few days after the end of the intake period, in order to determine the skeletal retention at the end, but for two of the subjects these determinations were also made at intervals during the intake period. In this way the entry of ^{85}Sr into the skeleton could be followed. Thus, "profile" curves of the radioactivity distribution in the body were made by using a crystal fitted with a slit collimator (Lillegraven, 1964). Scattered radiation was eliminated by recording only those pulses in the full-energy peak. Examination of the profile curves showed that the counting-rates observed in the measurements made up to 40 cm and beyond 100 cm from the top of the head were independent of the presence of radioactivity in the gastrointestinal tract.

The area under a profile curve in these regions was therefore used as a measure of the skeletal content; it was converted to microcuries from the results of a profile curve made at a time when the body content, measured with the four-crystal spectrometer, was essentially all in the skeleton. All the estimates of body and skeletal contents were corrected for radioactive decay to the end of the intake period. The half-life of ^{85}Sr was taken as 65·0 days.

RESULTS

In Fig. 1 the skeletal retention for subject G.R.H. is plotted semi-logarithmically as a function of time after the radioactive diet to demonstrate the graphical analysis into three exponential components. R_t/m (the retention per unit daily intake) has been plotted, rather than R_t, so that the intercepts of the

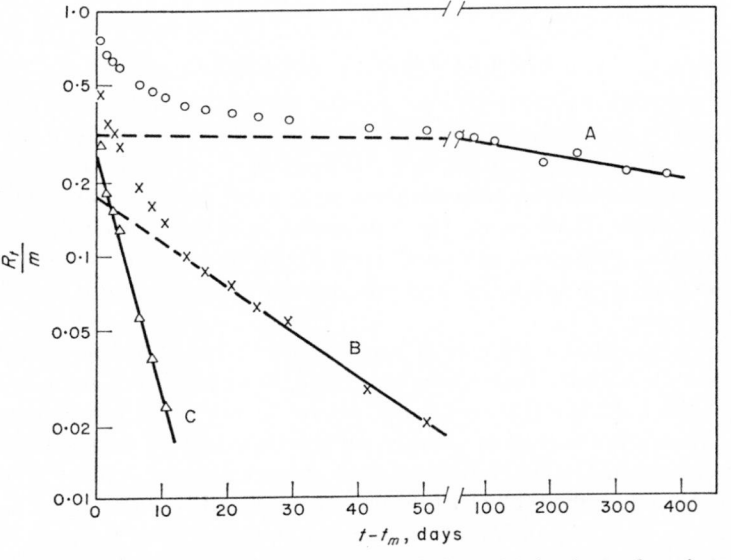

FIG. 1. The retention of ^{85}Sr in subject G.R.H. ($t_m = 21$ days) as a function of time after the end of the constant diet, showing the analysis of the data into three exponential functions. Note the change of scale at $t - t_m = 55$ days. A, 610-day component; B, 16·1-day component; C, 3·03-day component.

three straight lines at $t - t_m = 0$ are the values of fA_i. From the slopes of the lines the values of λ_i are obtained; the corresponding values of the half-lives are set out in Table I. For subjects G.E.H. and J.F.L., the values may be compared with those obtained from a similar analysis of the retention curve obtained after an intravenous injection of 0·5 μCi ^{85}Sr 4 years previously (Bishop *et al.*, 1960). The half-lives obtained from that study are included in parentheses in Table I, and they are quite similar to those found in the present study.

TABLE I

Half-lives in days of the three components for the five subjects

Subject	$\dfrac{0\cdot693}{\lambda_1}$	$\dfrac{0\cdot693}{\lambda_2}$	$\dfrac{0\cdot693}{\lambda_3}$
T.E.F.C.	2·6	24	860
G.E.H.	3·2	20	670
	(2·7)	(25)	(635)
G.R.H.	3·0	16	610
J.F.L.	2·0	17	630
	(1·7)	(12)	(630)
F.S.W.	2·6	19	980
Mean	2·68	19·2	750

From the values of fA_i and λ_i, the values of fa_i were calculated by using the definition (equation 3):

$$fA_i = \frac{fa_i}{\lambda_i}\,(1 - e^{-\lambda_i t}m).$$

The results are given in Table II. It is of interest to compare the values of $f\ (=\sum_i fa_i)$ with Harrison's estimates of this quantity in his paper at this Symposium (p. 161); he found a mean value of 0·148, with a range of 0·103 to 0·214. There is good agreement between the two sets of values, as determined by entirely different methods. The largest discrepancy (in the case of subject G.R.H.) is a little more than 25%.

TABLE II

Values of fa_i and of f for the five subjects

Subject	fa_1	fa_2	fa_3	$f=\sum_i fa_i$
T.E.F.C.	0·132	0·024	0·031	0·187
G.E.H.	0·047	0·015	0·017	0·079
G.R.H.	0·059	0·013	0·015	0·087
J.F.L.	0·109	0·051	0·024	0·184
F.S.W.	0·081	0·020	0·026	0·127
Mean	0·095	0·025	0·023	0·133

Finally, values of a_i follow immediately since:

$$a_i = \frac{fa_i}{\sum_i fa_i}$$

and these are set out in Table III, where the results obtained 4 years previously from the intravenous-injection study (Bishop *et al.*, 1960) are again shown in parentheses; they are in respectable agreement with the values from the present study.

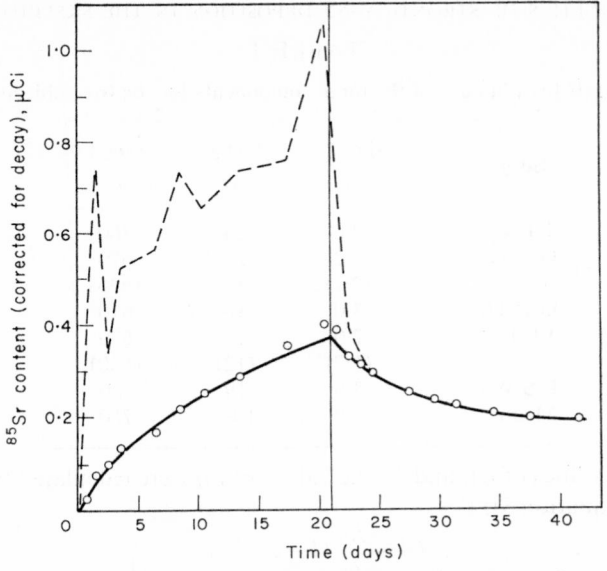

FIG. 2. Skeletal (continuous line) and total body (broken line) contents of ^{85}Sr during the period of the constant diet and for the 3 weeks after, for subject G.R.H. For $t \geqslant 21$ days the smooth curve was calculated from equation (3), and for $t \leqslant 21$ days the curve was calculated from equation (2).

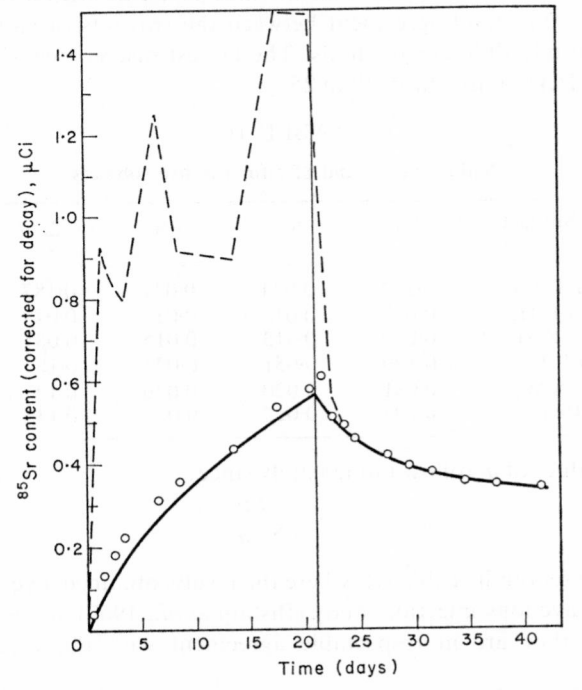

FIG. 3. As for Fig. 2, but results for subject F.S.W. (Skeletal content, continuous; whole body content, broken line).

For subjects G.R.H. and F.S.W. profile curves were determined at intervals during the 21-day exposure period and the results were used to calculate the bone content, which was corrected for radioactive decay to the end of the period of the radioactive diet. The values are plotted as a function of time in Figs. 2 and 3; the curves drawn through the experimental points for $t \leqslant 21$ days were calculated from equation (2) with the appropriate values of m, f, a_i and λ_i. For $t \geqslant 21$ days the curves were calculated from equation (3)., the constants of which were of course determined from the experimental retention values as shown in Fig. 1. Also included in Figs. 2 and 3 are the values of the total body contents (broken lines). From 2·5–3·5 days after the end of the constant diet, the total body contents were indistinguishable from the skeletal contents.

In any assessment of the differences between calculated and observed skeletal contents two points should be noted: (i) the profile counter also responded to ^{85}Sr in the blood, so that the skeletal content would have been over-estimated slightly; and (ii) entry into the skeleton is not an instantaneous process, as is assumed in the application of equation (2) so that the skeletal content must lag behind the calculated content.

With these reservations in mind, the excellent agreement between calculated and observed skeletal contents for G.R.H. (Fig. 2) is gratifying, but a little surprising. Possibly the two effects just mentioned cancel each other. The agreement is less good for F.S.W. (Fig. 3), where the measured content exceeds the calculated by about 0·05 μCi (10% of one day's intake) in the first 9 days, and rather less subsequently.

CONCLUSIONS

The entry of ^{85}Sr into the skeleton during chronic oral exposure seems to be characterized by three exponential components, the rate constants and relative compartment sizes of which may be determined from the retention after

TABLE III

Relative compartment sizes for the five subjects

Subject	a_1	a_2	a_3
T.E.F.C.	0·71	0·13	0·16
G.E.H.	0·59	0·19	0·22
	(0·64)	(0·13)	(0·23)
G.R.H.	0·68	0·15	0·17
J.F.L.	0·59	0·28	0·13
	(0·62)	(0·18)	(0·20)
F.S.W.	0·64	0·16	0·20
Mean	0·642	0·182	0·176

the end of the exposure. It is reasonable to attribute the fastest and largest component (amounting to 0·64 of the absorbed fraction; Table III) to the freely exchangeable Sr pool, consisting of plasma, extra-cellular fluid and (predominantly) bone surfaces. The average half-life of 2·68 days is consistent with the findings of Eisenberg and Gordan (1961), Cohn et al. (1962) and Harrison (1963). The second component (average half-life 19·2 days) amounts to about 0·18 of the absorbed fraction and it is attributed to a closer binding of Sr with the apatite crystals of the bone (Bishop et al., 1960).

The third component also amounts to about 0·18 of the absorbed fraction, and its very long half-life (750 days) is probably a consequence of the penetration of the Sr into the apatite crystal lattice. Bone resorption is then the only means by which Sr is released from the skeleton.

I am grateful to Mr A. L. Lillegraven and Miss J. I. Mason for much assistance and to Dr G. E. Harrison for many helpful discussions.

REFERENCES

Bishop, M., Harrison, G. E., Raymond, W. H. A., Sutton, A., and Rundo, J. (1960). *Int. J. Radiat. Biol.* **2**, 125–142.

Carr, T. E. F., Harrison, G. E., Loutit, J. F., and Sutton, A. (1962). *Nature, Lond.* **194**, 200–201.

Carr, T. E. F., Harrison, G. E., Loutit, J. F., and Sutton, A. (1965). *Proc. Nutr. Soc.* **24**, 120–126.

Cohn, S. H., Spencer, H., Samachson, J., and Robertson, J. S. (1962). *Radiat. Res.* **17**, 173–185.

Eisenberg, E. and Gordan, G. S. (1961). *J. clin. Invest.* **40**, 1809–1825.

Harrison, G. E. (1963). *In* "Diagnosis and Treatment of Radioactive Poisoning", pp. 119–129. Vienna, International Atomic Energy Agency.

Lillegraven, A. L. (1964). *J. appl. Phys.* **35**, 1974–1982.

Lillegraven, A. L., and Rundo, J. (1965). *Acta Radiol.* **3**, 369–383.

Rundo, J. (1958). *2nd Int. Conf. peaceful Uses atom. Energy*, **23**, 101–112.

Rundo, J., and Lillegraven, A. L. (1966). *Br. J. Radiol.* **39**, 676–685.

Rundo, J., Lillegraven, A. L., and Mason, J. I. (1964). *In* "Proceedings of the 1st European Bone and Tooth Symposium", pp. 159–165. Oxford, Pergamon Press.

An Attempt to Quantitate the Short-term Movement of Strontium in the Human Adult

T. E. F. CARR

*Medical Research Council Radiobiological Research Unit,
Harwell, Didcot, Berks*

SUMMARY

Results previously reported (after measurements on five subjects fed for 3 or 4 weeks on a diet containing known amounts of stable Sr and ^{85}Sr) are used in conjunction with a simple model to quantitate the movement of Sr in the adult human. Calculations are made of absorbed, retained and excreted fractions of the dietary Sr as well as Sr removed from bone and either excreted or returned to the bone. Estimates are also given of the Sr balance of the subjects over the periods concerned and of approximate Sr pool sizes and equilibration times.

INTRODUCTION

The continuous feeding of a diet of known specific activity has been used by several authors to estimate the various parameters of Ca and Sr metabolism, in particular, bone resorption and bone accretion rates (Gran, 1960; Adams and Carr, 1965; Nordin *et al.*, 1964; Holtzman, 1965). Similar estimates have been made from experimental data previously reported by this laboratory in which five normal adult men consumed a constant diet of known ^{85}Sr specific activity for three or four weeks.

Much of the procedure used in these experiments has been described previously (Carr *et al.*, 1962a, 1962b, 1965; Harrison, 1963). A controlled intake of food with known stable Sr and ^{85}Sr content was given to each subject and measurements of both isotopes were made in urine and faeces. The ^{85}Sr content of samples was measured in a well-scintillation γ-ray counter, and the stable Sr content was measured by flame spectrophotometry. Skeletal retention of ^{85}Sr in each subject was estimated by body radioactivity measurement during and after the period of ingestion of the diet (Rundo *et al.*, 1964; Rundo and Lillegraven, 1966).

THEORETICAL CONSIDERATIONS

A relatively simple model consisting of three compartments, is visualized for the movement of Sr in the body (Fig. 1). Sr absorbed through the gut supplies the first compartment which consists of the plasma and part of the extracellular fluid. There is a loss of Sr from this pool by urinary and endogenous faecal excretion.

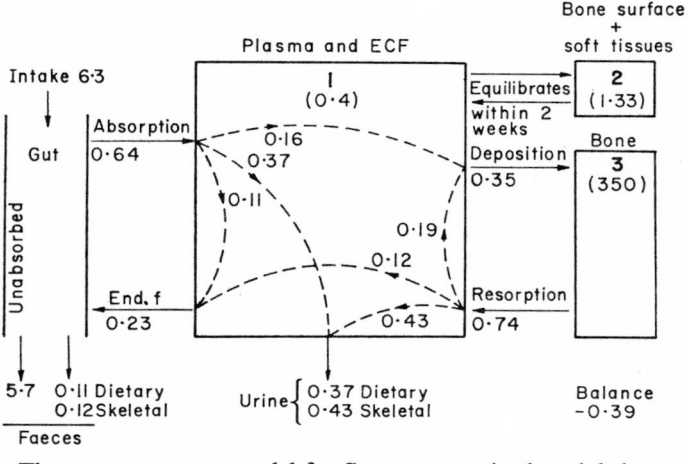

FIG. 1. Three-compartment model for Sr movement in the adult human. Quantities shown as mg Sr/week for the third week on the experimental diet of subject G.E.H.

The second compartment, consisting of soft tissue and bone surfaces, equilibrates with the first compartment within 2 weeks. The third compartment, entirely in bone, is known to be complex, but is treated as a single compartment over the time concerned in these experiments. Both the second and third compartments are in equilibrium with the first, and Sr movement into and out of the third compartment, considered as bone formation and resorption, can then be estimated.

When this model is applied to the observations it is necessary to make certain assumptions: (a) that the specific activity of the plasma (S.A.P.) is the same as that of the urine (S.A.U.); (b) that the ratio (f/U) of endogenous faecal excretion (f) to urinary excretion is known; (c) that Sr resorbed from bone enters the plasma and is completely mixed with this pool before any returns to bone; (d) that no significant resorption of labelled bone occurs during the period of observation.

It is possible to calculate the rate and magnitude of strontium movements from the following equations:

$$\text{Dietary Sr retained} = \frac{^{85}\text{Sr retained}}{\text{S.A.D.}} \tag{1}$$

$$\text{Plasma Sr retained} = \frac{^{85}\text{Sr retained}}{\text{S.A.U.}} \tag{2}$$

$$\text{Resorbed Sr re-incorporated} = \text{Plasma Sr retained} - \text{dietary Sr retained} \tag{3}$$

$$\text{Dietary Sr in urine} = \frac{^{85}\text{Sr in urine}}{\text{S.A.D.}} \tag{4}$$

$$\text{Dietary Sr in } f = \text{Dietary Sr in urine} \times f/U \tag{5}$$

$$\text{Resorbed Sr in urine} = \text{Total Sr in urine} - \text{dietary Sr in urine} \tag{6}$$

$$\text{Resorbed Sr in } f = (\text{Sr in urine} \times f/U) - \text{dietary Sr in } f \tag{7}$$

$$\text{Total resorbed Sr} = \text{Resorbed Sr re-incorporated} + \text{resorbed Sr in urine} + \text{resorbed Sr in } f \tag{8}$$

$$\text{Sr balance} = \text{Dietary Sr retained} - (\text{resorbed Sr in urine} + \text{resorbed Sr in f}) \tag{9}$$

RESULTS

The daily excretion of ^{85}Sr in the urine of one subject (G.E.H.) during the period of the diet is shown in Fig. 2. Although the daily variation in output is appreciable, the mean weekly output after the second week shows only a

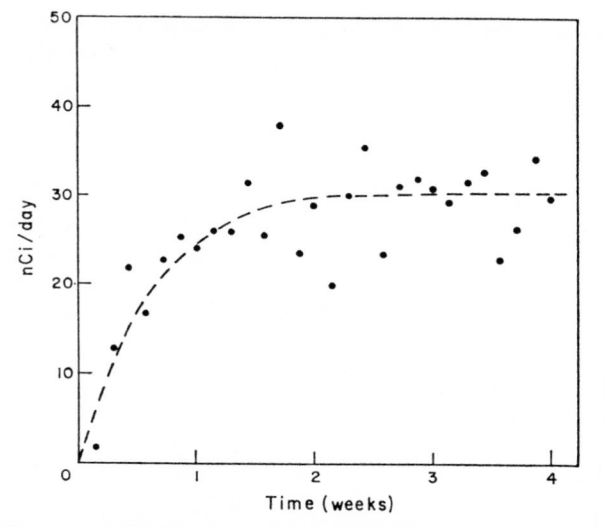

FIG. 2. Daily output of ^{85}Sr in urine by subject G.E.H. during period of ingestion of the marker.

slight rise. The daily urinary output of ^{85}Sr from the other subjects gave curves which were similar in shape to that in Fig. 2, differing only in the magnitude of the urinary excretion.

The stable Sr assays in urine were made on weekly bulked samples, and the weekly mean specific activities for the five subjects are given in Fig. 3. Again after the first two weeks the values are approximately constant at about half the specific activity of the diet.

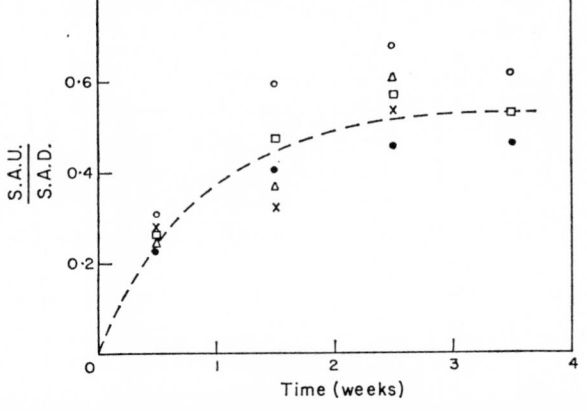

FIG. 3. Specific activity of the urine (S.A.U.) as a fraction of the specific activity of the diet (S.A.D.) for: ○, T.E.F.C.; △, F.S.W.; ●, G.E.H.; □, J.F.L.; ×, G.R.H.

The values of skeletal retention of ^{85}Sr, derived from body radioactivity measurement by Rundo and Lillegraven (1966) are shown in Fig. 4 for subject G.E.H. during the ingestion of the tracer. After the first week, these values

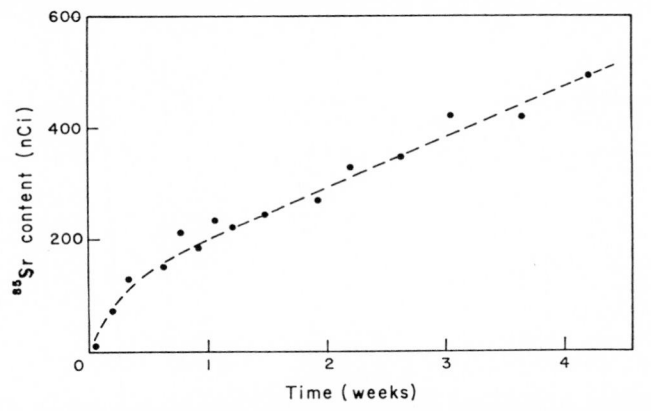

FIG. 4. Estimated skeletal retention of ^{85}Sr in subject G.E.H. during the ingestion of the marker, derived from body radioactivity measurements.

show an approximately constant rise with time and apart from the first week, a linear relationship has been assumed over this period. Corresponding values were derived from similar curves for the other subjects and were used in the calculations.

The mean ratio of endogenous faecal excretion to urinary excretion of Sr is required for each subject, so that estimates of endogenous faecal excretion can be made from the urine during the period of ingestion. It was not possible to measure this directly during the intake of ^{85}Sr, so the ratio has been derived from the total faecal and urinary output of ^{85}Sr after the end of the intake period and when the gut was clear of unabsorbed ^{85}Sr (G. E. Harrison, this book, p. 161).

The intake of stable Sr was measured and the output was determined in both urine and faeces, so that estimates of the stable Sr balance could be made for each subject. The weekly balances are given in Table I for each subject during the dietary period.

TABLE I

Stable strontium balances from intake minus excretion (mg Sr/week)

| | | Week | | |
Subject	1	2	3	4
T.E.F.C.	$-4{\cdot}3$	$-0{\cdot}8$	$-0{\cdot}3$	$-1{\cdot}9$
G.E.H.	$-4{\cdot}2$	$-1{\cdot}1$	0	$-1{\cdot}3$
G.R.H.	$-1{\cdot}5$	$-2{\cdot}0$	$-0{\cdot}1$	—
J.F.L.	$-2{\cdot}2$	$-2{\cdot}2$	$+0{\cdot}3$	$-1{\cdot}5$
F.S.W.	$-2{\cdot}2$	$+0{\cdot}1$	$-0{\cdot}3$	—

Table II gives the basic weekly data for subject G.E.H. from which the estimates of Sr movement in the body are derived. Values for the other

TABLE II

Intake, urine and retention of strontium and strontium-85 in subject G.E.H.
Values are weekly, intake of ^{85}Sr was 3·5 μCi/week and f/U was 0·29

| | | Week | | |
	1	2	3	4
Stable Sr in diet, mg	6·3	6·3	6·3	6·4
Stable Sr in urine, mg	0·99	0·88	0·80	0·83
^{85}Sr in urine, nCi	124	198	202	212
^{85}Sr retention, nCi	204	90	90	90
Specific activity of diet (S.A.D.), nCi/mg	555	555	555	547
Specific activity of urine (S.A.U.), nCi/mg	125	225	253	256

subjects are not given but the estimates of Sr movements are given in later Tables. All values have been normalized to an intake of 0·5 μCi ⁸⁵Sr/day. Values of the parameters defined by equations (1) to (8) were calculated and are given in Table III.

TABLE III

Retention, excretion and resorption of strontium in subject G.E.H.
derived from data in Table II (mg Sr/week)

	Week			
	1	2	3	4
Dietary Sr retained	0·37	0·16	0·16	0·16
Plasma Sr retained	1·63	0·40	0·35	0·35
Resorbed Sr reincorporated	1·26	0·24	0·19	0·19
Dietary Sr in urine	0·22	0·36	0·37	0·38
Dietary Sr in f	0·065	0·11	0·11	0·11
Resorbed Sr in urine	0·77	0·52	0·43	0·45
Resorbed Sr in f	0·22	0·15	0·12	0·13
Total Sr resorbed	2·25	0·91	0·74	0·77

TABLE IVa

Absorption and retention of dietary strontium for subject G.E.H.

		Week				
		1	2	3	4	Mean
Absorption	% dose/week	10·4	9·9	10·0	10·4	10·2
	mg Sr/week	0·66	0·63	0·64	0·66	0·65
Retention	% dose/week	5·8	2·6	2·6	2·6	2·6
	mg Sr/week	0·37	0·16	0·16	0·16	0·16

TABLE IVb

Mean absorption and retention of strontium per week for five subjects

	Absorption		Retention	
Subject	% dose	mg Sr	% dose	mg Sr
T.E.F.C.	22·5	1·55	5·0	0·34
G.E.H.	10·2	0·65	2·6	0·16
G.R.H.	11·6	0·98	3·4	0·35
J.F.L.	17·0	1·11	3·3	0·22
F.S.W.	12·6	1·15	3·9	0·26

Note: The mean values for retention exclude the first week.

From the data in Tables II and III the absorption and retention can be found in milligrams stable Sr per week or as percentage of intake per week; these are given in Table IVa for subject G.E.H. The mean values for absorption are based on all four weeks; the mean values for retention exclude the first week. In Table IVb these mean values are given for all five subjects.

The total resorbed Sr for all five subjects is given in Table V, and has been calculated as for subject G.E.H. in Table III. After the second week, the total resorbed Sr is 0·74 mg/week and is very similar for the other subjects.

TABLE V

Total strontium resorbed for five subjects (mg Sr/week)

Subject	Week			
	1	2	3	4
T.E.F.C.	4·07	0·97	0·74	0·94
G.E.H.	2·25	0·91	0·74	0·77
G.R.H.	2·40	1·92	0·77	—
J.F.L.	3·53	1·20	0·80	0·96
F.S.W.	3·91	1·85	0·78	—

TABLE VIa

Loss of resorbed strontium, retention of dietary strontium and resultant balance, subject G.E.H. (mg Sr/week)

	Week			
	1	2	3	4
Loss of resorbed Sr	0·99	0·67	0·55	0·58
Retention of dietary Sr	0·37	0·16	0·16	0·16
Balance	− 0·62	− 0·51	− 0·39	− 0·42

TABLE VIb

Strontium balance for five subjects (mg Sr/week)

Subject	Week			
	1	2	3	4
T.E.F.C.	− 0·51	− 0·39	− 0·22	− 0·38
G.E.H.	− 0·62	− 0·51	− 0·39	− 0·42
G.R.H.	− 0·76	− 1·12	− 0·39	—
J.F.L.	− 1·21	− 0·75	− 0·42	− 0·55
F.S.W.	− 0·65	− 0·90	− 0·20	—

The amount of resorbed Sr lost in urine and endogenous faeces, in the process of resorption and re-incorporation, must to some extent be compensated by the dietary Sr retained. If this compensation is complete a condition of balance will exist; if not, then either a positive or negative balance will occur. The weekly loss of Sr by excretion and the weekly gain of dietary Sr retained (see equation (9)) together with the balance for subject G.E.H. is given in Table VIa. Weekly balances for all five subjects are given in Table VIb. All subjects were in negative balance throughout the experiment and by the third week it was somewhat less than 0·5 mg Sr/week.

The estimated skeletal retention of ^{85}Sr and the urinary output over the period on the diet, Figs. 2 and 4, show that the second compartment reaches equilibrium with the first within two weeks. Thus an estimate of the size of the second compartment can be made. It will be equal to the excess of the total plasma Sr retained in the first two weeks over that in the third or fourth weeks (see Table III). The values for the size of the second compartment are given for all subjects in Table VII.

TABLE VII

Size of second compartment for five subjects (mg Sr)

T.E.F.C.	G.E.H.	G.R.H.	J.F.L.	F.S.W.
3·06	1·33	1·48	1·81	3·05

By using all the previous data it is possible to quantitate the weekly movement of Sr in the subjects. An example is given in Fig. 1 for subject G.E.H. during the third week of the dietary intake.

DISCUSSION

The method used in this paper to quantitate Sr metabolism has previously been used by other authors, principally to measure bone formation and resorption rates with Ca isotopes. In earlier experiments where retention was only measured at the end of the intake period (Gran, 1960; Adams and Carr, 1965), it was necessary to make the plateau of specific activity in plasma long compared to the rise time, so that short-term exchange could be neglected.

The advantages of the foregoing calculations are due to the estimates of ^{85}Sr skeletal retention available during the period of the intake. The estimates of ^{85}Sr retention from body radioactivity measurement exclude the effect of gut loading of the isotope and make it unnecessary to use total faecal measurements to estimate retention as dose minus excretion.

The estimate of the f/U ratio used during the period of the diet was based on a value found following the cessation of the diet. However, the f/U found after the intake period might not hold during the intake, as the diet was different in the two periods. The use of an intravenous marker for Sr to measure the f/U during the intake was precluded for technical reasons.

The various theories (Aubert *et al.*, 1963) on the origin of endogenous faecal excretion are not directly concerned in this measured f/U; it is only the loss of Sr by this route that is required. The only way in which the origin of the endogenous faecal loss affects the calculations is in the estimate of absorption. The values in Tables IVa and IVb are for net absorption. Since the endogenous faecal Sr results from Sr secreted into the lumen of the gut, but not re-absorbed, the relative position of secretion and absorption sites along the length of the gut will influence any correction made to this net absorption.

The total Sr resorbed in all subjects is very similar during the third week; some part of the variation could be due to skeletal size and age of the subjects. That the Sr deposition rate was less than the resorption rate (Tables VIa and VIb) for all subjects indicates negative balance on the diet employed. This state of negative balance is further upheld by the balances (Table I) obtained from total Sr intake and output, though it is well known that these are difficult to measure unless the imbalance is very large. The experimental diet involved was low in Sr compared to the normal diets of the subjects; in fact for subjects G.E.H., J.F.L. and T.E.F.C., it was about half the normal intake on an unrestricted diet.

Strontium compartments

The size of the first compartment (plasma plus extra-cellular fluid), cannot be estimated from this experiment, but for subject G.E.H. it is known from other experiments when intravenous injections of Sr were given (Bishop *et al.*, 1960; Harrison *et al.*, 1966). These indicate a volume of 13·5 l or 0·4 mg Sr (plasma contains 30 μg Sr/l).

The second compartment, the size of which can be found by these calculations, equilibrated within the first 2 weeks, but without detailed curve analysis its time constant cannot be found. This compartment is usually identified with a labile exchange process in bone, but may involve further movement in extra-cellular fluid and soft tissue as well as surface bone phenomena.

The third and last compartment considered in the model is the bulk of the bone Sr. Since an attempt is made to estimate deposition and resorption rates over the period concerned, it must be assumed that significant resorption of labelled bone does not occur. This must occur in the long run, but the size of this compartment is very large compared to the rates found. The size of this compartment for subject G.E.H. would be about 350 mg Sr with a resorption rate of 0·74 mg Sr/week and a deposition rate of 0·35 mg Sr/week during the third week of the diet. The negative balance that is found is probably due to a reduction in deposition rate rather than an increase in resorption rate. If this is so, then under balanced conditions on a higher intake of Sr, the bone resorption and deposition rates would be about 0·74 mg Sr/week for subject G.E.H. (weight 71 kg).

By using this figure of 0·74 mg Sr/week with plasma levels of 30 μg Sr/l and 100 mg Ca/l, the daily deposition and resorption for Ca would be 0·35 g Ca/day (5 mg Ca/day/kg), which is well within the accepted normal limits. The bone resorption rate of 0·74 mg Sr/week and a total skeletal Sr content of about 350 mg would give a skeletal turnover of 10%/year, also in good agreement with other estimates (Bryant and Loutit, 1964).

I gratefully acknowledge the help given by A. Sutton in the preparation of this paper and the technical assistance of Miss Hilda Shepherd and Mr K. B. Edward.

REFERENCES

Adams, P. J. V., and Carr, T. E. F. (1965). *In* "Proceedings of the 2nd European Symposium, 1964" (Richelle, L. J., and Dallemagne, M. S., eds), pp. 145–155. Liege, The University.

Aubert, J. P., Bronner, F., and Richelle, L. J. (1963). *J. clin. Invest.* **42,** 885–897.

Bishop, M., Harrison, G. E., Raymond, W. H. A., Sutton, A., and Rundo, J. (1960). *Int. J. Radiat. Biol.* **2,** 125–142.

Bryant, F. J., and Loutit, J. F. (1964). *Proc. R. Soc.* **159B,** 449–465.

Carr, T. E. F., Harrison, G. E., Loutit, J. F., and Sutton, A. (1962a). *Nature, Lond.* **194,** 200–201.

Carr, T. E. F., Harrison, G. E., Loutit, J. F., and Sutton, A. (1962b). *Br. med. J.* ii, 773–775.

Carr, T. E. F., Harrison, G. E., Loutit, J. F., and Sutton, A. (1965). *Proc. Nutr. Soc.* **24,** 120–126.

Gran, F. C. (1960). "Studies on calcium and strontium-90 metabolism in rats." Norwegian Monographs on Medical Science. Oslo, University Press.

arrison, G. E. (1963). *In* "Diagnosis and Treatment of Radioactive Poisoning", pp. 119–129. Vienna, International Atomic Energy Agency.

Harrison, G. E., Carr, T. E. F., Sutton, A., and Rundo, J. (1966). *Nature, Lond.* **209,** 526–527.

Holtzman, R. B. (1965). *Radiat. Res.* **25,** 277–294.

Nordin, B. E. C., Smith, D. A., and Nisbet, J. (1964). *In* "Medical uses of ⁴⁷Ca", pp. 47 and 75–85. 2nd Panel Report (Technical Report Series No. 32). Vienna, International Atomic Energy Agency.

Rundo, J., and Lillegraven, A. L. (1966), *Br. J. Radiol.* **39,** 676–685.

Rundo, J. Lillegraven, A. L., and Mason, J. I. (1964). *In* "Proceedings of the 1st European Bone and Tooth Symposium", pp. 159–165. Oxford, Pergamon Press.

A Comparison between Calcium-45 and Strontium-85 Absorption, Excretion and Skeletal Uptake

J. SHIMMINS,

Western Regional Hospital Board, Regional Physics Department,
9–13 West Graham Street, Glasgow, C.4

D. A. SMITH,

University Department of Medicine
Gardiner Institute, Western Infirmary, Glasgow

B. E. C. NORDIN

M.R.C. Mineral Metabolism Research Unit
Leeds General Infirmary, Leeds

AND L. BURKINSHAW

M.R.C. Environmental Radiation Research Unit,
Leeds General Infirmary, Leeds

SUMMARY

The absorption, faecal and urinary excretion and bone mineralization rates (M) of ^{85}Sr and ^{45}Ca have been compared by following the continuous feeding of the isotopes at different levels of dietary Ca. The net absorption of ^{85}Sr was found to be less than that of ^{45}Ca at all levels of Ca intake. The net absorption of both isotopes fell as the Ca intake was increased. The urinary excretion and skeletal retention of both isotopes also decreased with increasing Ca intake. A discrimination factor of 3·34 in favour of ^{85}Sr compared to ^{45}Ca was found in the kidneys when simultaneous blood and urine samples were compared. The bone-mineralization rates calculated from ^{85}Sr and ^{45}Ca were not found to differ significantly.

The bone-mineralization rates were also estimated by the method of Bauer *et al.* by following the simultaneous intravenous administration of both isotopes and the data combined with those of Dow and Stanbury. The ^{85}Sr bone-mineralization rate was found to correlate with the ^{45}Ca bone-mineralization rate but there were large individual discrepancies

INTRODUCTION

Discrimination between Ca and Sr in the transport of these ions across biological membranes has been recognized for some time. This discrimination has been shown to occur in transport across the gut (Bailey *et al.*, 1960; Spencer *et al.*, 1960; Samachson, 1963; Nordin *et al.*, 1964b), in re-absorption in renal tubules (Harrison *et al.*, 1955; MacDonald *et al.*, 1957; Barnes *et al.*, 1961) and in placental transfer (Hodges *et al.*, 1950). The discrimination between Ca and Sr at the bone surface has been investigated in several ways. The transfer of Sr and Ca tracers into bone powder has been followed (Harrison *et al.*, 1959; Boyd *et al.*, 1959; Neuman *et al.*, 1963), and the uptake of these tracers *in vivo* in animals (Bauer *et al.*, 1955; Comar *et al.*, 1956; Likins *et al.*, 1959), and in man (Dow and Stanbury, 1960; Nordin *et al.*, 1962; Bronner *et al.*, 1963) has also been investigated.

It is clear that there is discrimination against the transfer of Sr as compared with Ca across biological membranes, though why it occurs is not understood. Whether discrimination occurs at the bone surface has not been clearly established.

We report in this paper two series of studies in man. In the first, we have investigated the effect of different stable Ca intakes on the retention and faecal and urinary excretion of ^{45}Ca and ^{85}Sr during continuous administration of these isotopes (Nordin *et al.*, 1964a).

In the second, we have administered ^{45}Ca and ^{85}Sr intravenously by a single injection and estimated the bone-mineralization rate with both isotopes, and we have combined these results with those of Dow and Stanbury (1960).

EXPERIMENTAL METHODS

All the patients were studied in a metabolic ward, and were undergoing Ca-balance studies as well as isotope investigation, the urine and faeces being collected in 7-day aliquots.

Continuous-feeding studies

^{45}Ca and ^{85}Sr were administered simultaneously twice daily with the morning and evening meals. Doses, consisting of 0·05 μCi of ^{45}Ca and 0·25 μCi of ^{85}Sr, were given in 2·5 to 5 mg of Ca as CaCl$_2$. Three weeks were allowed for equilibration before estimations were begun (Nordin *et al.*, 1964a). The isotopes and stable Ca were estimated in the 7-day urine and faecal collections; simultaneous plasma and urine samples were collected for the estimation of stable Ca, ^{45}Ca and ^{85}Sr.

The net absorption of each isotope was taken to be the oral dose less the faecal activity averaged over 2–3 weeks. Skeletal retention was taken to be the oral dose less the faecal and urinary activities averaged over the same period.

Bone-mineralization rate (M) with ^{85}Sr as tracer was determined by dividing the skeletal retention of ^{85}Sr by the average plasma ^{85}Sr specific activity. The value with ^{45}Ca as tracer was found by dividing the skeletal retention of ^{45}Ca by the weekly urine ^{45}Ca specific activity.

Intravenous studies

Doses consisting of 5 μCi of ^{45}Ca and 30 μCi of ^{85}Sr were administered simultaneously. Both isotopes were estimated weekly in the urine and faeces. The plasma specific activities were estimated at 10 and 30 min, 6, 12 and 24 h, and 2, 3, 5, 7, 10, 12, 14, 21 and 28 days after administration.

The M value was determined by the calculation of Bauer *et al.* (1957) applied at 7 and 14 days.

The plasma and urine stable Ca were estimated by a modification of the AutoAnalyzer technique (McFadyen *et al.*, 1965). Faecal Ca was estimated in an aliquot of a solution of the ashed sample by titrating against EDTA with ammonium purpurate as an indicator.

^{85}Sr was determined in plasma and in urine by counting 5-ml samples in an automatic well scintillation counter. Faecal ^{85}Sr activity was determined by counting an aliquot of homogenized faeces in a large well scintillation counter.

Sufficient ammonium oxalate was added to the plasma and urine samples to precipitate all their Ca as the oxalate. This precipitate was then filtered off on to a disc of glass-fibre filter paper, which was counted in a thin-window gas-flow Geiger counter. For each sample, the counts in the Geiger counter due to ^{85}Sr were estimated by counting a standard source of ^{85}Sr, firstly in solution in the well scintillation counter and secondly, as a precipitate in the Geiger counter. The ratio of these standard counts multiplied by the count-rate of the sample in the well scintillation counter gave the count-rate of the ^{85}Sr in the Geiger counter. By subtracting this count-rate from the total count-rate in the Geiger counter, the count-rate due to ^{45}Ca was found.

Faecal ^{45}Ca was similarly estimated by precipitating the Ca from an aliquot of the solution of ashed faeces used to estimate the stable Ca.

Each isotope sample was counted for the time taken to record 5,000 counts. Therefore, as the count-rate due to the isotope fell, the standard error of the counts due to the isotope increased. The least value of this standard error was 1·5% and this increased to 7% when the count-rate due to the isotope fell to its lowest value.

Thirty-two patients were given ^{85}Sr by continuous feeding and 19 were given ^{45}Ca. Six patients were given both ^{45}Ca and ^{85}Sr by intravenous injection. The clinical details are shown in Tables I and III.

The results of these investigations are shown in Figs. 1 to 6, and Tables II and III.

TABLE I

Clinical details of patients given isotopes by continuous feeding

Case	Age	Sex	Weight (kg)	Diagnosis	Bone status
Patients given ^{45}Ca and ^{85}Sr					
P.C.	55	M	48	Steatorrhoea	Osteoporosis
J.M.	59	M	52	Diarrhoea	Normal
R.T.	50	M	72	Osteoporosis	Osteoporosis
A.D.	49	M	52	Steatorrhoea	Normal
J.L.	76	F	37	Osteoporosis	Osteoporosis
T.R.	64	M	51	Osteoporosis	Osteoporosis
F.A.	60	F	57	Osteoporosis	Osteoporosis
R.W.	63	M	57	Rheumatoid arthritis (steroids)	Osteoporosis
J.L.	63	F	66	Backache	Normal
A.M.	54	M	56	Osteoporosis	Osteoporosis
F.D.	26	M	56	Osteoporosis	Osteoporosis
J.P.	59	F	50	Rheumatoid arthritis	Osteoporosis
T.C.	48	M	45	Steatorrhoea	Osteomalacia
Patients given ^{45}Ca					
H.L.	65	F	67	Osteoporosis	Osteoporosis
M.S.	53	F	43	Rheumatoid arthritis	Rheumatoid arthritis (steroids)
I.K.	43	F	62	Osteoporosis	Osteoporosis
A.S.	46	F	62	Osteoporosis	Osteoporosis
A.H.	67	F	38	Steatorrhoea	Steatorrhoea
F.S.	70	F	46	Renal failure	Renal failure
Patients given ^{85}Sr					
A.McS.	50	M	47	Steatorrhoea	Osteoporosis
M.G.	43	F	90	Arthritis	Normal
J.McN.	58	M	80	Myocardial infarction	Normal
A.M.	80	F	52	Osteoporosis	Osteoporosis
M.R.	57	F	58	Osteoporosis	Osteoporosis
M.H.	58	F	41	Gastroenteritis	Osteoporosis
J.I.	68	F	61	Osteoporosis	Osteoporosis
E.B.	60	F	35	Steatorrhoea	Osteomalacia
F.M.	60	F	58	Osteoporosis	Osteoporosis
M.McM.	57	F	46	Osteoporosis	Osteoporosis
M.McE.	72	F	43	Osteoporosis	Osteoporosis
A.McL.	78	F	48	Osteoporosis	Osteoporosis
D.V.	68	F	61	Osteoporosis	Osteoporosis
M.McL.	72	F	61	Oxteoporosis	Osteoporosis
W.Y.	60	F	57	Osteoporosis	Osteoporosis
E.W.	75	F	40	Renal failure	Osteomalacia
C.W.	31	M	82	Calculi	Normal
E.S.	59	F	69	Calculus	Osteoporosis
K.C.	34	M	70	Hypertension	Normal

TABLE II

Bone-mineralization rates (mg/day/kg) measured by continuous feeding of strontium-85 and calcium-45

Case	Bone-mineralization rate measured by ^{85}Sr	Bone-mineralization rate measured by ^{45}Ca
P.C.	5·0	9·7
J.M.	4·8	10·5
R.T.	0	7·0
A.P.	13·4	13·0
T.R.	33·0	8·0
T.R.	22·0	8·7
F.A.	18·5	7·2
F.A.	12·4	14·0
R.W.	8·3	4·6
R.W.	8·8	3·9
R.W.	7·9	5·3
J.L.	5·5	4·9
J.L.	0·9	7·2
J.L.	6·6	7·3
A.M.	12·5	7·4
A.M.	3·2	8·4
A.M.	10·5	10·2
F.D.	15·6	4·5
F.D.	22·0	8·0
J.P.	21·0	9·5
J.P.	11·6	8·7
T.C.	13·0	10·0
Mean	11·7	8·1
Standard error	1·7	0·6

TABLE III

Patients given intravenous calcium-45 and strontium-85
M values are expressed in mg/day

Case	Age	Weight (kg)	Diagnosis	^{85}Sr M	^{85}Ca M
J.R.	64	50	Osteoporosis	220	234
H.G.	53	72	Osteoporosis	176	306
A.P.	70	44	Osteoporosis	60	358
F.T.	62	48	Renal failure	103	141
M.McK.	69	46	Osteoporosis	138	153
E.T.	64	54	Osteoporosis	170	181

DISCUSSION

The interest in these data lie in the comparison between [85]Sr and [45]Ca. At all levels of Ca intake in the continuous feeding studies, the net absorption of radiostrontium is less than that of radiocalcium (Fig. 1). If the discrimination

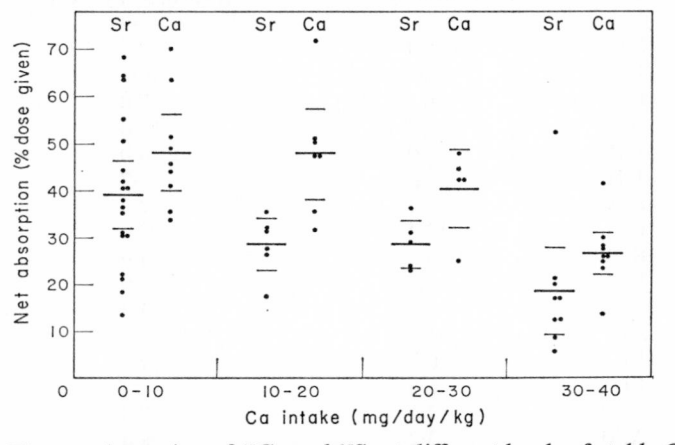

FIG. 1. The net absorption of [45]Ca and [85]Sr at different levels of stable Ca intake during the continuous feeding of the isotopes. The mean and 2 × standard error ranges are shown.

against Sr is simple competitive inhibition, it would be expected that the discrimination would increase with increasing Ca intake. However, there is no change in discrimination with increasing Ca intake, the average discrimination value being 0·69. Bailey *et al.* (1960) and Samachson (1963) found a discrimination factor of 0·5. Spencer *et al.* (1960) found a factor varying from 0·45 to 0·40 by a single oral dose of the isotopes. Nordin *et al.* (1964b) found a factor of 0·55 by using a single oral dose and measuring the level of [45]Ca and [85]Sr in the plasma 2 h later.

Discrimination by kidney

The urinary excretion of [85]Sr and [45]Ca during continuous feeding of isotope at different Ca intakes is shown in Fig. 2. The urinary excretion of [85]Sr is the same or slightly higher than that of [45]Ca at all levels of Ca intake, and the excretion of both isotopes decrease with increasing intake. Since the renal excretion of [45]Ca and [85]Sr depends in part upon the plasma levels of these isotopes, discrimination by the kidney can only be shown by comparing simultaneous plasma and urine samples. This has been done and the results are shown in Fig. 3. When compared in this way, the urinary excretion of [85]Sr is 3·34 times higher than that of [45]Ca. This agrees well with the factor of

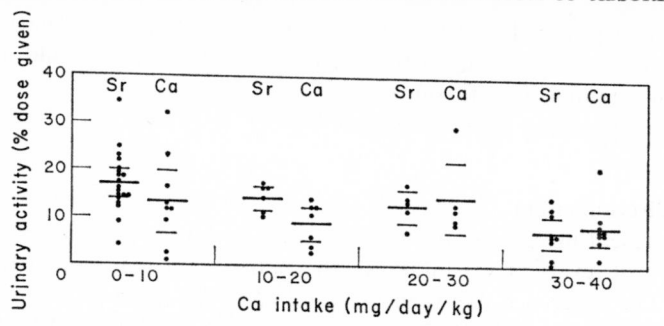

FIG. 2. Urinary excretion of ^{45}Ca and ^{85}Sr at different levels of stable Ca intake during the continuous feeding of the isotopes. The mean and $2 \times$ standard error ranges are shown.

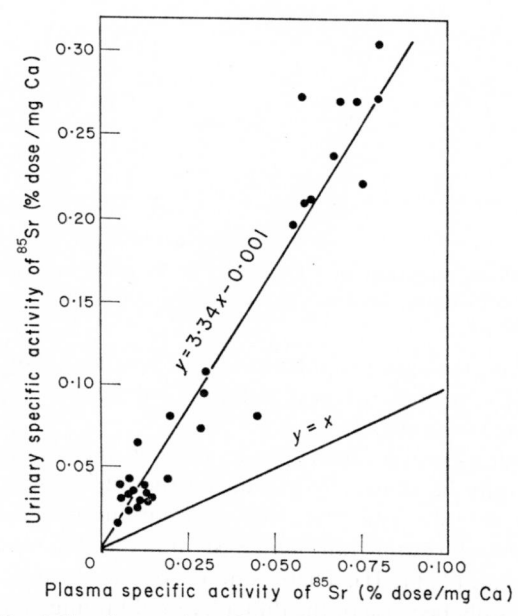

FIG. 3. The discrimination against the re-absorption of ^{85}Sr by the renal tubules obtained from the measurement of simultaneous plasma and urine samples: correlation coefficient, $r = 0.92$.

3·5 found in six normal subjects by Barnes *et al.* (1961). Discrimination might arise in the glomerulus during filtration of plasma water or during re-absorption of Ca and Sr by the renal tubule. If the protein binding of Sr is the same as Ca then the above factor will be determined by tubular re-absorption assuming that all the smaller complexes are freely filterable. If the binding of Sr is different from that of Ca, then the renal discrimination factor would be affected by filtration. If the protein binding of Sr is less than that of Ca, as has been suggested by Samachson and Lederer (1958), this would itself tend to cause discrimination in favour of Sr.

The protein binding of Sr is discussed below.

Skeletal discrimination

In the present investigation, the skeletal retention of [85]Sr is lower than that of [45]Ca at all levels of Ca intake in the patients on continuous feeding of isotopes. The differences between [45]Ca and [85]Sr are factors of 0·67, 0·37, 0·67 and 0·59 at intakes of 10, 10–20, 20–30 and 30–40 mg/day/kg (Fig. 4). These

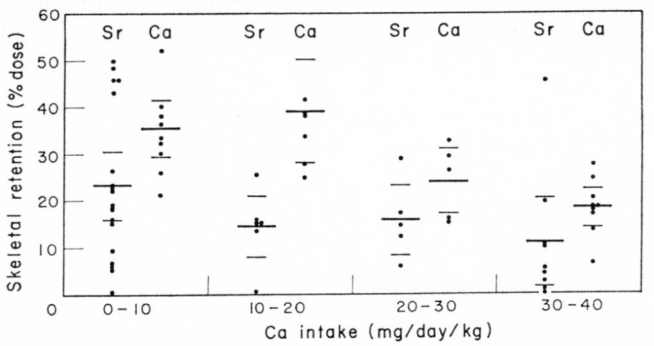

FIG. 4. The skeletal retention of [45]Ca and [85]Sr at different levels of stable Ca intake during the continuous feeding of the isotopes. The mean and 2 × standard error ranges are shown.

figures demonstrate that the proportion of [85]Sr retained shows little change compared with [45]Ca, with increasing Ca intake. The retention of both isotopes fell with increasing Ca intake.

Because of discrimination against Sr at the gut and kidney, skeletal discrimination can only be shown by measuring the mineral-transfer rate. The skeletal uptakes of [45]Ca and [85]Sr with the continuous-feeding technique show a higher mineralization rate with [85]Sr, but the difference between the two is not significant (Table II). There is no correlation between the [45]Ca and [85]Sr M values possibly because of the difficulty in obtaining an average plasma specific activity for [85]Sr as the isotopes were administered twice daily. Because of this difficulty, we have used the method of Bauer et al. (1957) to calculate M between 7 and 14 days, following the simultaneous, intravenous administration of [45]Ca and [85]Sr. This method of calculating M assumes that no isotope returns from the bone by resorption of labelled material, and that the plasma has the same specific activity as the rapidly exchanging Ca pools. The M values of our six patients are shown in Table III. The only comparable data are those of Dow and Stanbury (1960), which are in such a form that it is possible to calculate the mineral-transfer rates in the same way by using the

Bauer bone-mineralization rate between 7 and 14 days. This has been done and the combined data are shown in Fig. 5. There is a correlation between the

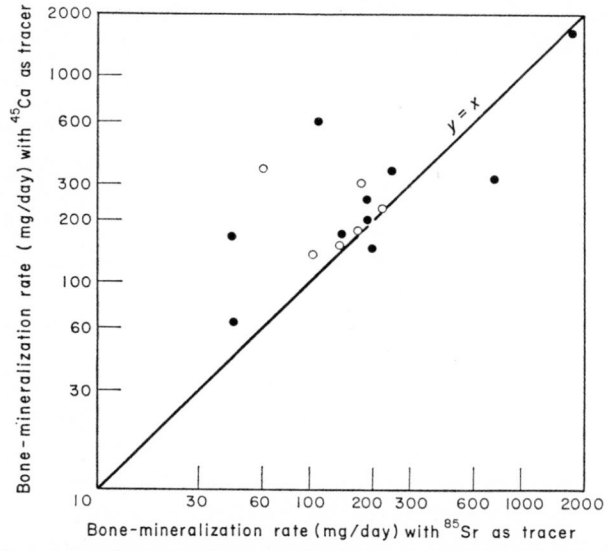

FIG. 5. The bone-mineralization rates in six patients in the present study (○) and ten patients studied by Dow and Stanbury (1960) (●). The M value was estimated by the method of Bauer *et al.* (1957) following the simultaneous intravenous administration of ⁴⁵Ca and ⁸⁵Sr.

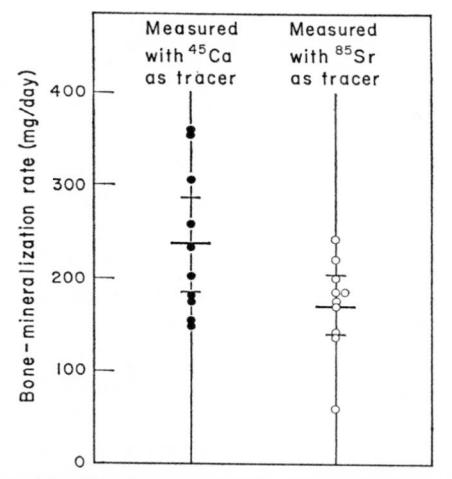

FIG. 6. The bone-mineralization rates in normal and osteoporotic subjects estimated from the simultaneous intravenous administration of ⁴⁵Ca and ⁸⁵Sr. The mean and 2 × standard error ranges are shown.

measurement of transfer of Ca into bone by using ^{45}Ca and ^{85}Sr as tracers if all the cases are considered ($r = 0.89$), but there are large individual discrepancies. Since the M values do not have a normal distribution, only the normal and osteoporotic subjects are considered in Fig. 6.

The ^{85}Sr M values are significantly lower than the ^{45}Ca values in this group. Bronner et al. (1963) have found no significant correlation between ^{45}Ca and ^{85}Sr mineral-transfer rates in man. It is evident therefore that the discrimination varies substantially in different individuals.

As it is the ionic Ca and Sr that is transferred into bone, a difference in the rate of transfer of mineral from blood to bone, obtained by using ^{45}Ca and ^{85}Sr as tracers, could be due to the protein binding of ^{45}Ca being different from the protein binding of ^{85}Sr. Harrison et al. (1955) found no difference between the protein binding of Ca and Sr in two patients, but Samachson and Lederer (1958) added ^{45}Ca and ^{85}Sr to plasma in vitro and showed that the protein binding of ^{85}Sr was less than ^{45}Ca.

If the Ca and Sr are equally bound to protein, then it is permissible to calculate mineral-transfer rate into bone by using plasma specific activity of ^{85}Sr and ^{45}Ca. If the protein binding of Sr is less than the protein binding of Ca, the estimated bone-mineralization rates with ^{85}Sr are too high, and the difference we have found between the bone-mineralization rates measured with ^{45}Ca and ^{85}Sr would be even greater than is shown in Fig. 6.

A difference in the transfer rate of Sr and Ca into bone has been described in rabbits by Kshirsagar et al. (1966). They continuously fed young rabbits on a diet containing stable Sr, and found that the skeletons of these animals contain a lower Sr/Ca ratio than the plasma and concluded that there is a discrimination factor of about 1·6 in favour of Ca in the transfer from blood to bone. However, this discrimination factor is only valid if the whole skeleton is labelled with Sr and is in equilibrium with plasma, and this has not been clearly demonstrated to be so in these studies. Other workers have found no difference in the rate of skeletal uptake of these isotopes in animals (Bauer et al., 1955; Comar et al., 1956; MacDonald et al., 1957).

In vitro experiments with bone powder in a bathing fluid have shown a discrimination in favour of the transfer of Ca ions into the bone powder in several studies (Harrison et al., 1959; Boyd et al., 1959; Neuman et al., 1963).

The uptake of isotope in bone powder is, of course, due to exchange. The discrimination appears to depend upon many factors, such as temperature, pH, ionic concentration of Ca and phosphate, and the concentration of other ions, and there seems to be no good reason why this will not occur in vivo. Boyd et al. (1959) in their work on bone powder have postulated that under certain conditions the limiting factor determining exchange of Ca and Sr into bone is the release of stable Ca from the bone. Neuman (1964) also suggested two different lattice positions for Ca ions, one of which is not available for exchange of Sr ions.

There is therefore general agreement on discrimination against Sr compared with Ca *in vitro*, when exchange governs the movement of these ions. There are, however, large individual discrepancies *in vivo*, and no consistent discrimination against the uptake of Sr by the skeleton has been found. The variation in individual values suggests that [85]Sr is not a satisfactory tracer for Ca.

REFERENCES

Bailey, N. T. J., Bryant, F. J., and Loutit, J. F. (1960). AERE R–3299. London, H.M.S.O.

Barnes, D. W. H., Bishop, M., Harrison, G. E., and Sutton, A. (1961). *Int. J. Radiat. Biol.* **3,** 637–646.

Bauer, G. C. H., Carlsson, A., and Lindquist, B. (1955). *Acta physiol. scand.* **35,** 56–66.

Bauer, G. C. H., Carlsson, A., and Lindquist, B. (1957). *Acta med. scand.* **158,** 143–150.

Boyd, J., Neuman, W. F., and Hodge, H. C. (1959). *Arch. Biochem. Biophys.* **80,** 105–113.

Bronner, F., Aubert, J. P., Richelle, L. J., Saville, P. D., Nicholas, J. A., and Cobb, J. R. (1963), *J. clin. Invest.* **42,** 1095–1104.

Comar, C. L., Wasserman, R. H., and Nold, M. M. (1956). *Proc. Soc. exp. Biol. Med.* **92,** 859–863.

Dow, E. C., and Stanbury, J. B. (1960). *J. clin. Invest.* **39,** 885–903.

Harrison, G. E., Raymond, W. H. A., and Tretheway, H. C. (1955). *Clin. Sci.* **14,** 681.

Harrison, G. E., Lumsden, E., Raymond, W. H. A., and Sutton, A. (1959). *Arch. Biochem. Biophys.* **80,** 97–105.

Hodges, R. M., MacDonald, N. S., Nusbaum, R., Stearns, R., Ezmirlian, F., Spain, P. C., and McArthur, C. (1950). *J. biol. Chem.* **185,** 519–524.

Kshirsagar, S. G., Lloyd, E., and Vaughan, J. (1966). *Br. J. Radiol.* **39,** 131–140.

Likins, R. C., Posner, A. S., Kunde, M. L., and Craven, D. L. (1959). *Arch. Biochem. Biophys.* **83,** 472–481.

MacDonald, N. S., Noyes, P., and Lorick, P. C. (1957). *Am. J. Physiol.* **188,** 131–136.

MacFadyen, I. J., Nordin, B. E. C., Smith, D. A., Wayne, D. J., and Rae, S. L. (1965). *Br. med. J.* i, 161–164.

Neuman, W. F. (1964). *In* "Bone Bio-Dynamics", (Frost, H. M., ed.), pp. 393–408. London, Churchill.

Neuman, W. F., Bjornerstedt, R., and Mulryan, B. J. (1963). *Arch. Biochem. Biophys.* **101,** 215–224.

Nordin, B. E. C., Bluhm, M., and MacGregor, J. (1962). *In* "Radio-Isotopes and Bone", (McLean, F. C., *et al.*, eds), pp. 105–125. Oxford, Blackwell.

Nordin, B. E. C., Smith, D. A., and Nisbet, J. (1964a). *Clin. Sci.* **27,** 111–122.

Nordin, B. E. C., Smith, D. A., MacGregor, J., and Nisbet, J. (1964b) Technical Report No. 32, Second Panel Report, p. 124. Vienna, International Atomic Energy Agency.

Samachson, J. (1963). *Clin. Sci.* **25,** 17–26.

Samachson, J., and Lederer, H. (1958). *Proc. Soc. exp. Biol. Med.* **98,** 867–870.

Spencer, H., Li, M., Samachson, J., and Laszlo, D. (1960). *Metabolism,* **9,** 916–925.

Ratio of the Faecal to Urinary Clearance of Strontium in Man

G. E. HARRISON AND ALICE SUTTON

Medical Research Council Radiobiological Research Unit, Harwell, Berks

SUMMARY

Measurements of the ratio of the faecal to the urinary clearance of Sr are reported for five human adults on a controlled diet corresponding to a daily intake of 1·6 g Ca and 1 mg Sr.

Values of the ratio of the endogenous faecal to the urinary output of ^{85}Sr have also been obtained for two male adults following a single intravenous injection of the radioactive marker.

Results from the two experiments are compared and the importance of the ratio of the endogenous faecal to the urinary output of Sr in the assessment of oral absorption is discussed.

INTRODUCTION

It has been shown that the mean daily renal clearance of Sr is approximately constant and independent of the plasma level for a given subject (Harrison *et al.*, 1955; Barnes *et al.*, 1961). Experimental data for the urinary and faecal excretion as well as the plasma concentration of ^{85}Sr, following a chronic ingestion have been used to derive mean values for the ratio of the endogenous faecal excretion, *f*, to the urinary output, *U*, and the mean intestinal clearance rates of the tracer for five adult men. These values of f/U have been used to calculate the cumulative endogenous excretion and the oral absorption of ^{85}Sr for the subjects on a constant diet.

EXPERIMENTAL METHODS

Five male subjects from this Unit, aged 30 to 57 years, were on a constant diet which included milk labelled with ^{85}Sr for 21–32 days (Carr *et al.*, 1962). The diet remained unchanged, except that inactive milk was substituted for the labelled milk, for a further period of 1 week before the subjects reverted to a diet of choice. The daily output of [85]Sr in urine and faeces was measured during the first 5 or 6 weeks, and later it was determined in separate 14-day bulkings of urine and faeces up to about 17 weeks after the intake of [85]Sr had ceased. At intervals over the first few weeks, the plasma concentration of the marker was also measured so that the renal clearance of [85]Sr, defined as the ratio of the rate of excretion in urine to the concentration in contemporary plasma, could also be obtained.

The body retentions of [85]Sr in the five subjects have been measured by Rundo *et al.* (1964). The observed values for this retention at $t = 7.5$ days after the last [85]Sr ingestion, when exogenous tracer had been excreted, have been used to derive the body retentions at $t = 0$, 20 and 100 days from the excretion data. The oral absorption of [85]Sr has been derived for each subject on the assumption that f/U remained constant during the period on the labelled diet.

RESULTS

Graphical treatment

Semi-logarithmic graphs of the [85]Sr activity in the daily urine and faeces from $t = 0$, the time of the last ingestion, were made for each subject. After 30 days the daily urinary output was assumed to follow a single exponential function of time and a straight line of best fit was drawn through the experimental points for which t was >30 days. A second line was drawn through the experimental points for the faeces at $t > 30$ days parallel to that for the urine. The constant ratio for f/U represented by the ratio of the intercepts of the two parallel lines at $t = 0$ was assumed to hold over the whole experimental period.

Daily urine and faecal excretions of [85]Sr were fitted graphically to a two-term exponential expression, apart from the faecal excretion between $t = 0$ and $t = 7$, when part of the faecal output was exogenous. By integration of this two-term exponential expression from $t = 0$ to $t = t$, the total endogenous excretion E_0^t was obtained. Thus:

$$E_0^t = U_0^t(1 + f/U) = (1 + f/U)\left[\frac{C_1}{k_1}(1 - e^{-k_1 t}) + \frac{C_2}{k_2}(1 - e^{-k_2 t})\right] \qquad (1)$$

where U_0^t represents the cumulative urinary output from $t = 0$ to $t = t$, C_1/k_1 and C_2/k_2 are space constants and k_1 and k_2 are rate constants for the urinary output of the tracer.

If R_0 is the body retention of [85]Sr excluding the gut contents at $t = 0$ and $\sum U$ is the cumulative urinary excretion of the tracer during the ingestion of the [85]Sr labelled diet, the oral absorption may be written:

Absorption $= R_0 + \sum U(1 + f/U)$, provided that the ratio of f/U remains constant.

Derived values

A typical semi-logarithmic graph of the daily urinary and faecal excretion between $t=0$, the time of the last ^{85}Sr ingestion, and $t=110$ days is shown in Fig. 1. The two straight lines corresponding to the two exponential components are also shown.

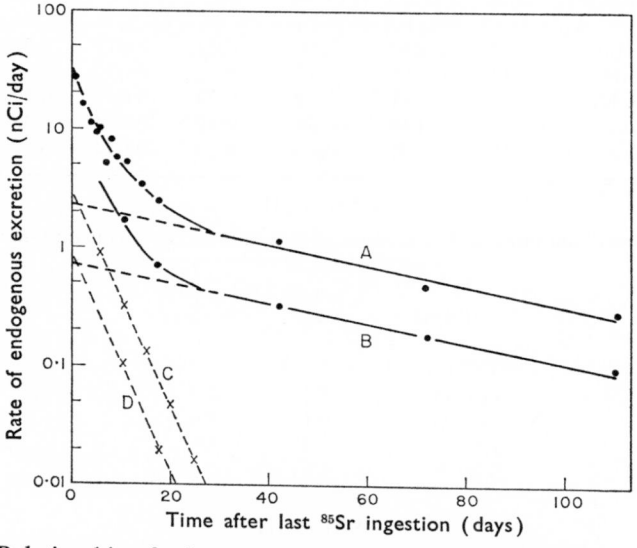

FIG. 1. Relationship of urinary and faecal excretion of ^{85}Sr with time: A, urine; B, faeces with C and D the first exponential components of A and B, respectively, $\times 0\cdot1$.

The mean values for f/U, derived as the ratio of the intercepts of these lines at $t=0$ from the graphs are shown in Table I, with the two rate constants k_1 and k_2 and the two space constants C_1/k_1 and C_2/k_2 (equation (1)). A result of interest is the almost constant value of the rate constant k_1 for the different subjects. The standard deviation for k_1 was 8% of the mean and that for f/U 12%. Differences between the space constants for the different subjects did not appear to be correlated with body weight.

Table II shows the mean renal and intestinal clearance of ^{85}Sr during the ^{85}Sr intake.

Table III shows the total excretion derived from equation 1 at $t=0$, 20 and 100 days after the last ^{85}Sr intake for the five subjects. The body retentions of ^{85}Sr calculated from the excretion data are compared with the appropriate interpolated values of the whole body retention measured by Rundo *et al.* (1964). The agreement is within the experimental error of the separate observations and it is concluded that this is substantial evidence for the correctness of the present calculations.

TABLE I

Space constants and rate constants for the urinary output of strontium-85 with mean values for f/U after a constant diet labelled with strontium-85

Subject	Days on ^{85}Sr	Urinary output Space constants (nCi) C_1/k_1	C_2/k_2	Rate constants (days^{-1}) k_1	k_2	f/U
T.E.F.C.	32	240	160	0·21	0·015	0·325
G.E.H.	32	118	122	0·21	0·020	0·29
G.R.H.	21	113	88	0·24	0·049	0·27
J.F.L.	28	194	88	0·19	0·014	0·36
S.R.W.	21	123	116	0·23	0·044	0·28

TABLE II

Renal and intestinal clearances of strontium-85 on a constant diet

Subject	Clearance (litre/day) renal	intestinal
T.E.F.C.	8·1	2·6
G.E.H.	8·1	2·3
G.R.H.	9·1	2·5
J.F.L.	15·4	5·5
S.R.W.	8·9	2·5

TABLE III

Excretion, body retention and absorption of strontium-85

Subject	Days after last ^{85}Sr ingestion	Cumulative excretion, E_0^t (nCi)	Body retention (nCi) calculated	observed	Cumulative urinary output $\sum U$ (nCi)	Absorption of ^{85}Sr (% dose)
T.E.F.C.	0	0	1000	1050	1836	21·4
	20	369	631	627		
	100	483	517	468		
G.E.H.	0	0	542	520	856	10·3
	20	202	340	320		
	100	245	227	245		
G.R.H.	0	0	405	390	620	11·4
	20	211	194	192		
	100	254	151	142		
J.F.L.	0	0	724	690	1313	17·9
	20	286	438	420		
	100	351	371	308		
S.R.W.	0	0	575	610	625	13·1
	20	242	333	333		
	100	304	271	254		

DISCUSSION

The constant value of f/U for a given subject derived from the present observations, may appear somewhat surprising since extensive recycling of both Ca and Sr is known to take place in the kidney (Chen and Neuman, 1955) and most probably in the intestine also. We conclude that, at least for subjects on a constant diet, the renal and intestinal recycling of Sr is such that f/U remains constant. As the renal clearance of Sr has been shown to be independent of the plasma concentration, we conclude that the faecal clearance is similar in this respect.

There was substantial diurnal variation in the faecal output of ^{85}Sr and to a lesser degree in the urinary output even for subjects with the most regular excretion patterns. In fact, one of the principal problems in any investigation such as the present is the choice of representative totals for the faecal and urinary excretions. Certainly over periods less than 14 days, the unknown and variable time of transit of food through the intestine makes the faecal totals unreliable. It was for this reason that the graphical treatment was adopted to obtain f/U.

The values of f/U (Table I) may be compared with mean values derived for subjects G.E.H. and J.F.L. following intravenous doses of ^{85}Sr (Bishop et al., 1960; Harrison et al., 1966). The daily urinary output following these intravenous doses could not be fitted to a two-term exponential over the whole period of the observations and the cumulative totals over 20, 40 or 100 days

TABLE IV

Cumulative urinary and faecal excretion of an intravenous dose of strontium-85

Subject	Days after dose	Cumulative excretion (% dose) urinary U	faecal f	f/U	Renal clearance (litres/day)
G.E.H.*	20	56·0	14·5	0·26	
	100	60·2	18·9	0·31	10·5
J.F.L.*	20	64·0	10·9	0·17	
	100	69·7	13·7	0·20	17·5
G.E.H.†	20	56·8	14·8	0·26	
	40	59·4	15·4	0·26	10·9

* Bishop et al. (1960).
† Harrison et al. (1966)

are given in Table IV. For G.E.H. there is good agreement between f/U derived from the oral and intravenous doses, but for J.F.L., for whom the renal clearance of Sr was always appreciably greater than for the others, there is an appreciable difference between f/U oral and f/U intravenous. This may have been due to a change in the Ca intake which has recently been shown to affect f/U (Samachson, 1966).

The mean daily endogenous faecal excretion of Ca in adults obtained independently by Bronner (1964) and Heaney and Skillman (1964) was 130 mg. If the plasma concentration of Ca was 100 mg/l, the endogenous faecal clearance of Ca for these subjects was about 1·3 litres/day, which is to be compared with a mean intestinal clearance of 3 litres/day for Sr for the present subjects. It would appear, therefore, that the relative endogenous faecal excretion, Sr to Ca, is about 2.

An interesting result is the two-fold range in the oral absorption of ^{85}Sr in the five subjects on a constant diet (C, 1·6 g; Sr, 1 mg; P, 2·2 g/day). All subjects were in negative Sr balance during the ^{85}Sr intake (Carr, see p. 139) and this could well have reduced the oral absorption of the tracer. Comparison of the derived values for the oral absorption of ^{85}Sr (Table III) with those obtained by Rundo (see p. 131) from different data show satisfactory agreement.

The skilled assistance of Miss Hilda Shepherd and Messrs K. B. Edwards and G. R. Howells is gratefully acknowledged.

REFERENCES

Barnes, D. W. H., Bishop, M., Harrison, G. E., and Sutton, A. (1961). *Int. J. Radiat. Biol.* **3**, 637–646.

Bishop, M., Harrison, G. E., Raymond, W. H. A., Sutton, A., and Rundo, J. (1960). *Int. J. Radiat. Biol.* **2**, 125–142.

Bronner, F. (1964). *In* "Mineral Metabolism" (Comar, C. L., and Bronner, F., eds.), Vol. II, Part A, p. 408. New York, Academic Press.

Carr, T. E. F., Harrison, G. E., Loutit, J. F., and Sutton, A. (1962). *Nature, Lond.* **194**, 200–201.

Chen, P. S., and Neuman, W. F. (1955). *Am. J. Physiol.* **180**, 632–636.

Harrison, G. E., Raymond, W. H. A., and Tretheway, H. C. (1955). *Clin. Sci.* **14**, 681–695.

Harrison, G. E., Carr, T. E. F., Sutton, A., and Rundo, J. (1966). *Nature, Lond.* **209**, 526–527.

Heaney, R. P., and Skillman, T. G. (1964). *J. Lab. clin. Med.* **64**, 29–41.

Rundo, J., Lillegraven, A. L., and Mason, J. I. (1964). *In* "Proceedings of the 1st European Bone and Tooth Symposium", pp. 159–165, Oxford, Pergamon Press.

Samachson, J. (1966). *Radiat. Res.* **27**, 64–74.

A Comparison of the Metabolism of Calcium and Strontium in Rabbit and Man

ELIZABETH LLOYD

*Medical Research Council Bone-Seeking Isotopes Research Unit,
The Churchill Hospital, Oxford*

SUMMARY

The reduced uptake and retention of radioactive Sr in the adult rabbit compared with man is the result of many dietary and physiological differences. The most important of these are probably the absence of dairy products from the rabbit's diet, the continuous feeding habits of the rabbit, and the presence of large quantities of bile in the gastro-intestinal tract. The renal tubular reabsorption of Ca in the rabbit was considerably lower than that reported for humans, which may in turn be linked with the higher concentrations of citrate found in rabbit blood.

The greatly increased rate of transfer of Sr and Ca ions from diet to bone in the rabbit compared with man provides us with an opportunity to study the uptake and release of a radioactive isotope in a time interval that is about ten to twenty times shorter than the corresponding time in man.

INTRODUCTION

Since bone-seeking isotopes became a hazard from fall-out, a vast amount of literature has been published on the uptake, retention and excretion of ^{90}Sr in animals. Similar studies with ^{45}Ca and more recently with ^{47}Ca have helped to correlate Sr metabolism with the basic mineral Ca. The information available on the normal metabolism of Sr and Ca in healthy man, however, is still incomplete. If the difference in the readily measurable quantities in man and in animal experiments could be related to known differences in physiological or other known factors, this might be expected to lead to the possible control of the mechanisms by which Sr and Ca are transported. No animal can adequately reproduce the same conditions as are found in man. The rabbit was used in the present study, being the smallest animal with similar bone structure to that of man. The work reported here, forms part of a wider investigation on the same animals that included autoradiographic and kinetic studies as well as a detailed study of plasma–bone interrelationships. These have been reported elsewhere (Lloyd, 1964a, 1964b, 1965; Kshirsagar *et al.*, 1966). Details of the method are included in the 1966 publication and are therefore omitted in the present report.

*

RESULTS

Table I shows the difference in the retention of radioactive Sr at different time intervals after intravenous injection in the adult rabbit and that reported by Bishop *et al.* (1960) for the normal human adult. The figures given here were derived from the excretion data reported for G.E.H. From these figures it can be seen that the retention by the rabbit is less at all times.

TABLE I

**Comparison of the retention of radioactive strontium in
adult man and rabbit following intravenous injection**

All values are expressed as the percentage of the injected dose

Time (days)	Man*	Rabbit
1	85·7	41·6
3		23·0
5	54·6	20·1
9		16·6
10	38·4	
15	32·4	
22	28·8	
30		10·7
50	24·4	
100	20·9	7·6
150	19·1	
200	18·3	
460		3·1

* These values are derived from excreta data for G.E.H. (Bishop *et al.*, 1960).

Table II shows the relative amounts of ^{90}Sr found in human plasma compared with mean values for the rabbit at different times after injection, assuming that the plasma volume is 4·5% of the body weight. Here it can be seen that rabbit plasma is cleared of ^{90}Sr much more rapidly than human plasma. This is reflected in the different excreta patterns seen in Table III. Much more radioactive Sr is excreted by the rabbit during the first day in both the urine and faeces. At later times smaller quantities are excreted in the

TABLE II

Comparison of radioactive strontium in the plasma of adult man and rabbit following intravenous injection

All values are expressed as the percentage of the injected dose
in the total plasma volume

Time	Man*	Rabbit
10 min		20·0
$\frac{1}{2}$ h	20·9	
1 h	15·5	
$1\frac{1}{2}$ h		6·85
2 h	11·5	
4 h		2·95
5 h	7·2	
10 h	4·7	
20 h	3·2	
24 h		0·45
2 days 2 h	1·7	
3 days		0·044
9 days		0·013
30 days		0·0022
100 days		0·0008

* These values are obtained from those reported for G.E.H. (Bishop *et al.*, 1960) by assuming plasma volume = 4·5% of body weight.

TABLE III

Comparison of the daily excretion of radioactive strontium in adult man and rabbit following intravenous injection

All values are expressed as the percentage of the injected dose.

Time (days)	Man* Urine	Man* Faeces	Rabbit Urine	Rabbit Faeces
1	14·3	—	35·0	23·4
2	10·2	0·07	8·2	5·9
3	6·9	0·82	2·4	1·6
4	4·3	2·92	1·2	0·7
5	4·6	1·19	0·6	0·5
6	2·33	4·34	0·6	0·6
7	2·52	—	0·4	0·3
8	1·89	2·34	0·3	0·2
9	1·33	0·45	0·3	0·2
Total	48·37	12·13	49·0	33·4

* These values are those reported for G.E.H. by Bishop *et al.* (1960).

rabbit urine, but the endogenous faecal excretion is comparatively more important in the rabbit: the faecal/urine ratio is 0·7, compared with 0·25 in man.

In order to elucidate the reasons for these differences in the uptake and retention of a radioactive tracer, it was felt that it might be instructive to compare the average daily intake and excreta of the stable elements of Sr and Ca in the two species under conditions of normal dietary intake. Typical values are given in Table IV. Here it can be seen that the normal daily intake of Ca in the rabbit is about the same as that in man, even though the rabbit weighs only 3–5 % of the human adult. Moreover, this high intake is necessary for health maintenance. Experiments in our laboratory showed that animals on 0·3 mg Ca/day were in negative Ca balance.

TABLE IV

Comparative metabolism of the stable elements of strontium and calcium in man and rabbit

	Man Ca (g)	Man Sr (mg)	Man mg Sr / g Ca	Rabbit Ca (g)	Rabbit Sr (mg)	Rabbit mg Sr / g Ca
Diet	1·2*	1·6*	1·3*	0·92	3·7	4·0
Absorbed	0·45†	0·30†	0·67	0·42	0·5	1·2
Urine	0·25†	0·21†	0·84	0·14	0·3	2·1
Faeces (Total)	0·95	1·4	1·5	0·78	3·4	4·4
End. Faeces	0·20‡	0·90	0·45	0·28	0·2	0·72
Adult Bone	1200*	380*	0·32	26	13·5	0·52
Plasma/litre	0·10	0·04	0·4§	0·14	0·11	0·8

* Loutit (1962).
† Dolphin and Eve (1963).
‡ Comar (1963).
§ Harrison et al. (1955).

Because of the herbivorous nature of the rabbit diet, the intake of Sr is higher than that of the normal human diet. This higher intake of Sr by the rabbit is reflected both in a higher absorption and higher excretion of Sr. The Sr/Ca ratios are also higher in the rabbit for all the samples measured. However, the Observed Ratio for the rabbit (0·12) expressed as:

$$\left(\frac{\text{Sr/Ca in bone}}{\text{Sr/Ca in diet}}\right)$$

is about half the value for man (0·25).

The discrimination in passing from diet to bone involves (1) discrimination in the gastro-intestinal tract; (2) urinary discrimination; and (3) discrimination between blood and bone.

Only (1) and (2) are discussed here, as blood/bone discrimination has already been reported in full elsewhere (Kshirsagar et al., 1966).

STRONTIUM/CALCIUM RELATIONSHIPS IN ABSORPTION AND ENDOGENOUS FAECAL EXCRETION

The higher discrimination by the gastro-intestinal tract of the rabbit compared with man is almost certainly the result of many factors.

The discrimination by the gut depends not only on the absolute level of Sr and Ca present, but also on the hydrogen ion concentration of the contents of the lumen and the absence or presence of other compounds in the diet that may compete for these minerals, which in turn may accelerate or inhibit their absorption. However, perhaps the most significant differences in the gastro-intestinal absorption, observed for the two species, are concerned with (1) the absence of dairy products in the rabbit diet (Lengemann (1957) has shown that milk, added to a grain diet given to rats, increases the absorption of both Sr and Ca, but the *discrimination* in absorption is decreased). (2) The fact that the rabbit is a continuous feeder and depends on bulk for emptying the stomach contents, since it has little power of contraction, which in turn, probably results in much smaller fluctuations in pH than are found in man whose eating habits are much more erratic. (3) Very high production of bile in the rabbit, which may be important for facilitating the absorption of large amounts of Ca and Sr by the rabbit. It may also be an important factor in the high endogenous faecal excretion, although Lengemann (1963) suggested that in dogs and rats it is responsible for less than 10% of the endogenous material excreted into the gut. (4) Presence of a large caecum in the rabbit, although this may affect absorption, it is much more likely to be important in the secretion from the blood back to the gut, since most of the absorption would be expected to take place closer to the stomach.

The gastro-intestinal tract of the rabbit seems therefore, both by virtue of its size in proportion to body weight and the presence of large quantities of bile, to be ideally suited for the efficient transfer of Ca and Sr ions both into and out of the blood.

URINARY EXCRETION

Although the amount of Ca absorbed by the rabbit is of the same order as that absorbed in man, the urinary excretion is somewhat less in the rabbit. However, when one considers the greatly reduced glomerular filtration rate and the smaller urine volume, the concentration of Ca in rabbit urine seems surprisingly high (Kennedy, 1965).

The concentration of Ca in rabbit urine in the present study was about ten times as high as that found in human urine. The amount of Sr excreted in rabbit urine is greater than that excreted in human urine; however, if the amount excreted is considered in relation to the amounts of Sr absorbed into the blood stream from the diet, then the Sr excreted in human urine represents about 70 % of the absorbed Sr, compared with about 60 % in the rabbit.

The clearance rate of Sr by the kidney, relative to Ca, was found to be 2·3–2·4 in the present study, whereas values reported for man vary between 3 and 4·5 (Barnes *et al.*, 1961).

The amounts of Ca and Sr excreted by the kidney depend not only on the glomerular filtration rate, but the fraction of the element that is in diffusible form in the blood stream, and the degree of reabsorption in the kidney tubules. The rate of glomerular filtration in the adult rabbit is about 9 ml/min (Smith, 1951). This value is not very different from the values obtained for man when expressed in terms of volume filtered per unit weight.

The percentage of ultrafilterable Ca may be about the same in each species: values between 45 % and 80 % were reported for rabbits by Updegraaf *et al.* (1927). Similarly, values between 40 % and 70 % have been reported for man, although when care was taken to reproduce the *in vivo* conditions in man, regarding temperature and pH, a value of 52 % was obtained (Munday and Mahy, 1964). The amount of Ca filtered by the kidney can be obtained by multiplying the volume of plasma filtered by the concentration of Ca in the ultrafilterable fraction. Assuming a total plasma concentration of 14 mg Ca/100 ml, as observed in the present studies, and taking 67 % as the ultra-filterable fraction, the amount of Ca filtered is estimated to be about 1,220 mg/day. Since 140 mg Ca were excreted daily, the tubular reabsorption is about 89 % compared with the figure of 99 % commonly reported for man, i.e. the amount excreted in rabbit urine is 11 % of that filtered at the glomerulus compared with only 1 % in man.

The reason for this lower tubular re-absorption in the rabbit is probably the result of a higher plasma citrate content (McLean and Hastings, 1935) since the complexed fraction of the plasma is thought to be poorly re-absorbed (Neuman and Neuman, 1958).

This may also account for the lower Sr/Ca discrimination by the rabbit kidney, since the infusion of sodium citrate into dogs (Della Rosa *et al.*, 1961) produced not only large increases in urinary excretion of both Sr and Ca but also caused the disappearance of the renal discrimination process.

ACKNOWLEDGMENTS

I am indebted to Dr Janet Vaughan, who has been a constant source of encouragement through this work; also to Dr S. Kshirsagar who made the Sr measurements and assisted at various other stages of the research.

I am also grateful to Dr J. H. Marshall and Dr R. E. Rowland who read the manuscript and made helpful suggestions.

Finally I should like to thank Mr Michael Davies and Miss Fae Schofield for excellent technical assistance.

REFERENCES

Barnes, D. W. H., Bishop, M., Harrison, G. E., and Sutton, A. (1961). *Int. J. Radiat. Biol.* **3,** 637–646.

Bishop, M., Harrison, G. E., Raymond, W. H. A., Sutton, A., and Rundo, J. (1960). *Int. J. Radiat. Biol.* **2,** 125–142.

Comar, C. L. (1963). *In* "Transfer of Calcium and Strontium across Biological Membranes", pp. 405–416. Academic Press, New York.

Della Rosa, R. J., Smith, F. A., and Stannard, J. N. (1961). *Int. J. Radiat. Biol.* **3,** 557–578.

Dolphin, G. W., and Eve, I. S. (1963). *Physics Biol. Med.* **8,** 193–203.

Harrison, G. E., Raymond, W. H. A., and Tretheway, A. C. (1955). *Clin. Sci.* **14,** 681–695.

Kennedy, A. (1965). *J. comp. Path.* **75,** 69–74.

Kshirsagar, S. G., Lloyd, E., and Vaughan, J. (1966). *Br. J. Radiol.* **39,** 131–140.

Lengemann, F. W. (1957). *Proc. Soc. expl. Biol. Med.* **94,** 64–66.

Lengemann, F. W. (1963). *In* "Transfer of Calcium and Strontium across Biological Membranes", pp. 85–96. Academic Press, New York.

Lloyd, E. (1964a). *In* "Proceedings of the 1st European Bone and Tooth Symposium, pp. 93–101. Pergamon Press, Oxford.

Lloyd, E. (1964b). *In* "Assessment of Radioactivity in Man II", pp. 329–343. International Atomic Energy Agency, Vienna.

Lloyd, E. (1965). *In* "Proceedings of the Second European Symposium on Calcified Tissues, Deuxieme Symposium Européen, a l'Institut de Therapeutique Experimentale de l'Universite de Liege", pp. 381–390.

Loutit, J. F. (1962). "Irradiation of Mice and Men". University of Chicago Press, Chicago.

McLean, F. C., and Hastings, A. B. (1935). *J. biol. Chem.* **108,** 285–321.

Munday, K. A., and Mahy, B. W. J. (1964). *Clinica chim. Acta,* **10,** 144–151.

Neuman, W. F., and Neuman, M. W. (1958). "The Dynamics of Bone Mineral". University of Chicago Press, Chicago.

Smith, W. H. (1951). "The Kidney: Structure and Function in Health and Disease". Oxford University Press, Oxford.

Updegraaf, H., Greenberg, D. M., and Clark, C. W. (1927). *J. biol. Chem.* **71,** 87–117.

The Role of Oxidative Phosphorylation in Calcium and Strontium Absorption from the Gastro-intestinal Tract

D. M. TAYLOR

Department of Biophysics, Institute of Cancer Research, Sutton, Surrey

SUMMARY

Orally administered tracer doses of ^{45}Ca and of ^{85}Sr given to suckling rats are almost totally absorbed from the gastro-intestinal tract. The high absorption and lack of discrimination contrast with the pattern observed in adult rats and in other animals. A study of the influence on Ca and Sr absorption of known inhibitors of oxidative phosphorylation suggests that the metabolically dependent block which normally restricts the passage of Ca across the small intestine is inhibited by lactose (and possibly other substances) present in milk diet. Similar mechanisms may operate for Sr and Ba.

INTRODUCTION

Previous studies of the absorption of alkaline earth metals from the gastro-intestinal tract of rats of various ages (Taylor *et al.*, 1962) showed that suckling rats absorbed almost all of orally administered tracer doses of ^{45}Ca and ^{85}Sr and more than 80% of similar doses of ^{140}Ba and ^{226}Ra.

One possible explanation for the high absorption of, and lack of discrimination between, alkaline earth metals in the very young rat is that the Ca-specific duodenal active transfer mechanism (Schachter *et al.*, 1960) may be able also to transfer Sr and other alkaline earth metals. However studies of the transfer of ^{45}Ca and ^{85}Sr *in vitro* across everted intestinal sacs prepared from rats of various ages have failed to show any evidence of active transfer of Sr, Ba or Ra (Taylor, 1964). The more likely explanation is therefore that the permeability of the small intestine to all alkaline earth metals is increased by the action of some factor, or factors, present in the milk diet, for example lactose and/or lysine, by some mechanism not yet fully understood.

Fournier and Dupuis (1963) have shown that the metabolic inhibitors fluorocitrate and fluoride increase the serum Ca levels in Ca-deficient rats to the same extent as lactose. Wasserman (1964) has suggested that there is a metabolically dependent block to the passage of Ca across the small intestine and that lactose increases Ca absorption by inhibition of this block. Such a metabolic block would appear to involve oxidative phosphorylation, and a study was therefore made of the effect of various known inhibitors of oxidative phosphorylation on Ca and Sr absorption in the intact animal.

METHODS

The experimental methods were similar to those previously described by Taylor *et al.* (1962). The animals used were female rats of the highly inbred August strain, aged 6–8 weeks, raised on M.R.C. Diet 41 B, which contains about 0·8 % of Ca. After fasting overnight, the rats were given the appropriate inhibitor, followed 15 min later by a single oral dose of ^{47}Ca and ^{85}Sr, or ^{45}Ca and ^{85}Sr or ^{85}Sr only. The inhibitors were administered either orally or by intraperitoneal injection. The animals were killed 6–7 h after dosing, and absorption was estimated from the amount of each isotope retained in the body less that retained in the gastro-intestinal tract. The retention of the various nuclides in the femur was also measured.

RESULTS

The absorption of Ca and Sr by treated and untreated rats is shown in Table I. It will be seen that all the inhibitors examined, except arsenate, significantly increase the absorption of both elements. The sites of action in the oxidative phosphorylation chain of the various inhibitors are shown in Fig. 1. This is derived from data obtained mainly by studies of isolated mitochondrial systems *in vitro*, and it has been assumed that the inhibitors act similarly *in vivo*.

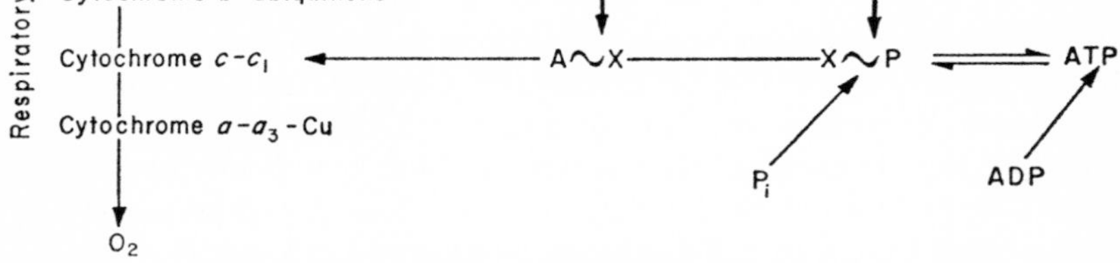

FIG. 1. The sites of action of inhibitors of oxidative phosphorylation.

TABLE I

The effect of metabolic inhibitors on the absorption of calcium and strontium

Inhibitor	No. of rats	Ca (% ± SEM)		Sr (% ± SEM)		Sr/Ca ratio	
		Femur	Total absorbed	Femur	Total absorbed	Femur	Total absorption
None	9	1·92±0·11	53·9±1·8	0·92±0·04	25·8±1·0	0·52±0·01	0·48±0·02
Lactose: 0·84 mM, orally	6	2·85±0·06	73·8±1·1	2·92±0·09	65·9±4·8	1·02±0·02	0·85±0·09
Phloridzin: 0·1 mM, intraperitoneally	7	2·74±0·11	76·7±2·0	2·86±0·23	71·7±5·4	1·04±0·07	0·94±0·05
0·1 mM, orally	10	2·81±0·08	73·7±1·2	1·89±0·06	50·2±0·6	0·67±0·02	0·69±0·04
2,4-Dinitrophenol: 0·01 mM, orally	10	2·81±0·10	75·5±0·7	1·70±0·05	41·4±1·9	0·64±0·03	0·55±0·02
0·005 mM, intraperitoneally	5	..	..	1·58±0·08	43·8±3·1	..	..
Barbital: 0·05 mM, intraperitoneally	9	2·72±0·07	73·0±1·8	1·68±0·17	43·7±3·9	0·62±0·02	0·64±0·02
Arsenate: 0·01 mM, orally	4	1·47±0·23	42·7±5·0	0·99±0·15	31·5±3·9	0·66±0·03	0·73±0·05
0·0025 mM, intraperitoneally	11	..	..	0·73±0·09	22·7±2·4	..	..
Oligomycin: 25 μg, intraperitoneally	9	2·06±0·08	60·5±2·0	1·11±0·05	30·8±1·9	0·54±0·02	0·51±0·01
50 μg, intraperitoneally	4	..	..	1·27±0·15	34·6±5·9	..	..
100 μg, intraperitoneally	4	..	..	1·01±0·12	28·9±5·6	..	..

The failure of arsenate to influence the absorption of either Ca or Sr, and the slight effects observed with oligomycin at each of the three dose levels studied, suggests that the mechanism of the metabolic block to Ca absorption requires the participation of a non-phosphorylated high-energy intermediate.

DISCUSSION

Comparison of the data for Ca and Sr show that for all inhibitors tested, except arsenate and oligomycin, the degree of enhancement of Ca absorption is constant at about 1·4 times the normal level, but that the increase in Sr absorption is more varied, ranging from about 1·7 to 3 times the control value. The constancy of the degree of stimulation of Ca absorption may indicate that, in the untreated animal, Ca transfer is already proceeding at a high rate and that it is stimulated maximally by a lower degree of inhibition of oxidative phosphorylation than is required for maximal stimulation of Sr transfer. This opinion is supported by the observation that 0·1 mM phloridzin, administered orally, increases Sr absorption by a factor of two, compared with a threefold increase when the same dose is injected intraperitoneally; the enhancement of Ca absorption is similar in each case. Although no measurements have been made of the relative inhibition of oxidative phosphorylation produced under these two circumstances, it seems reasonable to suppose that the oral dose would lead to a lower degree of inhibition due to incomplete absorption, or breakdown of the drug in the intestine.

Comparison of the Sr/Ca ratios listed in Table I shows that most of the inhibitors examined result in an increase in the Sr/Ca ratio. However, in several of the experiments this is increased only from the control value of about 0·5 to about 0·6 and, when allowance is made for the fact that the inhibitors used are also likely to reduce or abolish the preferential excretion of Sr through kidney, the differences are not very significant except for phloridzin and lactose, for which the Sr/Ca ratio is close to unity.

CONCLUSIONS

Measurement of the Sr/Ca ratio under these experimental conditions does not yield direct information on Ca/Sr discrimination at the cellular level. In a recent paper, Marcus and Wasserman (1965) have shown that lactose and lysine alter Ca/Sr discrimination by increasing overall absorption and not by influencing the rate constants for Ca and Sr absorption. The data presented in this paper do not fit the mathematical treatment proposed by Marcus and Wasserman sufficiently well to allow any definite conclusion to be drawn about whether the various inhibitors tested act in a similar way to that proposed for lactose.

Even though the overall effect of the various inhibitory substances appears to be similar for Ca and Sr, the actual mechanism of the metabolic block may differ for the two elements. Vasington (1966) has shown in mitochondria that the uptake of Ca and Sr appears to involve different mechanisms, and Marcus and Wasserman have suggested that the processes of intestinal absorption, renal re-absorption and mitochondrial accumulation of Ca and Sr may be similar.

No detailed studies of the effects of inhibitors on the absorption of other alkaline earth metals have yet been made. However, preliminary studies with ^{140}Ba show that 0·1 mM phloridzin, given intraperitoneally, increased the absorption in the fasting rat from $19·8 \pm 2·0\%$ to $81·7 \pm 2·6\%$ and that oral administration of 0·005 mM dinitrophenol significantly increased the uptake of ^{140}Ba in the femur.

The data presented here support the Wasserman hypothesis that there is a metabolically dependent block to Ca absorption in the ileal cell or the mucosal membrane surfaces, and suggest that there is a similar block to Sr and possibly Ba absorption. The very high absorption observed in very young animals would appear to result from almost complete inhibition of this block by lactose, or other factors, present in the milk diet.

REFERENCES

Fournier, P. L., and Dupuis, Y. (1963). *C. r. hebd. Séanc. Acad. Sci. Paris,* **256,** 2238–2241.

Marcus, C. S., and Wasserman, R. H. (1965). *Am. J. Physiol.* **209,** 973–977.

Schachter, D., Dowdle, E. B., and Scheniker, H. (1960). *Am. J. Physiol.* **198,** 263–268.

Taylor, D. M. (1964). Abstracts of the 6th International Congress of Biochemistry, New York. VIII. 104.

Taylor, D. M., Bligh, P. H., and Duggan, M. H. (1962). *Biochem. J.* **83,** 25–29.

Vasington, F. D. (1966). *Biochem. biophys. Acta,* **113,** 414–416.

Wasserman, R. H. (1964). *Nature, Lond.* **201,** 997–999.

Discussion

NORDIN: You have measured the loss from the gut, by using a carrier-free tracer. Is it possible that you have determined flux rather than transport—in other words, have you been measuring only the isotopic exchange of carrier-free material. Do you think that the results might have been different if the tracer had contained a reasonable amount of carrier?

TAYLOR: We know very little about the part played by the active transport mechanism in the duodenum. Our results suggest that in certain circumstances this mechanism has a relatively minor role and that manipulation of the metabolic block lower down in the intestine may be more significant.

COMAR: There is an active transport system, but it is generally of small importance compared with the rest of the diffusion processes in the intestine.

Studies on the Dynamics of Strontium Metabolism under Condition of Continual Ingestion to Maturity

MARVIN GOLDMAN AND R. J. DELLA ROSA

*Radiobiology Laboratory, University of California,
Davis, California*

SUMMARY

Over 100 beagles were reared on diets containing constant $^{90}Sr/Ca$ from intra-uterine time to adulthood at 1·5 years. On a diet containing 1% Ca and Ca/P ratio of 1·5, skeletal burdens of ^{90}Sr approximated to ten times the total daily intake level. During intra-uterine growth, discrimination via the placenta resulted in $^{90}Sr/Ca$ ratios in fetal bone of 0·1 that of the maternal diet. Discrimination during nursing changed the observed ratio of skeletons of weaned pups to 0·2 of the maternal diet. When fed the diet directly, the pups rapidly equilibrated to an observed ratio of about 0·4, with maximal rate of body burden accretion seen at 3–4 months of age. By 8–10 months of age, the adult skeletal mass was achieved and was uniformly labelled as to $^{90}Sr/Ca$. Following removal of ^{90}Sr from the diet, at 1·5 years of age, the skeleton lost ^{90}Sr at a rate proportional to the integrated effect of remodelling and exchange with the greatest effect seen in the trabecular skeletal regions, e.g. epiphysis, sternum. Teeth reflected a minimal rate of change. Total body monitoring of ^{90}Sr bremsstrahlung indicated varying individual retention functions with the mean retention following a power function exponent of about $-0·4$ corresponding to an effective half period of about 5–6 years.

INTRODUCTION

The quantitative dynamics of Sr metabolism are being studied in a colony of several hundred beagles continually fed ^{90}Sr from the onset of fetal ossification to young adulthood (1·5 years of age). The possible long-term somatic consequences of chronic ^{90}Sr ingestion will be compared with those resulting from injections of ^{226}Ra during young adulthood. Comparison of ^{226}Ra effects in the beagle with the information available from humans exposed to ^{226}Ra earlier in the century will provide a basis of inter-species comparisons of bone-seeking radionuclide damage that will be utilized in

TABLE I

**Effects of continual strontium-90 ingestion during the growth period
of the beagle and its relation to radium-226 toxicity**

Dose Level	No. Beagles*	Diet† (μCi ^{90}Sr/g Ca‡)	Average μCi ^{90}Sr fed per day
^{90}Sr ingestion series (in utero to 540 days old)			
D5	35	3·33	12
D4	35	1·11	4
D3	35	0·37	1·3
D2	35	0·123	0·4
D1	35	0·021	0·07
D0·5	40	0·007	0·03
D0	65	0·000	
Subtotal	280		

			Average μCi ^{90}Sr injected/dog
^{90}Sr injection series (single intravenous injection at 540 days old)			
S4	15	33 μCi ^{90}Sr/kg body weight	330
S2	15	3·7 μCi ^{90}Sr/kg body weight	37
Subtotal	30		

		μg ^{226}Ra/kg body weight per injection	Average μg ^{226}Ra injected/dog
^{226}Ra injection series (eight semi-monthly intravenous injections starting at 435 days old)			
R5§	35	1·25	100
R4§	35	0·42	33
R3§	35	0·14	11
R2§	35	0·047	4
R1§	35	0·008	0·6
R0·5§	40	0·003	0·2
R0§	65	0·000	
Subtotal	280		

* Equal sexes within practical limits. For dosimetry verification and colony-health evaluation, five dogs from each level are killed as needed.

† Based on average daily consumption of 300–400 g food containing average of 1% Ca; Ca:P ratio of 1·5.

‡ Within limits of normal variations of dietary Ca content.

§ Based on ^{90}Sr comparison and assuming 25% retention of total injected dose (30 days).

extrapolating from the chronic beagle ^{90}Sr data to an evaluation of possible hazards to young humans exposed to ^{90}Sr. Of special significance to man is that the studies to be described are in a colony of relatively long-lived animals exposed during the period of rapid skeletal growth and development. In this report, emphasis will be given to comparing Sr dynamics in the rapidly growing skeleton with those in the adult.

METHODS AND MATERIALS

The basic ^{90}Sr and ^{226}Ra experimental design is summarized in Table I. Note that although the ^{226}Ra is administered on a body-weight basis for a short period, the ^{90}Sr is fed *ad libitum* in a diet of constant ^{90}Sr/Ca. The stable Sr concentration was about 1400 μg Sr/g Ca. The methods of dietary preparation and control have been described earlier (Andersen, 1960a). Sr and Ca are routinely determined in biologic samples by ion exchange–flame photometry (Della Rosa *et al.*, 1963b) or by X-ray emission spectrometry (Goldman and Anderson, 1965), and ^{90}Sr measurements are performed by β-counting of portions of thermally ashed samples. Quantitative *in vivo* bremsstrahlung ^{90}Sr + ^{90}Y determinations are routinely made in a whole body counter with a 20×10 cm NaI (Tl) crystal (Goldman *et al.*, 1964). Baseline data on beagle skeletal-growth values were obtained by studying the mineral quality and quantity of skeletons taken from ninety-two non-radioactive beagles from birth to senescence (Andersen, 1960b). In addition, special radiological determinations of ^{90}Sr, Ca and P of ashed whole dogs, bones, plasma, hair, soft tissues, urine and feces were performed on experimental dogs.

RESULTS AND DISCUSSION

Fetal strontium dynamics

Fetal ossification was found to begin at mid-gestation and proceeded rapidly during the final 30 days of gestation. Ca was deposited in the skeleton at a rate of about 0·15 mg/day/g fetal weight. At birth, the skeletal weight was about 10% of the body weight, which in the beagle corresponded to about 1–2 g Ca in the 25–50 g skeleton of a 200–300 g pup (Goldman, 1962).

When non-radioactive dams were fed a ^{90}Sr diet, from mid-gestation to whelp, their plasma ^{90}Sr rapidly equilibrated to a ^{90}Sr/Ca ratio of about 0·4 that of their diet. Placental and gastro-intestinal discrimination, and preferential urinary excretion of Sr, influenced ^{90}Sr uptake by the fetus such that the ^{90}Sr/Ca ratio in newborn bone was about 10% that of the dams' diet (Fig. 1). Bremsstrahlung production by newborn pups was less efficient than in the adult owing to a larger fraction of ^{90}Sr–^{90}Y β-energy lost from bone (Parmley *et al.*, 1962) and its relatively lower effective atomic number (Evans, 1962). Of the total Sr present in the pup at birth, only a very small fraction

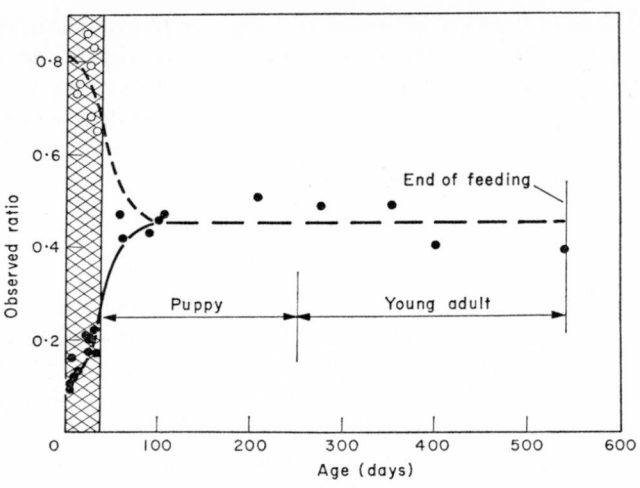

FIG. 1. Observed ratios in growing beagles:

$$OR = \frac{\mu Ci \ ^{90}Sr/g \ Ca \ in \ body \ or \ bone}{\mu Ci \ ^{90}Sr/g \ Ca \ in \ milk \ or \ diet} \ .$$

The cross-hatched area on the left represents suckling: broken lines $\dfrac{\text{pup body}}{\text{milk}}$;

continuous line, $\dfrac{\text{pup body}}{\text{maternal diet}}$.

was considered to be of maternal skeletal stores origin. However, in cases of prior maternal contamination, such sources can be of importance and should not be neglected (R. J. Della Rosa, unpublished data).

Strontium-90 derived by nursing

The 35–40 days from birth to weaning were characterized by an exponential growth rate in which the total skeletal mineral was increased ten-fold. At weaning, the skeleton was about 11% of the body weight. (The value decreased with age as the soft-tissue mass increased, such that at old age— 10 years—skeletons generally were only about 7% of the total body weight.)

The dams continued to afford considerable protection to the nursing pups; however, by weaning, the effect of the maternal milk diet had produced a net increase in ^{90}Sr/Ca of bone observed ratio from 0·1 to 0·2 that of the ^{90}Sr/Ca fed to the dam. (Observed ratio (OR) is the ratio of Sr/Ca in sample to Sr/Ca in precursor.) The amount of fetally deposited ^{90}Sr was considered to be a negligible fraction of the weanling body burden. Discrimination against ^{90}Sr by the suckling pup was not very marked. If one considers the maternal milk as the sole dietary source of mineral, the OR (^{90}Sr/Ca weaned skeleton to ^{90}Sr/Ca maternal milk) was about 0·8 (Della Rosa *et al.*, 1963a). Similar values have been found in miniature swine (McClellan *et al.*, 1962) and infants (Lough *et al.*, 1963). At weaning, the beagle skeleton is about 150 g, containing about 350 μg Sr/g Ca.

Adolescent uptake of strontium-90

The greatest increase in skeletal mineralization occurred during the 3-month period immediately following weaning and is illustrated by the Ca and P accretion curves in Fig. 2. The increase in total mineral was such that the residual "milkbone" soon became a small fraction of the skeletal mineral,

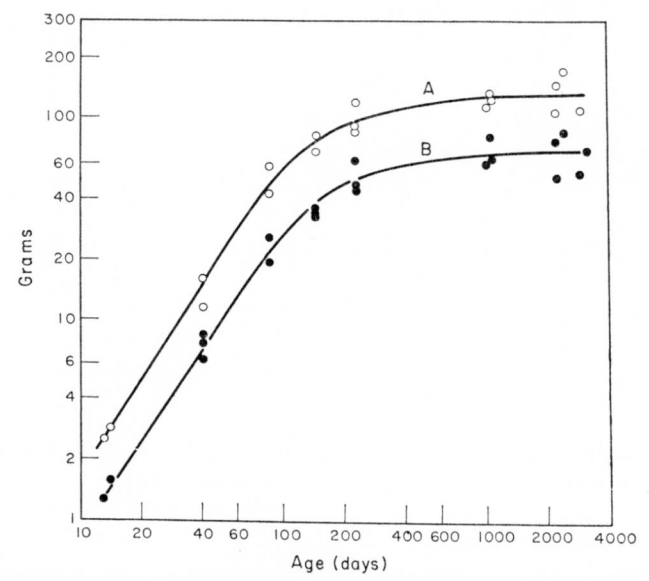

FIG. 2. Total Ca (A) and P (B) in the skeletons of growing and mature beagles.

illustrated by the rate at which the ^{90}Sr/Ca ratio of the total skeleton changed once the maternal discrimination influences were removed. After weaning to the solid diet, the young pups rapidly increased their ^{90}Sr burden and the OR increased to about 0·4–0·5 bone to diet (Fig. 1). With increasing age, ^{90}Sr discrimination did not change markedly; however, a slight diminution in the value of the ratio was suggested beyond a year of age.

The relative effect of age on uptake of a single ^{85}Sr and ^{47}Ca containing meal is summarized in Fig. 3 (Della Rosa *et al.*, 1965) in which the maximum mineral retention paralleled the period of maximal skeletal growth (Fig. 2). Although puberty occurs at 8–9 months of age and is accompanied by a change in mineral dynamics to an adult pattern, there does not seem to be any corresponding change in ⁹⁰Sr discrimination. A ten-fold quantitative difference was noted between weanling dogs and adults, with respect to percentage of mineral retention at a 1% dietary Ca level with a Ca/P ratio of about 1·5. When the Ca/P ratio was maintained constant, doubling of the Ca content of the diet halved the retention of Sr and Ca with little effect on discrimination.

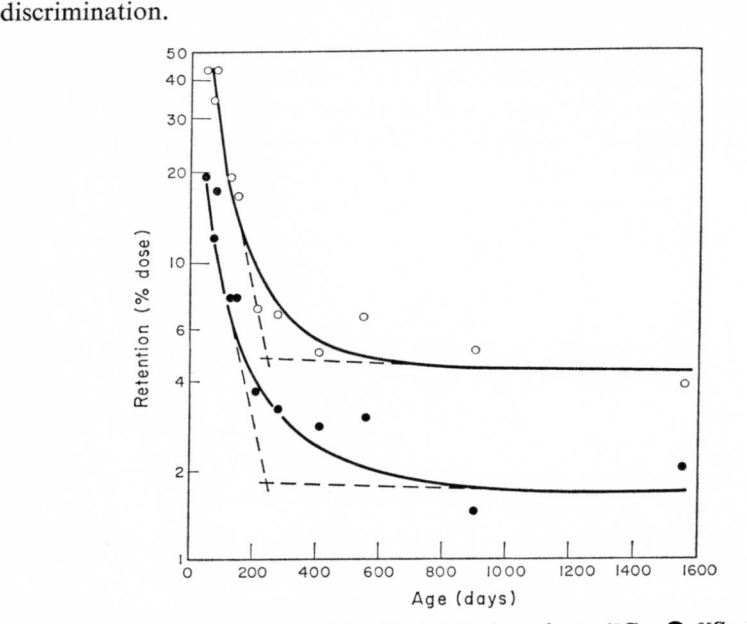

FIG 3. Total body retention of doubly labelled meal: ○, ⁴⁷Ca; ●, ⁸⁵Sr. Each point is the average retention 3–9 days after ingestion of a single meal.

Strontium-90 kinetics in uniformly labelled skeletons

Following 1·5 years on ⁹⁰Sr-containing diets, beagles' skeletons were considered to be uniformly labelled with respect to the ⁹⁰Sr/Ca ratio of bone. The body burden approximated to ten times the average daily intake of ⁹⁰Sr, and represented about 2% retention of the total ⁹⁰Sr fed. *In vivo* counting of ⁹⁰Sr bremsstrahlung indicated that 90% of the maximum body burden (1·5-year burden) was usually achieved by 8–10 months of age, a finding in agreement with the results of single-feeding studies shown in Fig. 3.

The ⁹⁰Sr burdens were realized at a slightly more rapid rate in females than in males. The age at which 50% of body burden maxima were reached was 104 days in the females and 124 days in the males (Goldman *et al.*, 1965). In addition, a trend toward litter grouping was noted with siblings generally following parallel accretion curves. With an average of about four pups per litter, there is an opportunity to investigate genetic and sex influences on uniformly labelled beagle skeletons. These factors are under study and may add significantly to our understanding of "skeletal biologic variability".

During the ^{90}Sr feeding period, plasma ^{90}Sr concentrations show little diurnal variation (Della Rosa, *et al.* 1964a), with only a 10–20% increase about 4 h after daily feeding in adults. The highest ^{90}Sr level of feeding corresponds to a specific activity of 2·4 μCi/mg Sr.

The plasma Sr content was about 5 μg/100 ml. After the ^{90}Sr was removed from the diet, the plasma ^{90}Sr concentration dropped rapidly, and was solely dependent on skeletal mineral remodelling and diffusion as its source (Fig. 4). (The levels were sufficiently high, so any additional ^{90}Sr ingested as a consequence of world-wide fallout contamination of dietary constituents could be neglected.)

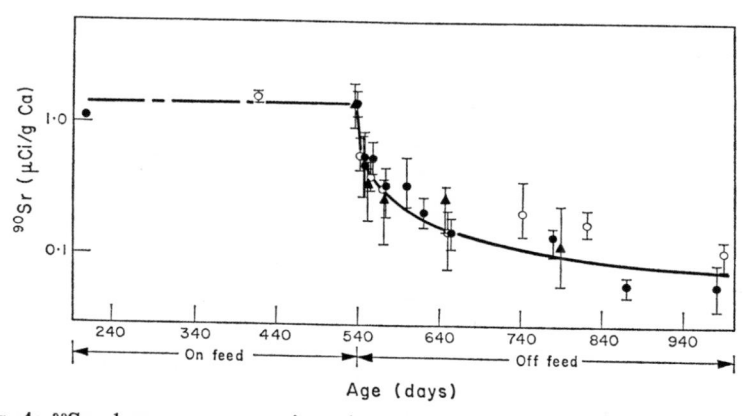

FIG. 4. ^{90}Sr plasma concentrations (normalized to 3·33 μCi ^{90}Sr/g Ca diet: ○, D3 dose level (6 dogs); ▲, D4 dose level (11 dogs); ●, D5 dose level (9 dogs). ^{90}Sr feeding was stopped at 540 days.

Over 100 beagles have been studied for up to 4 years to determine the dynamics of ^{90}Sr turnover in the total body, individual bones, plasma, urine, feces, and hair. The greatest relative loss of ^{90}Sr is from those areas of bone which are most active metabolically. This is demonstrated in Fig. 5, in which the relative radioactivity concentrations of bone and teeth are compared. The mineral pool of teeth is thought to lose its label with the slowest rate and therefore may constitute an almost "constant marker" for the initial skeletal ^{90}Sr concentration. The ^{90}Sr/Ca ratio in bones and teeth from dogs killed during the feeding period are essentially equal, but at later times the effect of post-labelling remodelling was graded and apparently related to the more trabecular bones.

During the two months immediately after ^{90}Sr feeding, the urine-to-fecal ratio rapidly increased from 0·6 to a value approaching unity (Figs. 6 and 7). Excretion measurements have been performed in some animals for 2·5 years

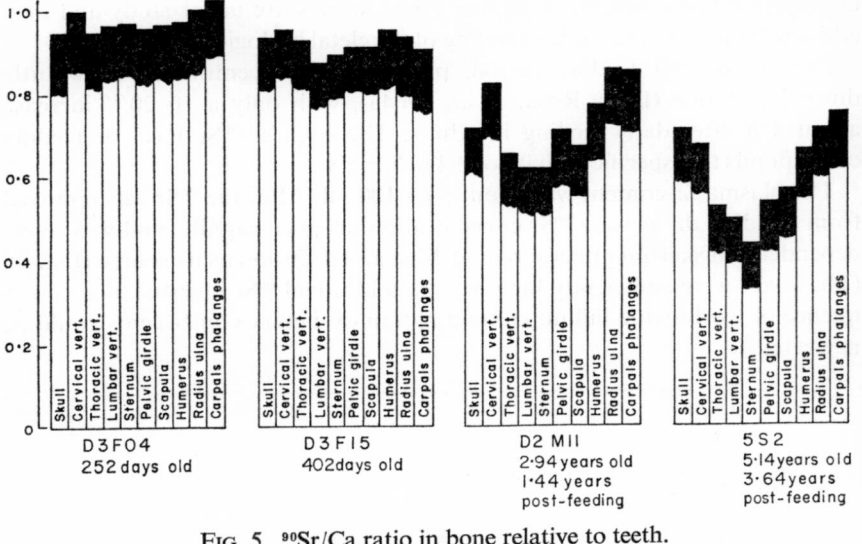

FIG. 5. ^{90}Sr/Ca ratio in bone relative to teeth.

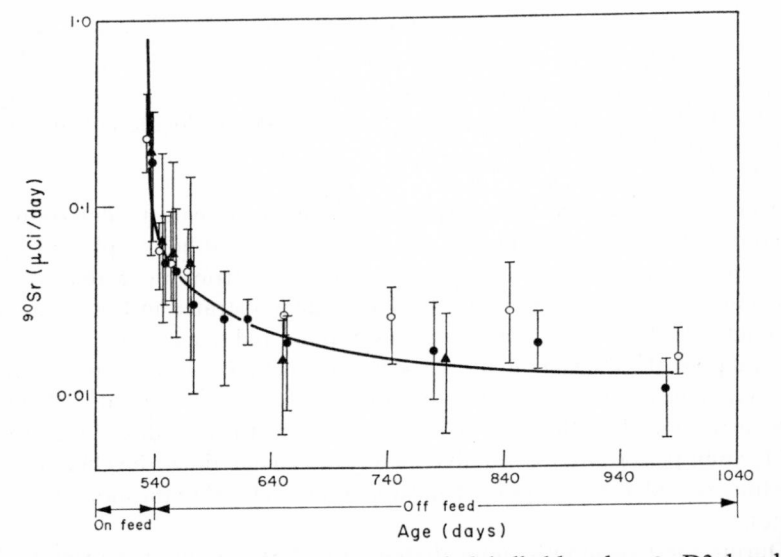

FIG. 6. Urinary ^{90}Sr excretion from uniformly labelled beagles; ○, D3 dose level (6 dogs); ▲, D4 dose level (11 dogs); ●, D5 dose level (9 dogs). ^{90}Sr feeding was stopped at 540 days.

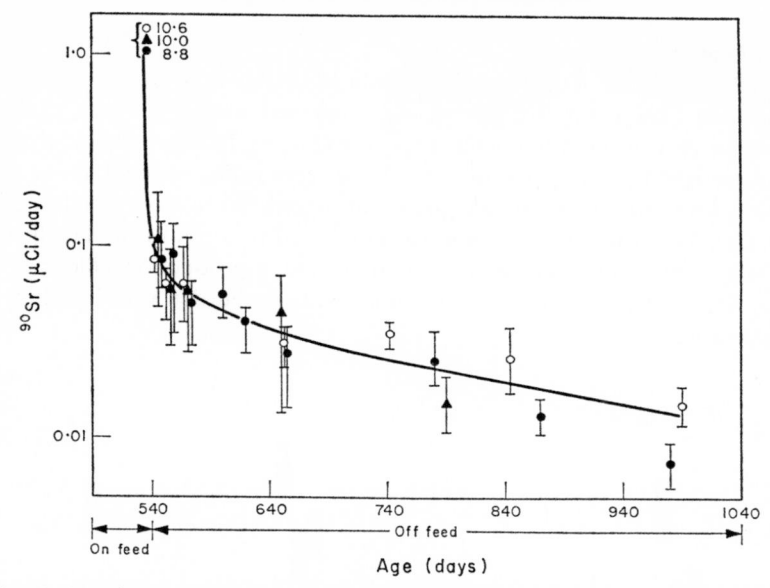

FIG. 7. Fecal ⁹⁰Sr excretion from uniformly labelled beagles: ○, D3 dose level (6 dogs); ▲, D4 dose level (11 dogs); ●, D5 dose level (9 dogs). ⁹⁰Sr feeding was stopped at 540 days.

after feeding. If one accounts for the hair growth period, a pattern was observed that was similar in time to that in plasma (Fig. 8) (Della Rosa *et al.*, 1964b, 1966). Hair may provide a useful bioassay some time after the incident, in cases of ⁹⁰Sr accidents.

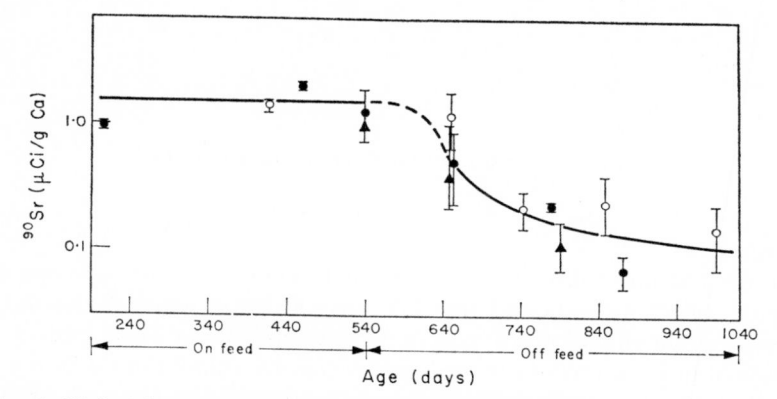

FIG. 8. Hair ⁹⁰Sr concentration from uniformly labelled beagles: ○, D3 dose level (6 dogs); ▲, D4 dose level (11 dogs); ●, D5 dose level (9 dogs). ⁹⁰Sr feeding was stopped at 540 days.

Acute exposure experiments

Further evidence of the role of metabolically active bone in Sr dynamics was seen when experimental dams or dogs injected with ^{90}Sr were examined following short term radionuclide exposure (Fig. 9). In these studies, little ^{90}Sr was deposited in teeth; therefore, the bone/teeth ratios were all above unity, with the trabecular bone again demonstrating greatest relative specific activity (^{90}Sr/Ca). When the injected dog was compared to a uniformly labelled one with the same body burden, the bremsstrahlung-production efficiency was about 8 % lower, suggesting that the injected body burden was distributed in less dense bone.

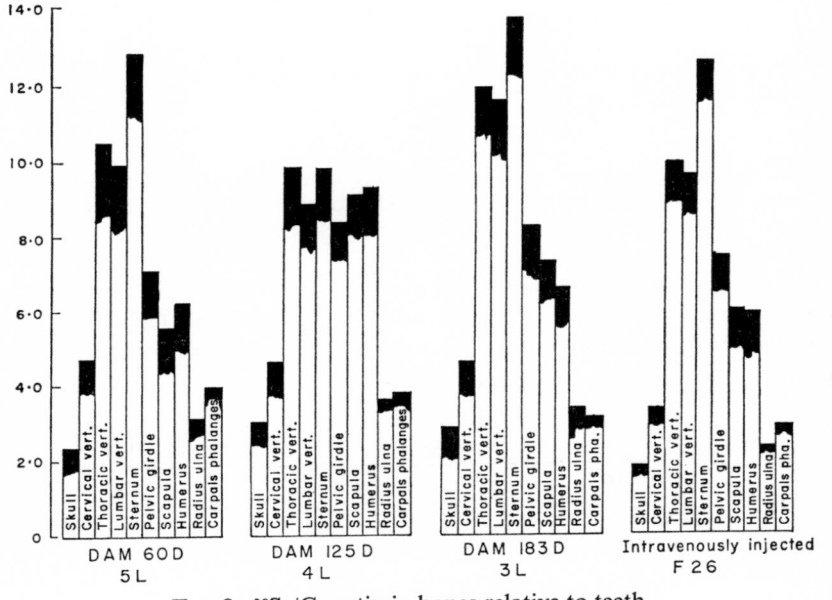

FIG. 9. ^{90}Sr/Ca ratio in bones relative to teeth.

In the ^{90}Sr injected dogs, *in vivo* bremsstrahlung measurements and total excreta collections indicated that ^{90}Y was efficiently retained and did not follow the ^{90}Sr elimination dynamics. Conversely, after a single feeding, ^{90}Sr deposited in bone was not immediately in secular equilibrium with the ^{90}Y. Similar findings on ^{90}Y metabolism were reported (Arnold *et al.*, 1955). A typical ^{90}Sr + ^{90}Y bremsstrahlung spectrum obtained from a beagle is reproduced in Fig. 10.

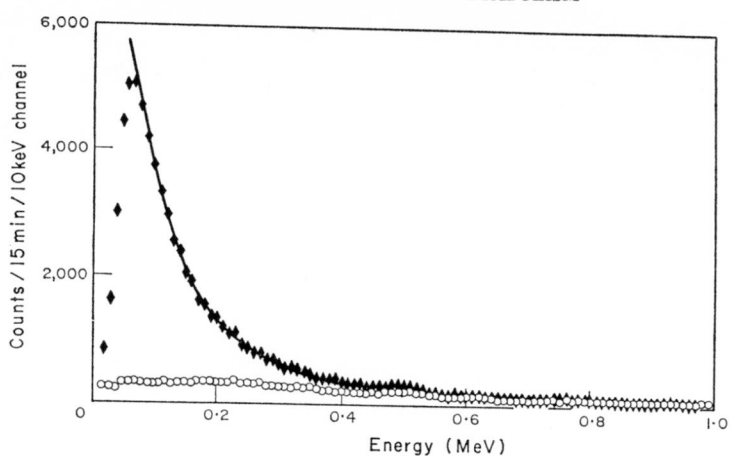

FIG. 10. Bremsstrahlung from a beagle (D3F15: 402 days old) with a 13·93 μCi ^{90}Sr body burden ($\blacklozenge$) and that of a comparable control ($\bigcirc$).

CONCLUSIONS

Analysis of over 2,000 ^{90}Sr + ^{90}Y bremsstrahlung measurements during and following ^{90}Sr feeding suggests an accretion and retention scheme, which for the beagle is outlined in Fig. 11. The various individual patterns of growth mentioned above are shown as are two general patterns of ^{90}Sr loss. The integrated effects of skeletal remodelling, diffusion and plasma clearance suggested a 5-year interval to deplete the body burden to 50% of its maximum value. It is not expected, however, that at very long times the same rate of loss will be operative. A least-squares fit of various retention models of Sr loss

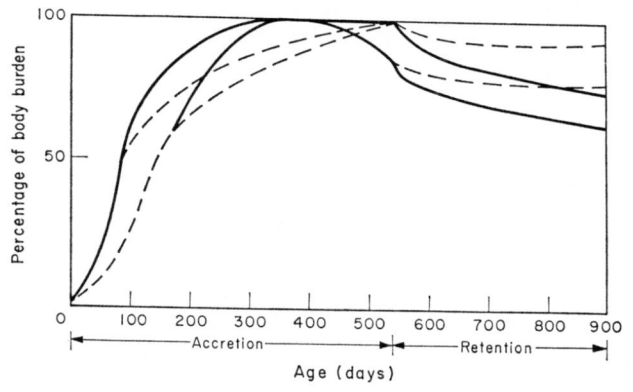

FIG. 11. Slow (broken lines) and fast (continuous lines) patterns of beagle skeletal growth demonstrated by lifetime ^{90}Sr body-burden estimates during growth (accretion) and adulthood (retention). ^{90}Sr feeding was stopped at 540 days.

TABLE II

Strontium-90 retention (years after continual feeding)

$t_{\frac{1}{2}}$ fit to single exponential model (number whole body counts); $-b \pm$ s.d. from $R = At^{-b}$ model fit by method of least squares to same data and standard deviation of estimate

Female No.	To 1·0 year $t_{\frac{1}{2}}$ (N) $-b\pm$s.d.	To 1·5 years $t_{\frac{1}{2}}$ (N) $-b\pm$s.d.	To 2·0 years $t_{\frac{1}{2}}$ (N) $-b\pm$s.d.	Male No.	To 1·0 year $t_{\frac{1}{2}}$ (N) $-b\pm$s.d.	To 1·5 years $t_{\frac{1}{2}}$ (N) $-b\pm$s.d.	To 2·0 years $t_{\frac{1}{2}}$ (N) $-b\pm$s.d.
D2F26	756 (7) 0·650±0·066			D2M29	974 (7) 0·500±0·133		
D2F21		1118 (10) 0·477±0·030		D2M30	1107 (7) 0·444±0·059		
D2F25	1287 (6) 0·372±0·182			D2M28	1161 (7) 0·421±0·036		
D2F06			1670 (7) 0·368±0·026	D2M22		1238 (10) 0·426±0·042	
D2F16		1693 (13) 0·313±0·027		D2M24		1268 (10) 0·420±0·026	
D2F15		1956 (12) 0·268±0·037		D2M04			1648 (7) 0·384±0·086
D2F17		2150 (12) 0·255±0·041		D2M11		1779 (6) 0·308±0·111	
D2F02			2831 (7) 0·224±0·112	D2M10			2002 (7) 0·304±0·052
				D2M31	3979 (7) 0·126±0·036		
x	*1022*	*1729*	*2251*	*x*	*1805*	*1428*	*1825*
D3F26	821 (8) 0·589±0·086		1399 (17) 0·402±0·021	D3M27	1389 (8) 0·363±0·071		
D3F25	874 (8) 0·559±0·030		1430 (15) 0·399±0·019	D3M22			1616 (15) 0·358±0·024
D3F16			1758 (15) 0·324±0·020	D3M10			1713 (10) 0·337±0·036
D3F19			2238 (8) 0·281±0·032	*x*	*1389*		*1665*
D3F20							
D3F02							
x	*848*		*1706*				
D4F15		993 (9) 0·555±0·071		D4M09		1246 (11) 0·435±0·083	
D4F12		1191 (9) 0·452±0·028		D4M04		1383 (10) 0·392±0·038	
D4F11		1415 (9) 0·383±0·026		D4M25	1551 (7) 0·319±0·023		
D4F08		1455 (11) 0·371±0·040		D4M06		1731 (10) 0·314±0·032	
D4F03		1637 (10) 0·326±0·018		D4M10		1850 (11) 0·289±0·039	
D4F14		1680 (9) 0·333±0·062		D4M05		1959 (10) 0·275±0·016	
x		*1395*		*x*	*1551*	*1634*	
D5F30	950 (9) 0·513±0·065			D5M31	887 (8) 0·551±0·048		
D5F29	1155 (8) 0·423±0·040			D5M08			1342 (16) 0·419±0·048
D5F06	1580 (13) 0·314±0·022			*x*	*887*		*1342*
D5F01			1593 (13) 0·359±0·016				
x	*1228*		*1593*				

from uniformly labelled beagles indicated that at this time a single exponential or a power function model were equally good representations of the rate of loss. As the metabolically slower compact bone becomes more dominant in the long term ^{90}Sr turnover, it might be reasonable to expect a lessening of the excretion slope if the residual burden is less accessible to remodelling. As discussed by Marshall (1964), a power function model is an effective means of expressing these rates. The exponent, $-b$, in the $R = At^{-b}$ model, at 3·5 years of age had an average value of about 0·4. Table II summarizes the slope constants fit to the retention data for 2 years following uniform labelling. The values for uniformly labelled beagles indicated a somewhat slower rate of ^{90}Sr loss than was found in beagles following a single injection of ^{85}Sr (Decker *et al.*, 1964). Whether chronic skeletal irradiation during bone growth and

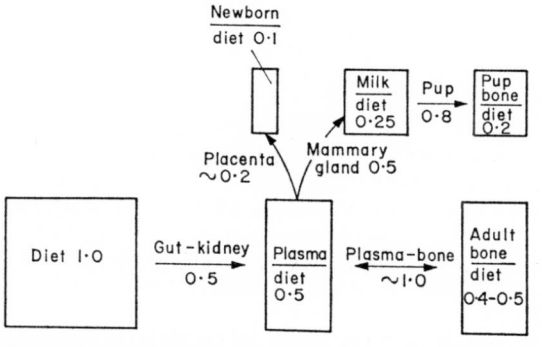

FIG. 12. Discrimination against Sr relative to Ca in beagles continually ingesting ^{90}Sr. OR values are ratio of Sr/Ca in sample to that in precursor.

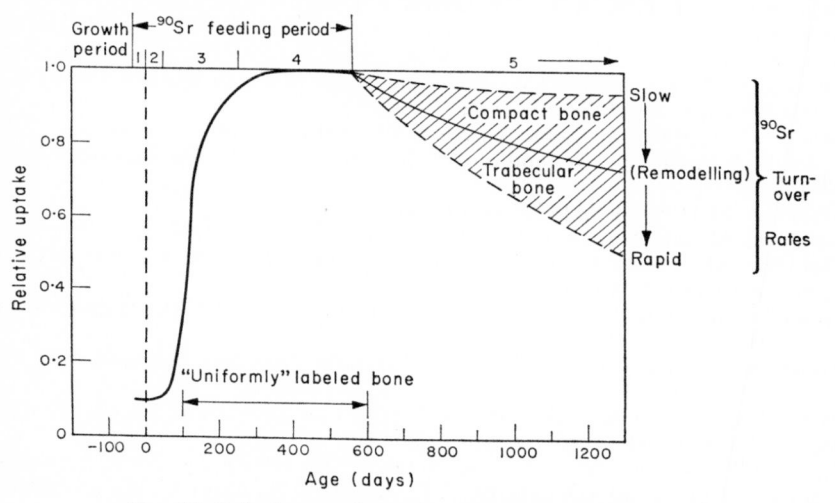

FIG. 13. Relative uptake and retention of ^{90}Sr in beagles.

development alters the "normal" ^{90}Sr dynamic picture at later times must await definitive microscopic studies. Bone alterations such as increased cortical thickness have been observed grossly in specimens from the highest radiation group (McKelvie, 1966, personal communication).

In conclusion, the ^{90}Sr metabolic picture for the growing beagle may be illustrated in Fig. 12. The growth pattern of uptake and retention resulting from these processes is summarized in Fig. 13. It would appear that the long-term dynamics of ^{90}Sr in the beagle are similar in some respects to that of ^{90}Sr in man, and in view of skeletal similarities will assist in furthering our knowledge of bone growth and development, as well as in establishing a realistic evaluation of the ^{90}Sr hazard.

The work described in this paper was supported by the U.S. Atomic Energy Commission.

REFERENCES

Andersen, A. C. (1960a). USAEC Report TID–11364, pp. 22–30.

Andersen, A. C. (1960b). USAEC Report TID–11364, pp. 45–62.

Arnold, J. S., Stover, B. J., and van Dilla, M.A. (1955). *Proc. Soc. exp. Biol. Med.* **90**, 260–263.

Decker, C. J., Kaspar, L. V., and Norris, W. P. (1964). *Radiat. Res.* **23**, 475–490.

Della Rosa, R. J., Gielow, F., and Peterson, G. (1963a). USAEC Report UCD–108, pp. 45–51.

Della Rosa, R. J., Nix, N., and Parcher, J. W. (1964a). UCD 472–110, pp. 42–46.

Della Rosa, R. J., Peterson, G., and Gielow, F. (1964b). USAEC Report UCD 472–110, pp. 54–59.

Della Rosa, R. J., Peterson, G., and Gielow, F. (1966). *Nature, Lond.* **211**, 777–779.

Della Rosa, R. J., Pool, R., and O'Sullivan, J. (1963b). USAEC Report UCD–108, pp. 66–74.

Della Rosa, R. J., Goldman, M., Andersen, A. C., Mays, C. W., and Stover, B. J. (1965). *Nature, Lond.* **205**, 197–198.

Evans, R. D. (1962). *In* "Some Aspects of Internal Irradiation" (Dougherty, T. F., Jee, W. S. S., Mays, C. W., and Stover, B. J., eds.), pp. 381–396. Pergamon Press, Oxford.

Goldman, M. (1962). USAEC Report UCD 106, pp. 45–53.

Goldman, M., and Anderson, R. P. (1965). *Analyt. Chem.* **37**, 718–721.

Goldman, M., Young, L. G., Powell, T. J., and Della Rosa, R. J. (1964). USAEC Report UCD 472–110, pp. 95–104.

Goldman, M., Powell, T. J., and Young, L. G. (1965). USAEC Report UCD 472–112, pp. 109–119.

Lough, S. A., Rivera, J., and Comar, C. L. (1963). *Proc. Soc. exp. Biol. Med.* **112**, 631–636.

McClellan, R. O., McKenney, J. R., and Bustad, L. K. (1962). *Life Sci.* **12**, 669–675.

Marshall, J. H. (1964). *J. theoret. Biol.* **6**, 386–412.

Parmley, W. W., Jensen, J. B., and Mays, C. W. (1962). *In* "Some Aspects of Internal Irradiation" (Dougherty, T. F., Jee, W. S. S., Mays, C. W., and Stover, B. J., eds.), pp. 437–453. Pergamon Press, Oxford.

Variation in "Turnover Rates" in Different Parts of the Skeleton in Relation to Tumour Incidence due to Strontium-90 Deposition

JANET VAUGHAN AND MARGARET WILLIAMSON

*Medical Research Council Bone-Seeking Isotopes Research Unit,
The Churchill Hospital, Oxford*

SUMMARY

As a result of an analysis of the sites of tumour incidence in rabbits given ^{90}Sr a study was made of "turnover rates" of ^{90}Sr in different bones and at different sites in the same bone.

It is concluded that bone sites with a slow "turnover rate" are at carcinogenic risk when the ^{90}Sr intake is low. When the intake is high, sites with a medium "turnover rate" are vulnerable. When "turnover rate" is rapid the risk is less.

INTRODUCTION

The rate of movement of Ca and other alkaline earth ions into and out of the skeleton, i.e. "turnover rate", is mediated by several complex factors which include: (1) the initial laying down of crystals on newly formed matrix; (2) ion exchange on crystal surfaces; (3) recrystallization; (4) resorption mediated by osteoclasts; and (5) osteolysis mediated by osteocytes modifying the matrix and resulting in loss of bone salts.

The skeletal turnover rate, as assessed in considering Ca kinetics, is the summation of many turnover rates for different bones and different parts of the same bone, since the relative importance of the factors listed varies in different parts of the skeleton at different ages.

Our attention was drawn to the possible importance of these different turnover rates when we came to analyse the tumour incidence in rabbits injected with different amounts of ^{90}Sr at different ages.

The tumour incidence observed is shown in Table I. The occurrence of jaw tumours is dependent upon the high uptake and retention of ^{90}Sr in the teeth and has been discussed elsewhere. It is not relevant to the present analysis (Rushton *et al.*, 1961). What is of interest is the high and unexpected incidence of squamous carcinoma of the external ear in rabbits injected when 2 days old with a high dose and in rabbits injected when 6 weeks old with a low dose of

H

[90]Sr. Further, osteosarcoma in the mid-diaphysis of the long bones appeared in this latter group only after an extremely long latent period (Kshirsagar *et al.*, 1965).

TABLE I

Tumour sites in rabbits injected with strontium-90

No. of rabbits	Age	Injection (μCi/kg)	Survival (months)	Osteosarcoma Meta-physis	Dia-physis	Jaw	Carcinoma ear
8	2 days	500	6–22	0	0	0	7
7	6–8 weeks	50–200	19–62	0	2	0	6
11	6–8 weeks	500–1,000	6	11	0	8	0
17	52 weeks	200–1,000	38	0	0	6	0

The relationship between the latent period and the injection-dose level in weaning rabbits is analysed in Table II. This shows that the lower the injected dose the longer is the latent period in this age group. The animals given 50 μCi/kg survived 5 years, whereas those given 500 μCi/kg survived 6 mo ths and developed tumours in a different site.

TABLE II

Relation of tumour latent period to initial injection dose strontium-90 to weanling rabbits

No. of rabbits	Injection (μCi/kg)	Mean survival days	Squamous carcinoma	Osteosarcoma
2	50	2000	2	1
2	100	1329	2	1
3	200	714	2	1
11	500–1,000	172	0	11

Differences in ^{90}Sr retention dependent on differences in turnover rate in different parts of the skeleton at different times seemed theoretically to be the most likely explanation of this observed tumour incidence in young animals. Two groups of young rabbits were therefore taken for further study: (*a*) rabbits aged 2 days injected with 500 μCi/kg ^{90}Sr; and (*b*) rabbits aged 6 weeks injected with 50 or 100 μCi/kg [90]Sr. Data, already available on a third group, rabbits aged 6 weeks injected with 600 μCi/kg ^{90}Sr, have been re-examined and are discussed.

EXPERIMENTAL METHODS

The retention of ^{90}Sr was measured by radiochemical techniques and related to both stable Sr and stable Ca content in the whole skeleton, in the tibia, femur and "ear bone". "Ear bone" for the present purpose is that portion of the petrous temporal bone that encloses the external, middle and inner ear (Kshirsagar et al., 1965). Radiation dose was also measured using a quantitative autoradiographic technique (Owen and Vaughan, 1959a). These measurements were made in all cases on the tibia and in some cases on the femur. Three points were chosen for measurement in the long bones in connection with the present analysis: (1) the top of the epiphysis; (2) the metaphysis; just below the epiphyseal plate (this site appears to move in relation to the epiphysis; the apparent movement is dependent on growth in length of the bone); and (3) the maximum point of autoradiographic blackening in mid-shaft. For ear bone, several measurements were made, and the highest dose-rate is recorded here. Some data on radiation-dose measurements were already available in 6-week-old rabbits given 600 μCi/kg (Owen and Vaughan, 1959a, b; Macpherson, et al., 1962). All the rabbits were of the same stock as those used in previous experiments; they were fed on a diet of hay, oats and cabbage. Except at the 3- and 5-year points in time, the results both for chemical and dose measurements recorded here are the mean of observations made on 3–5 rabbits. Intensive study in another connection (Kshirsagar et al., 1966) has shown that the behaviour of Sr and Ca in the femur and tibia is quantitatively the same. This has been confirmed again in the present series of animals. The figures for either the tibia or femur only are shown in the diagrams that follow.

The results have enabled an assessment to be made of the relative importance of the first four factors in determining turnover rates in the bones studied. The importance of factor (5), osteolysis, is not easily distinguished from factor (2), ion exchange on crystal surfaces, by the methods used.

RESULTS

Group 1. Rabbits aged 2 days injected with 500 μCi/kg ^{90}Sr

The rate of growth in the tibia is extremely rapid, and in the first 6 weeks of life the length increases from 1·6 cm to 5·7 cm. This growth-rate appeared unaffected by the presence of ^{90}Sr. The rate of growth as measured by ashed weight and the retention of ^{90}Sr ran parallel in the femur and ear bone for 9 days, after which the femur ashed weight increased more rapidly. There was a greater loss of ^{90}Sr from the long bones than from the ear bone (Fig. 1). At the end of 6 months retention of ^{90}Sr expressed as the percentage of the injected dose/g Ca in the whole skeleton was 1·0; in the femur it was 1·5 and in the ear bone 4·9. Autoradiographic measurements in the long bones showed

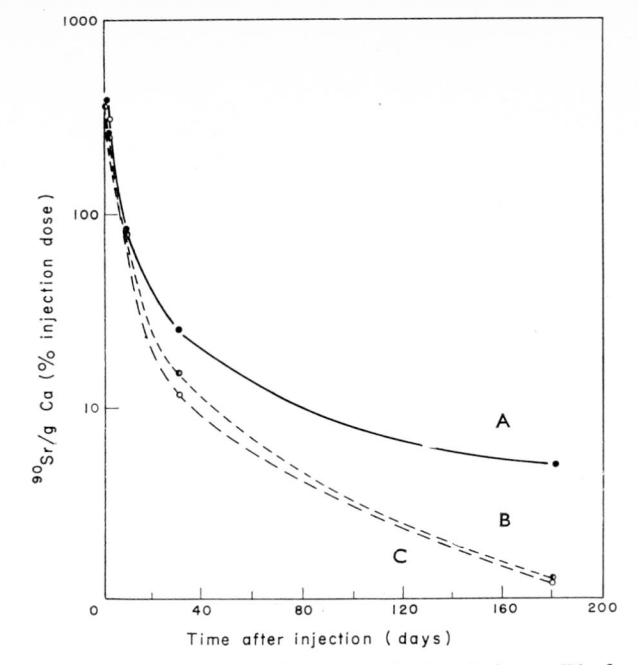

FIG. 1. Percentage of injected dose ^{90}Sr/g Ca in the skeleton (B), femur (C) and ear bone (A) of rabbits killed at different time intervals after an intraperitoneal injection (500 μCi/kg) when 2 days old.

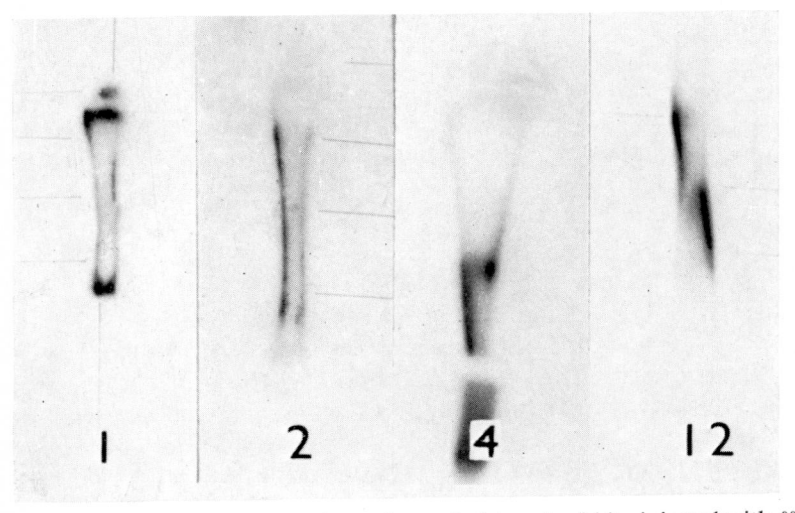

FIG. 2. Autoradiographs of thick sections of tibias of rabbits injected with ^{90}Sr. (500 μCi/kg) when 2 days old and killed 1 day, 9 days, 30 days and 6 months later Exposure time (in h) noted below. At 30 days, lower end is not shown; at 6 months, only mid-shaft exposed.

a re-distribution of ^{90}Sr throughout newly formed bone in the femur and tibia with little significant concentration after 30 days. The autoradiographic picture is shown in Fig. 2.

In examining the autoradiographs, it is essential to note the different exposure times required to give approximately similar blackening at different time intervals: 1 h is required at 1 day after injection; 2 h at 9 days; 4 h at 30 days; and 12 h at 6 months. The picture looks dramatic at 6 months, but had the section been exposed for 1 h only the result would have been a blank. Measurement of radiation dose both in the tibia and in the ear bone are shown in Fig. 3. The continuous lines represent the dose-rate at the three points in the

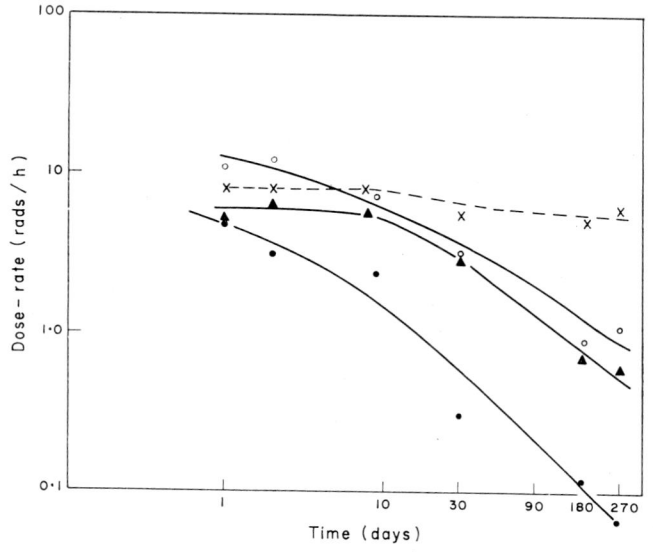

FIG. 3. Mean dose-rate in posterior wall of tibia and ear bone of rabbits killed at different time intervals after an intraperitoneal injection ^{90}Sr (500 μCi/kg) when 2 days old: ○, metaphysis below the epiphyseal plate; ×, ear bone; ▲, mid-diaphysis; ●, epiphysis.

tibia at which measurements were made, and the broken line the dose-rate in the ear bone up to 9 months after injection. It is at once apparent that the dose-rate falls at all points in the long bone but is well maintained in the ear bone. In Fig. 4 the measurements of accumulated dose at the same sites are shown, and it is again apparent that the accumulated dose is much greater in the ear bone reaching a figure of 35,600 rads 9 months after injection, whereas the maximum accumulated dose in the tibia at this time is only 12,300 rads. It is not therefore surprising that in this age group squamous carcinoma of the external ear is the tumour that kills.

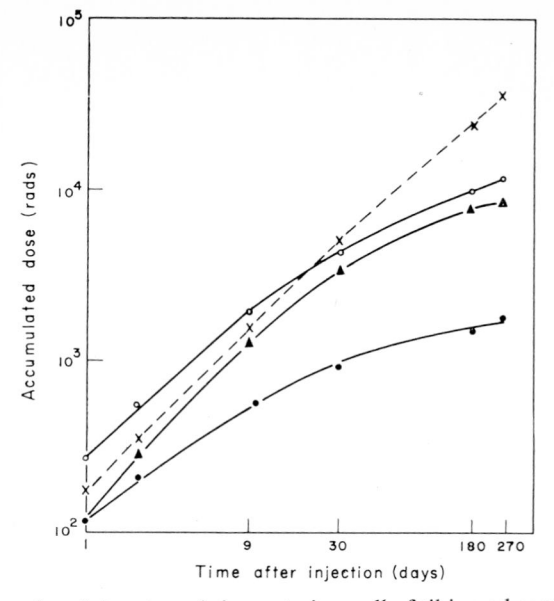

FIG. 4. Accumulated dose in rads in posterior wall of tibia and ear bone of rabbits killed at different time intervals after an intraperitoneal injection of ⁹⁰Sr (500 μCi/kg) when 2 days old: ×, ear bone; ○, metaphysis below the epiphyseal plate; ▲, mid-diaphysis; ●, epiphysis.

The explanation lies in the different turnover rates in the long bones and the ear bone in very young rabbits. The turnover rate in the long bones is rapid, largely due to the activity of factors (1), (2) and (4) throughout the bone, i.e. crystal formation, ion exchange and resorption. In the ear bone, however, factors (1), (2) and (3), though initially important, become less so, and factor (4) is hardly involved. Therefore a higher radiation dose is accumulated here than in the long bones.

Group 2: rabbits aged 6 weeks given 100 μCi/kg or 50 μCi/kg ⁹⁰Sr

In the rabbits given a low dose of ⁹⁰Sr when 6 weeks old there was no immediate skeletal damage, as evidenced by alteration in growth rate, but four out of four rabbits died with squamous carcinoma of the external ear, and two also had osteosarcoma arising on the endosteal surface of the mid-femoral diaphyses after 3–5 years. Chemical studies show a higher retention of injected ⁹⁰Sr in the ear bone than in the femur 6 months after injection, though the initial uptake was greater in the femur. One measurement of the ⁹⁰Sr retained 3 years after injection in the femur showed that it was approximately the same as the 6-month value (Fig. 5).

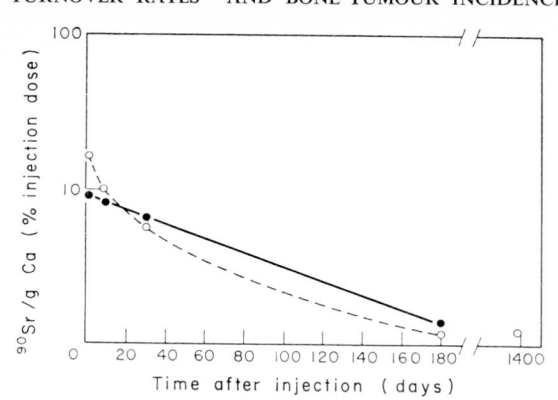

FIG. 5. Percentage injected dose, ^{90}Sr/g Ca in the tibia (broken line) and ear bone (continuous line) of rabbits killed at different time intervals after an intravenous injection (100 μCi/kg) when 6 weeks old.

Autoradiographs of thick sections of the upper end of the tibia at four time intervals up to 6 months after injection are shown in Fig. 6. These were all exposed to approximately the same degree of blackness for purposes of dose-rate measurement. The heavy uptake in the metaphysis directly after injection is apparent. Nine days later resorption has removed much of this. Thirty days after injection there is still some metaphyseal ^{90}Sr present, but at this time, and

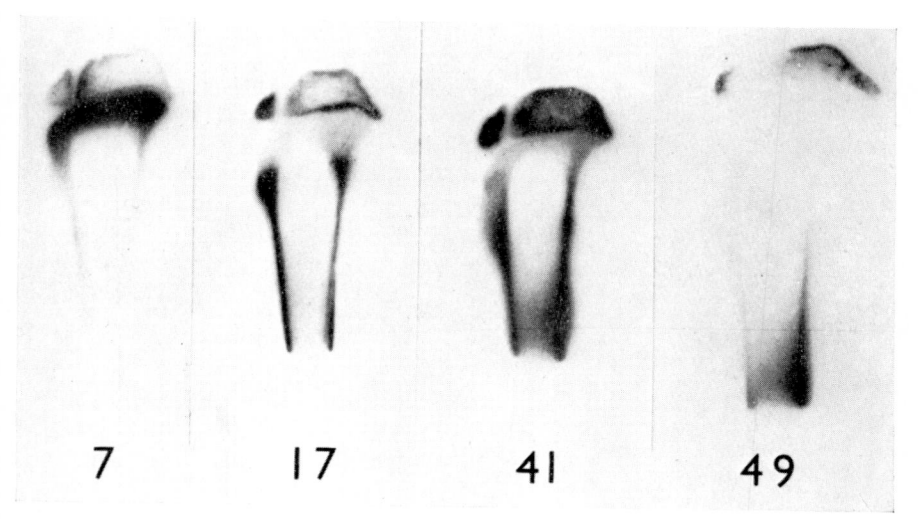

FIG. 6. Autoradiographs of thick sections of upper end of tibias of rabbits injected with ^{90}Sr (100 μCi/kg) when 6 weeks old and killed 1 day, 9 days, 30 days and 6 months later. Note changing distribution of ^{90}Sr. Exposure times (in h) noted below.

at 6 months, what is striking is the maintained uptake in mid-shaft. Measurements of dose-rate are shown in Fig. 7. The dose-rates both in the mid-shaft and the ear bone are initially low, but they remain relatively constant. The dose-rates in both the epiphysis and the metaphysis are initially high, but fall rapidly, with the result that when the accumulated dose is examined in Fig. 8 at 6 months after injection there is a higher accumulated dose in both ear bone

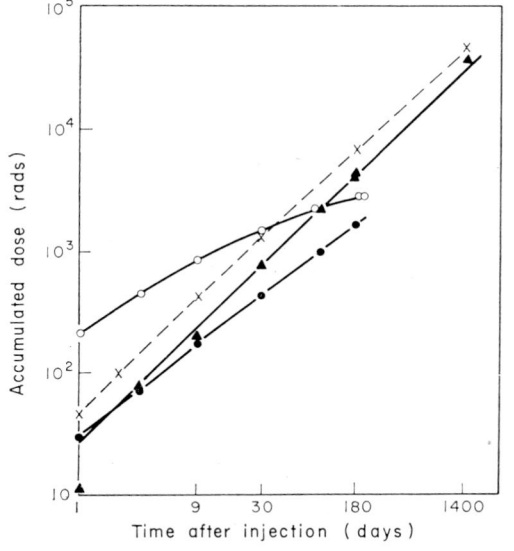

FIG. 7. Mean dose-rate in posterior wall of tibia and ear bone of rabbits killed at different time intervals after an intravenous injection of ^{90}Sr (100 μCi/kg) when 6 weeks old: ×, ear bone; ▲, mid-diaphysis; ○, metaphysis below the epiphyseal plate; ●, epiphysis.

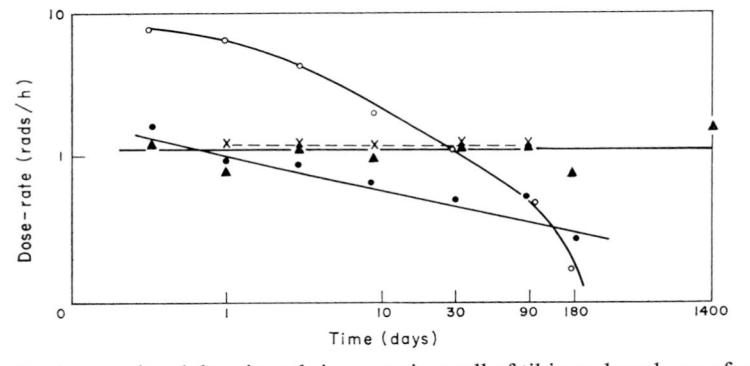

FIG. 8. Accumulated dose in rads in posterior wall of tibia and ear bone of rabbits killed at different time intervals after an intravenous injection of ^{90}Sr (100 μCi/kg) when 6 weeks old: ○, metaphysis below the epiphyseal plate; ●, epiphysis; ▲, mid-diaphysis; ×, ear bone.

and mid-shaft than elsewhere, whereas at 3 years after injection the accumulated dose in the ear bone and mid-shaft is about 40,000 rads. In rabbits
given 50 μCi/kg, the dose-rates are approximately halved and the same
accumulated dose is achieved in about 5 years.

Again the explanation of these findings lies in different turnover rates in
different sites in the long bone and in the ear bone.

At the time of injection crystal formation, ion exchange and recrystallization (factors (1)–(3)) were still active in the region of the epiphyseal plate and
metaphysis, but as the initial injection of ^{90}Sr was low, resorption (factor (4))
was not affected, and the normal remodelling that occurs in this site took place
so that no accumulated dose was built up. In the mid-shaft, however, and in
the ear bone the first three factors are probably not greatly out of balance
with one another. Factor (4), resorption, is not important at either site. They
are all effective at a slow rate. The result is a low turnover rate and therefore
there is a low dose-rate, but a build up of a high accumulated radiation dose.

Group 3: rabbits aged 6 weeks injected with 600 μCi/kg ^{90}Sr

In the third group of rabbits (weanlings injected with 600 μCi/kg) there
were no ear tumours, but all the rabbits died with osteosarcoma in the metaphyses of the long bones within 6 months. The tumours in the metaphyses can
be accounted for by the pattern of bone growth and its effect on the retention

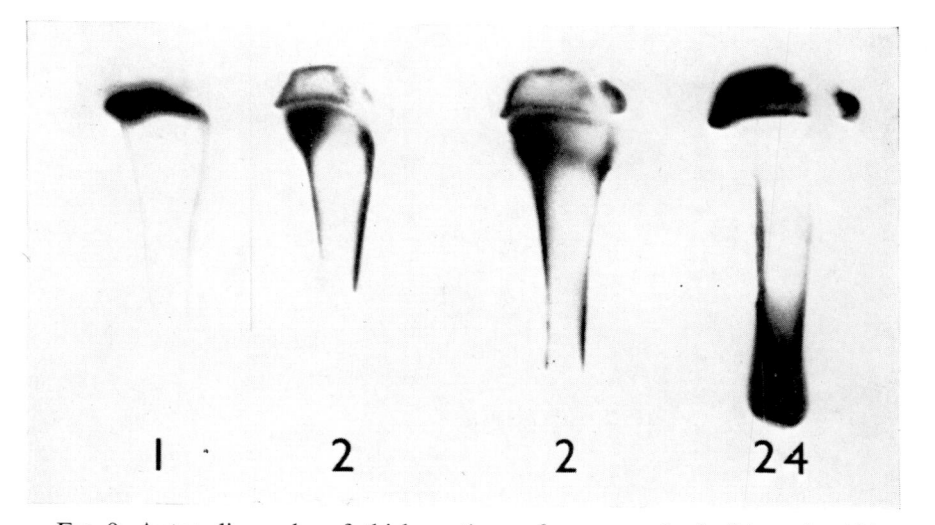

FIG. 9. Autoradiographs of thick sections of upper end of tibias of rabbits
injected with ^{90}Sr (600 μCi/kg) when 6 weeks old and killed 1 day, 9 days, 30 days
and 6 months later. Note retention of ^{90}Sr in metaphysis at 30 days. Exposure times
(in h) noted below.

of ^{90}Sr. This can be well seen in autoradiographs. The high rates of crystal formation, ion exchange and recrystallization (factors (1)–(3)) in the region of the epiphyseal plate and metaphysis of the long bones account in the growing bone for the high uptake of ^{90}Sr at this site directly after injection, as shown in Fig. 9. The resorption mechanism (factor (4)) however is damaged

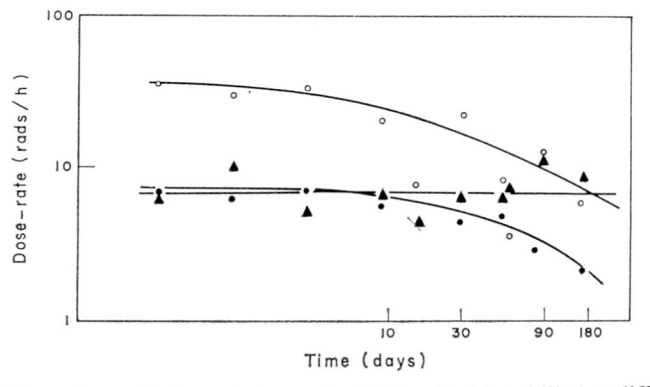

FIG. 10. Mean dose rate in posterior wall of tibia of rabbits killed at different time intervals after an intravenous injection of ^{90}Sr (600 μCi/kg) when 6 weeks old: ○, metaphysis beneath the epiphyseal plate; ▲, mid-diaphysis; ●, epiphysis.

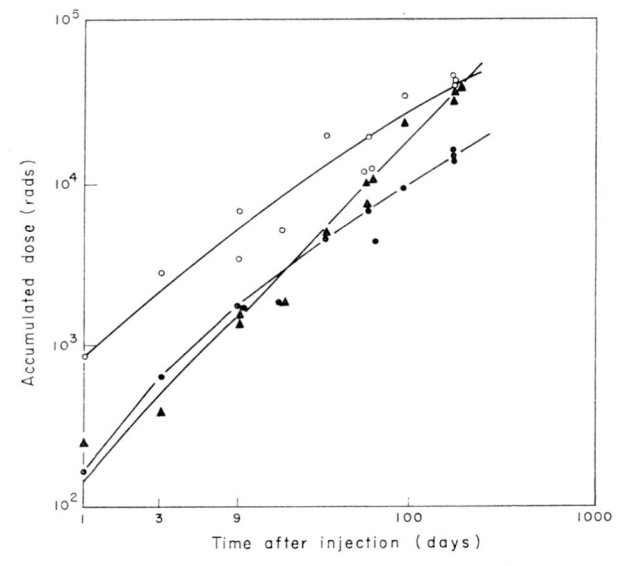

FIG. 11. Accumulated dose in rads in posterior wall of tibia of rabbits killed at different time intervals after an intravenous injection of ^{90}Sr (600 μCi/kg) when 6 weeks old: ○, metaphysis below the epiphyseal plate; ▲, mid-diaphysis; ●, epiphysis.

by the resulting high radiation dose following 600 μCi/kg, so that little ^{90}Sr is lost by remodelling and bone growth is somewhat inhibited. This is apparent particularly in the autoradiograph of the bone 30 days after injection. A high radiation dose therefore accumulates in this site before a significant dose has time to build up in the ear bone or in the mid-diaphysis. This is illustrated by the curves for radiation dose-rate and accumulated dose shown in Figs. 10 and 11. The accumulated dose in the region of the metaphysis at the time of injection is 40,000 rads 6 months later, i.e. it is already equivalent to that seen after 3 years in the mid-diaphysis and ear bone of rabbits given 100 μCi/kg.

DISCUSSION

These observations suggest that with a low intake of ^{90}Sr, bone sites with a slow turnover rate are at carcinogenic risk, whereas with a high intake sites with a medium turnover rate are at risk. In the sites of most rapid turnover, i.e. the long bones of a very young animal the risk is less.

A slow turnover rate has been recorded in the diaphysis of the long bones by other workers (Owen and Vaughan, 1959a, b; Rivera, 1964; Kshirsagar et al., 1966), but has not previously been measured in the ear bone. A similar low turnover rate in certain skull bones in man probably accounts in part for the high incidence of skull tumours in patients who have ingested radium (Dudley, 1960; Hasterlik et al., 1964).

It is concluded therefore that alteration in the relative balance of the many factors involved in turnover rate of the alkaline earths in different parts of the skeleton may be one of the parameters involved in determining the sites of carcinogenesis following the ingestion of certain bone-seeking isotopes.

Other important parameters, such as the relation of radiation dose to "sensitive tissue" are not for discussion here (Vaughan, 1965).

We are indebted to Sheila Macpherson for permission to use some of her unpublished data and for help in calculating the accumulated doses.

REFERENCES

Dudley, R. A. (1960). In "Radiation Damage in Bone", pp. 26–27. International Atomic Energy Agency, Vienna,
Hasterlik, R. J., Finkel, A. J., and Miller, C. E. (1964). Ann. N. Y. Acad. Sci. 114, 832–837.
Kshirsagar, S. G., Lloyd, E., and Vaughan, J. (1966). Br. J. Radiol. 39, 131–140.
Kshirsagar, S. G., Vaughan, J., and Williamson, M. (1965). Br. J. Cancer 19, 777–786.
Macpherson, S., Owen, M., and Vaughan, J. (1962). Br. J. Radiol. 35, 221–234.
Owen, M., and Vaughan, J. (1959a). Br. J. Radiol. 32, 714–724.
Owen, M., and Vaughan, J. (1959b). Br. J. Cancer, 13, 424–438.

Rivera, J. (1964). *Radiol. Hlth Data*, **5**, 98–99.
Rushton, M., Owen, M., Holgate, W., and Vaughan, J. (1961). *Archs. oral Biol.* **3** 235–246.
Vaughan, J. (1965). *Int. J. Radiat. Biol.* **9**, 513–543.

Discussion

QUESTION: Did you see any changes in the blood vessels in cases of high Sr dosage and significant retention in the epiphyses?

ANSWER: We have looked for damage of this kind by examination of serial sections of bone taken at different times. Occasionally we saw a few red cells that seemed to have escaped from the vessels, but nothing more.

Studies on Soft-tissue Dosage from Strontium-90

A. M. BRUES, H. AUERBACH, D. GRUBE
AND G. DeROCHE

Division of Biological and Medical Research,
Argonne National Laboratory, Argonne, Illinois

SUMMARY

Young adult male mice were injected intravenously with ^{85}Sr and the activities of blood and soft tissues were measured at serial sacrifices. After the first few hours, these activities declined at such a rate that the retention could be approximately described by the function $R = At^{-1\cdot3}$. The values for the testis, per g wet weight, were about 60–80% of those for whole blood; muscle and other soft tissues were generally about the same, except that kidney and intestine were somewhat higher for the first few hours and seminal vesicles (with contents) and salivary gland remained higher than blood for several days. Total integrated retention appeared to be about 0·4 days (as μCi days/g per μCi/g injected) for blood and about 0·2 days for testis and most soft tissues. Injection of ^{90}Sr (from which ^{90}Y was removed before injection) confirmed these findings, and measurements of ^{90}Y at sacrifice indicated an excess above the equilibrium ^{90}Sr concentration in liver and spleen. The absolute ^{90}Y excess declined with time, along with the decline in soft-tissue ^{90}Sr concentration, suggesting that its source was mainly Y produced endogenously by Sr decay in soft tissues. The ^{90}Y activity in blood remained consistently at about 10% of the ^{90}Sr activity.

INTRODUCTION

Because of the high relative concentration of alkaline earths in bone, relatively little attention has been given to the concentration of ^{90}Sr in the soft tissues and blood; and in contrast to voluminous data relative to its skeletal retention and metabolism, data on its fate elsewhere in the body are scattered, and those that exist relate largely to passage of Sr through the food chain, physiological discrimination against it in relation to Ca, and procedures aimed towards furthering its elimination.

Our attention was drawn to this problem through a series of observations that have been reported by Swedish investigators, which have indicated, along with related findings, that if male mice are treated with ⁹⁰Sr, their offspring conceived in the ensuing weeks show an increased incidence of failure of uterine implants, suggesting that lethal mutations have resulted from irradiation of the parental gonads (Lüning *et al.*, 1963). Since the alkaline earths have been shown under some circumstances to bind to DNA molecules (Åberg, this book, p. 261), the possibility had been seriously considered that genetic effects of ⁹⁰Sr might be attributed to energy of recoil of atoms of the nuclide that are so linked.

We previously reported (Brues *et al.*, 1963) a short series of experiments in which ⁹⁰Sr–⁹⁰Y was injected intraperitoneally in male mice, following which serial sacrifices were performed to determine the sources of testicular irradiation. Equating the radiation dosages to those that would be received from a 1 μCi/g injection of the equilibrium mixture, we concluded: (1) that the minimum testicular dose based on the concentration curve of radioactive material found in the testes, assuming dosage measured at a point surrounded by an infinite volume at that concentration, was about 6 rads, most of which was from the more energetic ⁹⁰Y daughter; (2) that in individual animals this dose might be exceeded by a factor of 4 or more, apparently owing to leakage of the injection solution into the scrotal sac or passage of the testis into the peritoneal space; and (3) that the concentration of ⁹⁰Sr and ⁹⁰Y in the adjacent skeletal structures, i.e. the pelvis and root of the tail, was of such magnitude that considerably larger total doses, extended over a longer period of time, might be expected to be received by the part of the testis nearest these structures, from skeletally deposited radionuclides. We then carried out some injections after placing silver phosphate glass dosimeters around the testes, and found after 2 to 3 days that doses up to 100 rads were recorded on some of the dosimeters. We proposed, at that time, that the major source of radiation dosage to the testis of the ⁹⁰Sr-treated mouse must be radioactivity deposited in the skeleton, which would not occur in a larger animal whose gonads were largely beyond the range of the ⁹⁰Y β-ray, or in a mouse treated with a soft β-ray emitting alkaline earth, such as ⁴⁵Ca.

EXPERIMENTAL METHODS

More recently, we have become interested in the whole question of the soft-tissue retention pattern of ⁹⁰Sr, and we wish to present a progress report on studies in this area. These experiments have included: (1) intravenous injection of ⁸⁵Sr followed by serial sacrifice and measurements of organ concentrations of radiostrontium; (2) intravenous injection of ⁹⁰Sr from which the ⁹⁰Y has been removed by solvent extraction within 2 h before injection, followed by organ analyses for ⁹⁰Sr and ⁹⁰Y; and (3) some preliminary attempts to measure radiation dosage in areas surrounding the testes by inserting fluorescing glass dosimeters (Kastner *et al.*, 1965). In all of these experiments we employed male CF-1 mice 2 to 3 months old, weighing between 28 and 38 g.

In point of fact, the entire pattern of irradiation would differ greatly in a human being absorbing ^{90}Sr by one of the usual routes, from that in the injected mouse: absorption of the ^{90}Y daughter accompanying the Sr would be negligible; β-radiation of gonads from the skeleton would be minimal or absent; the genetically effective dose would be due to β-radiation due to the Sr and its Y daughter lying within gonads and adjacent soft tissue and/or recoil of nuclide incorporated in chromatin. For physiological reasons the soft tissue retention could differ somewhat, but probably not greatly, from that in the mouse.

DISTRIBUTION OF STRONTIUM-85

Mice were injected intravenously with 5 μCi (in some cases, smaller amounts) of ^{85}Sr in 0·5 ml of saline solution, and sacrificed at intervals ranging from 6 min to 19 days. Just before sacrifice, the animals were bled by cardiac puncture under anesthesia, and the blood samples were weighed. Organs and tissues were then removed and weighed, including the entire left femur. The tissue and blood samples were counted in a well-type crystal counter against standards made up from the injection solution at the time of injection. The femur was later compared with the residual carcass in a whole-body mouse counter. All counts have been converted into terms of the injected mean body concentration, i.e. μCi/g tissue per μCi/g body weight injected.

The results are essentially as follows: the concentration in whole blood at about 30 min was equal to the injected mean body concentration and fell to 1/10 of that value at about 12 h; to 1/100 at about 3 days; and to 1/1000 at 20 to 25 days. Plotting this as a power function, the curve after a few hours has a slope of about $t^{-1·3}$ which corresponds to the slope of the excretion curve which is consistent with a retention function of $t^{-0·3}$ (Speckman and Norris, 1964). The curve is shown in Fig. 1.

Most of the soft tissues, including muscle, liver, spleen, and testes, rise slowly in the first 2–4 h and reach a level 0·6–0·8 that of whole blood (lower in liver), then decline parallel with the blood curve maintaining about the same relation to blood throughout the period of observation. Kidney and intestin representing routes of excretion, rise more rapidly and reach a level equal to that of blood within an hour. The seminal vesicles, which secrete calcium citrate, and the salivary gland, which maintains a high Ca concentration, parallel the blood specific activity at a slightly higher level. Lung rises to a level above that of blood and declines more slowly than blood, which we suspect is due to better retention in the cartilaginous tissues of the bronchi. Activities of various tissues, relative to blood, are tabulated in Table I.

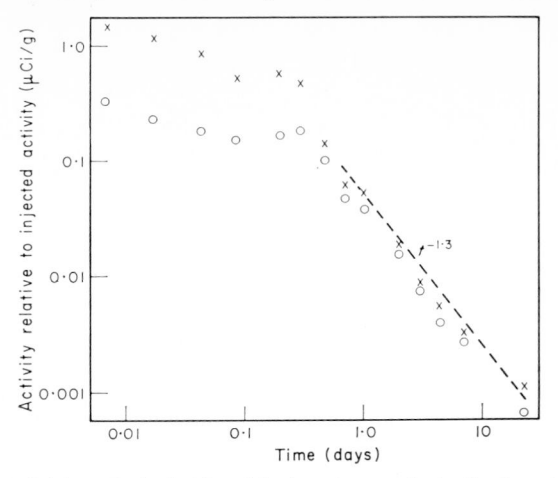

FIG. 1. [85]Sr activities of whole blood (×) and testes (○) of mice, at intervals after injection.

DISCUSSION

We attempted to estimate the whole tissue dose to the blood and testes by fitting three exponentials to their time-concentration curves. Blood shows an integrated retention of 0·4 μCi days per μCi. Testis, although it eventually reaches 80% of the blood activity, shows an integrated retention of 0·2 μCi days per μCi, within 10% error of estimate.

These figures represent the integrated retention, but since they refer only to Sr and since the Y must also be considered, they do not translate directly to radiation dosages. The second series of experiments, in which ^{90}Sr freed of ^{90}Y has been injected, show that the tissue concentrations of the two nuclides are in general the same, except for whole blood, in which the concentration of ^{90}Y is about 10% of that of ^{90}Sr, and liver (and probably also spleen) shows a relatively larger concentration of ^{90}Y than of [90]Sr. These values are shown in Table II.

From this experiment we may conclude that the endogenously formed ^{90}Y is cleared from the blood stream more rapidly than ^{90}Sr, and is taken up in the reticuloendothelial cells. Since the ^{90}Y that appears in excess in the liver declines at about the same rate as the ^{90}Sr in blood and soft tissue, we are led to believe that it arises mainly from ^{90}Sr decaying in soft tissue. The relative deficiency of ^{90}Y in the blood indicates that the Y escapes from its sites of formation more slowly than it is cleared from blood. The scope of the present data does not allow a more quantitative evaluation of the process.

Fluorescing dosimeters were inserted into the scrotal sac of mice before the mice were injected intravenously with ^{90}Sr that had been freed of [90]Y.

TABLE I

Tissue concentration of strontium-85, relative to blood concentration, as a function of time after intravenous injection of mice

Time after injection (h)	Blood ⁸⁵Sr (specific activity)*	Tissue activity/blood activity							
		Testes	Small intestine	Kidney	Spleen	Liver	Muscle	Salivary gland	Seminal vesicles
0·1	2·35	0·11	0·39	0·64	0·26	0·27	0·20	0·40	0·13
0·17	1·41	0·22	0·40	0·56	0·52	0·40	0·39	0·47	0·21
0·4	1·12	0·23	0·89	0·83	0·46	0·37	0·35	··	··
1·0	0·88	0·21	1·31	1·01	0·48	0·80	0·48	··	··
2	0·58	0·26							
3·6	0·46	0·43	1·13	1·10	0·44	0·38	0·77	1·00	0·29
11	0·14	0·70	0·96	0·83	0·44	0·39	0·61	··	··
17	0·060	0·78	··	··	··	··	··	··	··
24	0·052	0·73	··	0·88	0·47	0·47	··	··	··
29	0·063	0·82	1·13	1·02	0·69	0·36	0·72	1·72	1·41
48	0·018	1·12	0·81	0·91	0·62	0·43	0·75	··	1·71
72	0·0083	0·87	0·86	0·81	0·41	0·40	0·63	··	2·40
120	0·0055	0·71	0·86	0·67	0·80	0·54	1·03	1·32	3·03
168	0·0032	0·53	0·80	0·68	0·77	1·04	0·98	1·60	1·19

* μCi/g of blood per μCi/g body weight injected.

Note: Except for the blood values, which are given in absolute units, the tissue values given in this table are relative to blood of the same animal (i.e. blood = 1·0).

s.m.

TABLE II

Ratio of measured yttrium-90/strontium-90 at intervals after intravenous injection of strontium-90

Time after injection	Blood	Liver	Spleen	Testes	Muscle
3 days	0·09	2·3	1·0	0·9	1·2
	0·07	2·6	1·0	0·9	1·4
6 days	0·06	4·9	1·2	1·0	1·2
	0·10	3·5	1·5	1·6	1·2
12 days	0·09	4·2	2·1	1·0	1·2
21 days	0·11	4·3	1·7	0·6	0·9
38 days	0·07	1·3	*	*	*

* Total activity too low for significant estimate.

Note: β-Counts were made within 2 h after sacrifice by using absorbers to estimate ^{90}Sr and ^{90}Y separately. Yttrium values extrapolated to the time of sacrifice.

Because of the difficulty of locating these small glass rods at autopsy, black surgical silk sutures were tied around them before insertion. As they were serially removed, they were rinsed, blotted, and stored until the series of autopsies were completed. At that time they were examined routinely with a β-survey meter and were found to have retained a considerable amount of radioactivity, which had been bound to the sutures. One of these sutures was estimated to have fixed and retained for two weeks in the animal, 0·4 and 1·0 μCi/g ^{90}Sr and ^{90}Y, respectively, per μCi/g injected. (This unexpected observation raises the possibility that silk and other fibrous material might be used in some way for biological decontamination, but it seems very doubtful that this would have advantages over established methods.) The dosimetric values were high throughout the time-series, and the experiment, which we hoped to report at this time, is being repeated without the use of sutures.

REFERENCES

Brues, A. M., Auerbach, H., and DeRoche, G. (1963). Argonne National Laboratory Report ANL–6823, pp. 76–80.
Kastner, J., Roberts, D. R., and Prepejchal, W. (1965). *Am. J. Roentg.* **94**, 984–988.
Lüning, K. G., Frolen, H., Nelson, A., and Rönnbäck, C. (1963). *Nature, Lond.* **197**, 304–305
Speckman, T. W., and Norris, W. P. (1964). *Radiat. Res.* **23**, 461–474.

Some Aspects of Strontium-90 Metabolism in Miniature Swine Ingesting Strontium-90 Daily

R. O. McCLELLAN,* J. L. BEAMER AND P. L. SHELDON

Biology Department, Pacific Northwest Laboratory,
Battelle Memorial Institute, Richland, Washington

SUMMARY

The effect of two diets, varying in Ca and P content, on ^{90}Sr metabolism was studied in three generations of miniature swine chronically exposed to ^{90}Sr. The diets, which contained either 1·5% Ca and 0·8% P or 0·55% Ca and 0·50% P, as well as 25 μCi ^{90}Sr were fed to two groups of female miniature swine beginning at 11 months of age. When killed 7 months later, the high Ca animals contained 85 μCi ^{90}Sr and the low Ca animals 190 μCi ^{90}Sr. The body burden of ^{90}Sr of animals (F_1 generation) born to the original generation increased with age as they were continuously exposed to ^{90}Sr. The high Ca F_1 generation animals had a skeletal burden of 190 μCi ^{90}Sr when killed at 19 months of age, the low Ca animals contained 360 μCi ^{90}Sr. The newborn F_2 generation animals on the high Ca diet derived 28%, and the low Ca F_2 generation 39%, of their ^{90}Sr from the maternal skeleton that was present before conception. For the skeleton formed between birth and weaning (that originating from milk Ca and ^{90}Sr), 48% for the high Ca diet, and 54% for the low Ca diet, of the skeletal ^{90}Sr originated from maternal skeleton present before conception.

The comparative metabolism of ^{90}Sr and Ca was similar for both diets. The ^{90}Sr/Ca ratio of the milk was about 0·10 that of the diets. The ratio for the skeleton of newborn animals was about 0·15 that for the dam's skeleton. The ^{90}Sr/Ca ratio for skeleton formed during the suckling period was near unity, relative to the diet of dam's milk. After weaning and being placed on an adult diet, the ^{90}Sr/Ca ratio of the skeleton relative to that of the diet decreased to values of 0·10 to 0·13.

---

* Present address: Fission Product Inhalation Program, Lovelace Foundation Albuquerque, New Mexico.

INTRODUCTION

A long-term study to evaluate the toxicity of daily ingestion of ^{90}Sr in miniature swine has been in progress in our laboratory since 1958 (McClellan et al., 1962a, b). The miniature pig has been used in these studies because it is a long-lived omnivore similar in size and in some physiological features to man (Bustad and McClellan, 1966). To provide background information for the eventual establishment of a radiation dose–effect relationship it has been necessary to collect certain basic data on the metabolism of ^{90}Sr in miniature swine. Of particular interest is the animals' skeletal burden of ^{90}Sr when ingesting ^{90}Sr daily compared to similar data for man (Table I). The swine data are taken from our study in which animals are being fed 2 lb of pelleted feed (1·8 % Ca, 1·2 % P) per day and 25 μCi ^{90}Sr/day. The data for man are derived from the review by Snyder et al. (1964), who examined several methods for establishing the anticipated relationship between daily ^{90}Sr intake and the established maximum permissible body burden for ^{90}Sr, 2 μCi.

TABLE I

Relationship between daily strontium-90 intake and skeletal burden of strontium-90 for man and pig

Species	Daily Ca intake	^{90}Sr intake (μCi/day)	Skeletal ^{90}Sr burden (μCi)	Skeletal ^{90}Sr burden / Daily ^{90}Sr intake
Man*	~1 g	$5·7 \times 10^{-3}$	2	350
		18×10^{-3}	2	110
Pig†	~16 g	25	200–300	8–12

* Adapted from Snyder et al. (1964).
† Unpublished personal observations.

The estimated daily ^{90}Sr intakes that would result in a body burden of 2 μCi ^{90}Sr were derived by Snyder and co-workers by several methods, and agreed within a factor of three. The maximum and minimum estimates for man are shown in Table 1; both were derived by using a power-function model for expressing the retention of ^{90}Sr. The lowest value shown is almost three times the value of $2·2 \times 10^{-3}$ μCi ^{90}Sr/day derived from the recommendations of the ICRP (ICRP, 1950). The ratio of skeletal ^{90}Sr burden to daily ^{90}Sr intake for this latter value is about 900.

Earlier when faced with attempting to reconcile our swine data with the ICRP values for man we were completely bewildered by the marked difference. The publication of the analyses by Snyder and associates (1964) reporting a closer agreement with the swine data, led us to renew our efforts to establish a basis for the difference. The marked difference between the dietary Ca intake of the two species (Table I) and an awareness that differences in Ca intake can have a marked effect on Sr metabolism led us to explore this avenue. This paper discusses our initial investigations on the influence of dietary differences on the metabolism of ^{90}Sr in miniature swine.

MATERIALS AND METHODS

Six closely related 11-month-old female miniature swine were divided into two groups and placed on diets varying in Ca and P content. The low Ca diet (0·55% Ca, 0·5% P) consisted of ground barley, 28%; ground wheat, 24%; dehydrated alfalfa, 15%; soybean oil meal, 14%; blood meal, 13%; dried skim milk, 4%; mill run, 0·7%; corn distillers dried solubles, 0·6%; salt, 0·5%; oyster shell, 0·2%; bone meal 0·2%; and a trace of mineral mix. The high Ca diet (1·5% Ca, 0·8% P) had the Ca and P content increased by addition of dicalcium phosphate. Both diets were pelleted. The two groups were penned separately and each animal was fed individually in separate feeding compartments. The normal ration of 2 lb/day was given in two feedings, morning and afternoon. After allowing a week for adaptation the animals were started on 25 μCi ^{90}Sr/day given in a feed pellet with the morning aliquot of feed.

All animals were bred to the same normal male on their first estrus after being placed on ^{90}Sr feeding.

Before being placed on ^{90}Sr feeding, and at frequent intervals thereafter, the body burden of ^{90}Sr for all animals was determined by whole body bremsstrahlung counting with a 23-cm in diameter × 10 cm thick NaI (Tl) crystal and a multi-channel analyser (Case and McKenney, 1961).

Just before parturition the animals were placed in separate farrowing pens where they were held until their offspring were weaned at 6 weeks of age. The offspring born to the original female are referred to as the F_1 generation. For the period from 1 week post partum until the offspring were weaned at 6 weeks of age, the feed allowance of the animals was increased to 2·5 lb/day, although the ^{90}Sr intake remained at 25 μCi/day.

Immediately after birth, some offspring were killed. Most offspring, however, were allowed to remain with their dams until weaning at 6 weeks of age, when additional animals were killed to provide specimens for study.

At weaning, nine offspring from each diet were placed on 6·3 μCi ^{90}Sr and 0·5 lb/day of their respective ration. At 3 months of age, three animals from each diet were killed. The six surviving animals from each diet were placed on 12·5 μCi ^{90}Sr and 1 lb/day of their diet per day. At six months of age, three additional animals from each diet were killed.

The three surviving animals from each diet were placed on the adult intake of 25 μCi ^{90}Sr and 2 lb/day of feed. Starting at 6 months of age, the body burden of ^{90}Sr in the remaining F_1 generation animals was estimated at frequent intervals by whole-body counting. At 9·5–10 months of age, the F_1 generation animals were bred. From this point on the F_1 generation females were treated the same as the original females, their offspring were the F_2 generation. Milk samples were obtained from the F_1 generation females at the 4th and 6th week of lactation by hand milking after intravenous administration of 1 ml of oxytocin. Some of the F_2 generation animals were killed at birth, the remainder were killed when weaned at 6 weeks of age.

The original group of females were killed at just over 18 months of age. The six long-term F_1 generation animals were killed at just under 19 months of age.

The skeletons of the killed individuals were freed of attached flesh, separated into aliquots, dried at 120° C and then ashed at 600° C. The bone ash was then put into solution in conc. nitric acid, duplicate aliquots plated on stainless-steel planchets, and counted in proportional β-counters. Standard methods were used to calculate ^{90}Sr/g skeletal ash. Ca and P in the bone-ash solutions, as well as milk ^{90}Sr and Ca, were determined by standard techniques (Boltz, 1958; Yarboro, 1958).

RESULTS AND DISCUSSION

In Fig. 1 are plotted the changes in body burden of ^{90}Sr with age as influenced by the two experimental dietary Ca and P levels. The values at 6

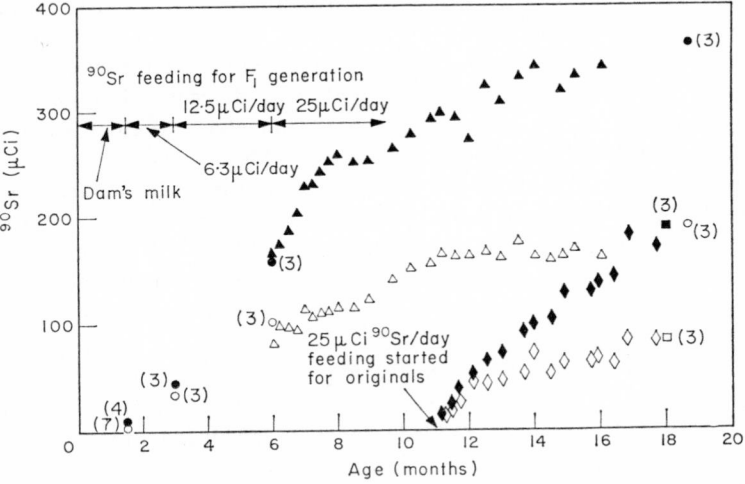

FIG. 1. Effect of diet on skeletal ^{90}Sr in miniature swine ingesting ^{90}Sr daily. Mean values. Key to points:

0·55% Ca, 0·50% P   1·5% Ca, 0·8% P

●	Skeletal analysis: exposure throughout life	○
■	Skeletal analysis: exposure starting at 11 months	□
▲	Whole-body counting: exposure throughout life	△
◆	Whole-body counting: exposure starting at 11 months	◇

TABLE II

Effect of age and diet on skeletal strontium-90 in miniature swine ingesting strontium-90 daily on an adult level of 25 μCi strontium-90/day

Results are quoted as mean ± standard deviation

⁹⁰Sr exposure	Age	Generation	1·5% Ca, 0·8% P				0·55% Ca, 0·5% P			
			No. of animals	Ashed skeleton weight (g)	Total μCi	Skeletal ⁹⁰Sr μCi/g ash	No. of animals	Ashed skeleton weight (g)	Total μCi	Skeletal ⁹⁰Sr μCi/g ash
1–3 days		F_1	9	24±7	0·25±·06	0·010±·002	8	21±7	0·49±·23	0·023±·004
Throughout life	6 weeks	F_2	5	25±5	0·34±·06	0·014±·001	5	22±6	0·83±·19	0·038±·002
	6 weeks	F_1	7	140±15	4·8±0·3	0·034±·004	4	140±30	8·1±1·0	0·059±·005
	3 months	F_2	8	180±8	12±1·9	0·064±·010	10	160±34	20±3·0	0·13±·02
	3 months	F_1	3	290±40	36±1·7	0·13±·02	3	230±56	44±11	0·20±·01
	6 months	F_1	3	850±90	100±9·0	0·12±·02	3	600±100	160±30	0·27±·08
	19 months	F_1	3	1,900±250	190±15	0·10±·02	3	1,600±200	360±7	0·22±·03
Starting at 11 months of age	18 months	Original	3	1,900±40	85±6·6	0·043±·003	3	1,900±270	190±8	0·10±·01

weeks, 3 months, 6 months, and 19 months were determined by radioanalysis of the skeletons of F_1 generation animals. The values at 18 months were obtained from the radioanalysis of the original animals' skeletons. Values obtained by whole-body counting of the originals from 11 to 18 months and the F_1 generation from 6 to 19 months are included. The data points for F_1 generation animals on both diets can be fitted reasonably well by an equation of the form $Y(\mu Ci\ ^{90}Sr) = b_0 - b_0 e^{-b_1 t}$. For the high Ca diet $b_0 = 446 \cdot 6$ and $b_1 = 0 \cdot 09$. For the low Ca diet $b_0 = 244 \cdot 8$ and $b_1 = 0 \cdot 07$.

The data obtained from radioanalysis of the skeleton are shown in Table II. It is of special interest to compare the values obtained for the F_1 and F_2 generation animals, since they have received the same basic treatment, except for differences in the ^{90}Sr exposure of their dams. The F_1 generation animals were born to original females which started ingesting ^{90}Sr (and labelling their skeleton) just before conception of the F_1 generation. The F_2 generation was born to the F_1 generation which having been exposed to ^{90}Sr throughout their life had skeletons relatively uniformly labelled with ^{90}Sr. The higher ^{90}Sr activity of the newborn F_2 generation animals' skeletons reflects the contribution of the ^{90}Sr mobilized from that portion of their dam's (F_1 generation) skeleton that had been formed before gestation. For the high Ca diet this represents 28 % of the ^{90}Sr in F_2 newborn skeleton; the similar value for the low Ca diet is 39 %. Pecher and Pecher (1941) were the first to note that radiocalcium and radiostrontium previously fixed in the maternal skeleton migrated to the fetus during the last days of pregnancy and to the suckling offspring. The difference between the above values and 100 % represents the ^{90}Sr passing directly from diet to maternal circulation to fetus or ^{90}Sr deposited in the maternal skeleton during the gestation period and then mobilized for transfer to the fetus. These values for the high and low Ca diets are 72 and 61 %, respectively, and are lower than the value observed by Wasserman, Comar and associates for the rat (Comar et al., 1955; Wasserman et al., 1957). They found that 88 % of the fetal Sr in the rat was of direct dietary origin.

A similar approach can be used to determine from the skeletons of the weanling animals what fraction of the dam's milk ^{90}Sr originates from skeleton formed before the beginning of gestation. For example from birth to weaning the low Ca F_1 skeleton increased by 119 g ash and 7·6 $\mu Ci\ ^{90}Sr$ for an activity of 0·064 $\mu Ci/g$. The low Ca F_2 skeleton increased by 138 g ash and 19·2 μCi ^{90}Sr for an activity of 0·14 $\mu Ci\ ^{90}Sr/g$. The difference between 0·14 and 0·064 $\mu Ci\ ^{90}Sr/g$ represents that portion of the ^{90}Sr in the dam's milk that originated in the maternal skeleton formed before gestation began.

This value for the high Ca diet is 48 % and for the low Ca diet is 54 % of the total ^{90}Sr transferred via the milk to the F_2 generation animal. The difference between these values and 100 %, i.e. 52 % for the high Ca diet and 46 % for the low Ca diet, represents that portion of the milk ^{90}Sr that originated rather directly from the maternal diet. This is in good agreement with work of

Comar et al. (1966) who found for cows maintained on diets containing 0·31 and 0·98 % Ca that about 50 % of the ^{90}Sr in the milk originated directly from the diet. When studying diets containing 1·6 and 1·7 % Ca in the cow and goat, respectively, Comar et al. (1961) found that essentially all of the milk calcium was coming directly from the diet.

In Table III the data for ^{90}Sr and Ca content of the skeleton are related to the ^{90}Sr and Ca content of the adult diet utilizing the observed ratio concept introduced by Comar et al. (1956). When values are expressed in this fashion taking into account the Ca content of the diet, the similarity in the values for the two diets is striking.

The low values for the newborn animal reflect the discriminatory processes operative between the dam's diet and the fetal skeleton. The differences between the newborn offspring and the maternal skeleton are indicative of the placental discrimination. The values for ^{90}Sr/Ca of the newborn skeleton (F_2 generation) are 0·13 and 0·18 those of the maternal skeleton (F_1 generation) for the high and low Ca diets, respectively. These values are lower than those found for humans, sheep, rabbits and rats which are in the range of 0·4 to 0·6 (Comar, 1963). The value of about 0·25 obtained for the beagle is in closer agreement with the swine data (Goldman and Della Rosa, this book, p. 181).

The low values for the 6-week-old weanling animals reflect the discriminatory processes operative between the dam's diet and the suckling animals' skeleton. The major discrimination occurs in transfer of ^{90}Sr and Ca from maternal diet to milk. A limited number of samples were obtained in this study and verify our earlier work (McClellan and Bustad, 1963) in which we found the ^{90}Sr/Ca of milk was 0·1 that of the diet. In this study similar values of 0·11 and 0·10 were obtained for the high and low Ca diets, respectively. The lack of any marked effect on the discrimination in milk transfer of Ca/Sr owing to alteration of dietary Ca and P (within physiological limits) is in agreement with the work of Comar and associates with cows and goats (Comar et al., 1961, 1966). The values of about 0·1 are in agreement with extensive data for cows, goats, dogs and humans (Lough et al., 1960; Comar et al., 1961, 1966; Goldman and Della Rosa, this book, p. 181). The swine data can be used to determine the extent to which discrimination occurs in the suckling offspring; this will be discussed later.

The values for the 3-, 6- and 19-month-old animals exposed throughout life more accurately reflect the true relationship between diet and the animal's skeleton, but still must be modified as noted later to obtain observed ratios that reflect the relationship for a given age period. The low values for the 18-month-old animals whose exposure started at 11 months reflects the incomplete labelling of their skeleton by ^{90}Sr.

In Table IV adjusted values are shown for ^{90}Sr/Ca discrimination between diet and skeleton for various age periods. The values for the period birth to 6 weeks were adjusted by taking into account the ^{90}Sr/Ca ratio of the animals'

TABLE III

Age-related changes in the relationship between strontium-90 and calcium in the diet and skeleton of animals ingesting two different diets

^{90}Sr exposure	Age	Generation	1·5% Ca, 0·8% P		0·55% Ca, 0·5% P	
			No. of animals	^{90}Sr/Ca skeleton / ^{90}Sr/Ca diet	No. of animals	^{90}Sr/Ca skeleton / ^{90}Sr/Ca diet
	1–3 days	F_1	9	0·015	8	0·012
		F_2	5	0·020	5	0·021
	6 weeks	F_1	7	0·051	4	0·032
Throughout life		F_2	8	0·093	10	0·070
	3 months	F_1	3	0·19	3	0·11
	6 months	F_1	3	0·17	3	0·15
	19 months	F_1	3	0·15	3	0·12
Starting at 11 months	18 months	Original	3	0·064	3	0·054

TABLE IV

Age-related changes in strontium-90–calcium discrimination in transfer from diet to skeleton of animals ingesting two different diets

Results expressed as $\dfrac{^{90}\text{Sr/Ca skeleton}}{^{90}\text{Sr/Ca diet}}$

Age period	Type of diet	Adult Diet: Ca and P	
		1·5% Ca, 0·8% P	0·55% Ca, 0·5% P
Birth to 6 weeks	Dam's milk	1·2	0·9
6 weeks to 3 months	Adult ration	0·30	0·22
3–6 months	Adult ration	0·18	0·17
6–19 months	Adult ration	0·13	0·10

actual diet, its dam's milk, and by considering the net increase in skeletal mass and ^{90}Sr content.

The values for later age periods were derived by taking into account the net increase in skeletal mass and ^{90}Sr during the age period under consideration. After 6 weeks of age, the ^{90}Sr/Ca relationship for the diet remained constant. By using this approach it can be seen that the actual ratios for ^{90}Sr/Ca of the newly formed skeleton relative to diet are generally higher than the values in Table III.

The changes in Ca/^{90}Sr discrimination with age observed here for the two diets are in accord with those we have previously noted for miniature swine

(McClellan *et al.*, 1962a, b; McClellan, 1964). Likewise, the data are in good agreement with data reported for man, dog and rat showing a decreased discrimination against ^{90}Sr relative to Ca in the young individual on a milk diet (Comar *et al.*, 1957; Lough *et al.*, 1962; Goldman and Della Rosa, this book, p. 181).

The values observed for the adult animals are generally lower than the value of about 0·25 which we and others have noted for the pig (Hogue *et al.*, 1961; McClellan *et al.*, 1962a, b; McClellan, 1964). Observed ratios (body/diet) of 0·2 to 0·5 have been noted for man and most other species investigated (Comar, 1963).

The data presented here generally fail to demonstrate any marked differences in the comparative metabolism of ^{90}Sr and Ca under the two dietary regimes. If one recognizes that the two diets are both physiologically normal with regard to Ca and P content, this is probably not surprising.

The data do demonstrate that, with normal diets, by lowering the Ca content, it is possible to increase the skeletal ^{90}Sr burden relative to the daily intake of ^{90}Sr for miniature swine, as would be expected from the work in laboratory species (Hogue *et al.*, 1961; Thompson, 1963; Palmer and Thompson, 1964).

ACKNOWLEDGMENT

The authors gratefully acknowledge the technical assistance of R. Keough for performing the calcium and phosphorus analysis, the biostatistical assistance of Dr John Thomas, the assistance of M. E. Kerr and R. L. McCluskey and the staff of the Experimental Animal Farm for care and handling of animals and processing of samples, and the members of the Biological Analyses Operation for analyses of samples. The encouragement and discussions of Drs L. K. Bustad and R. C. Thompson are most appreciated.

Work performed under Contract AT(45–1) 1350 between the U.S. Atomic Energy Commission and the General Electric Company and Contract (45–1) 1830 between the U.S. Atomic Energy Commission and the Battelle Memorial Institute.

REFERENCES

Boltz, D. F. (ed.) (1958). "Colorimetric Determination of Non-metals". Interscience, New York.

Bustad, L. K., and McClellan, R. O. (eds.) (1966). "Swine in Biomedical Research". Battelle Memorial Institute—Pacific Northwest Laboratory, Richland, Washington.

Case, A. C., and McKenney, J. R. (1961). Hanford Biology Research Annual Report (HW–69500, 1960), 28–30.

Comar, C. L. (1963). *In* "The Transfer of Calcium and Strontium Across Biological Membranes" (Wasserman, R. H., ed.), pp. 405–417: Academic Press, New York.

222   R. O. MCCLELLAN, J. L. BEAMER AND P. L. SHELDON

Comar, C. L., Russell, R. S., and Wasserman, R. H. (1957). *Science*, **126,** 485–492.
Comar, C. L., Wasserman, R. H., and Lengemann, F. W. (1966). *Hlth Phys.* **12,** 1–6.
Comar, C. L., Wasserman, R. H., and Nold, M. M. (1956). *Proc. Soc. exp. Biol. Med.* **92,** 859–863.
Comar, C. L., Wasserman, R. H., and Twardock, A. R. (1961). *Hlth Phys.* **7,** 69–80.
Comar, C. L., Whitney, I. B., and Lengemann, F. W. (1955). *Proc. Soc. exp. Biol. Med.* **88,** 232–236.
Hogue, D. E., Pond, W. G., Comar, C. L., Alexander, L. T., and Hardy, E. P. (1961). *J. Anim. Sci.* **20,** 514–517.
International Commission on Radiological Protection (1950). "Report of Committee II on Permissible Dose for Internal Radiation."
Lough, S. A., Hamada, G. H., and Comar, C. L. (1960). *Proc. Soc. exp. Biol. Med.* **104,** 194–198.
Lough, S. A., Rivera, J., and Comar, C. L. (1962). *Proc. Soc. exp. Biol. Med.* **112,** 631–636.
McClellan, R. O. (1964). *Nature, Lond.* **202,** 104–106.
McClellan, R. O., and Bustad, L. K. (1963). *Int. J. Radiat. Biol.* **6,** 173–180.
McClellan, R. O., McKenney, J. R., and Bustad, L. K. (1962a). *Life Sci.* **12,** 669–675.
McClellan, R. O., Clarke, W. J., McKenney, J. R., and Bustad, L. K. (1962b). *Am. J. vet. Res.* **23,** 910–912.
Palmer, R. F., and Thompson, R. C. (1964). *Am. J. Physiol.* **207,** 561–566.
Pecher, C., and Pecher, J. (1941). *Proc. Soc. exp. Biol. Med.* **46,** 91–94.
Snyder, W. S., Cook, M. J., and Ford, M. R. (1964). *Hlth Phys.* **10,** 171–182.
Thompson, R. C. (1963). *In* "The Transfer of Calcium and Strontium Across Biological Membranes" (Wasserman, R. H., ed.), pp. 393–404. Academic Press, New York.
Wasserman, R. H., Comar, C. L., Nold, M. M., and Lengemann, F. W. (1957). *Am J. Physiol.* **189,** 91–97.
Yarboro, C. L. (1958). *Analyt. Chem.* **30,** 504–506.

Metabolism of Strontium-89 and Calcium-45 Given Orally to Pigs 1, 3 or 7 Days after Whole-body Irradiation

M. C. BELL, R. S. LOWREY* AND G. WITHROW

Agricultural Research Laboratory of The University of Tennessee, Oak Ridge, Tennessee†

SUMMARY

Early plasma uptake of both ^{89}Sr and ^{45}Ca was reduced in young, rapidly growing pigs dosed at 1 or 3 days after exposure to whole-body ionizing radiation (450 rads). From 8 to 72 h after dosing, this effect was reversed. Data from pigs dosed 7 days after irradiation were not different from controls. No significant treatment differences were found in urinary and faecal excretion of ^{45}Ca, ^{89}Sr and nitrogen. ^{89}Sr retention in tibia of pigs dosed 1 day after whole-body irradiation was greater than in control pigs.

INTRODUCTION

Previous work at this laboratory (Lowrey and Bell, 1964) has shown that whole-body exposure to ionizing radiation at a level of 450 rads temporarily increased the absorption of ^{89}Sr by the young pig. This effect was evident in pigs dosed 1 day after irradiation, but was not present at a second dosing 21 days later. Irradiation showed no effects on oral ^{45}Ca nor on either isotope given intravenously. Typical whole blood cell and platelet depressions were found in all irradiated pigs. Structural changes were not found in intestinal histological preparations from these irradiated pigs. Growth and feed consumption were unaffected, and diarrhoea was not observed in contrast to pig data reported by Chambers *et al.* (1964). These differences may have been related to dose rate, since Chambers *et al.* used from 18 to 29 rads/min compared with 0·7 rads/min by Lowrey and Bell. Comar and Wasserman (1964) and Bond *et al.* (1965) have also summarized information on effects of irradiation on metabolism in porcine and other species. The present experiment was undertaken to define more thoroughly the influence of whole-body

* Present address: Georgia Coastal Plains Experiment Station, Tifton, Georgia.

† Operated by the Tennessee Agricultural Experiment Station for the U.S. Atomic Energy Commission under Contract AT–40–1–GEN–242

radiation on metabolism of ^{89}Sr and ^{45}Ca given during the first week after exposure to ionizing radiation.

EXPERIMENTAL PROCEDURE

Twelve barrows (average weight, 33 kg) of mixed breeding were allotted at random into four equal treatment groups. Three groups were dosed at 1, 3 or 7 days after irradiation, and the fourth group served as controls. The irradiated animals received a single dose of 450 rads at 0·5 rads/min from ^{60}Co while confined in metal crates. Groups were irradiated at different times to permit simultaneous dosing. The animals were placed in metabolism crates for an adjustment period of 7 days before dosing. Their feed intake was 400 g three times/day for the duration of the experiment. The pigs were dosed orally with 1·2 mCi ^{45}Ca and 1·0 mCi ^{89}Sr, and urine and faeces were collected daily for 6 days. A composite sample of urine was made from the six daily urine samples for ^{89}Sr, ^{45}Ca and nitrogen analysis. A composite was also made of the faecal samples for nitrogen determination. The animals were sacrificed 7 days after dosing and tibia and femur removed from each pig. Blood samples were taken 1, 2, 4, 8, 24, 48 and 72 hours post-dosing for plasma radiochemical analyses.

After the completion of the study with these pigs, a replicate group of 12 pigs was treated in the same manner except that feed consumption was only 1 kg/day. Other experimental procedures were identical to those described by Lowrey and Bell (1964). Duncan's multiple-range test was used to test differences between means.

RESULTS AND DISCUSSION

Blood plasma samples taken 1, 2 and 4 h after dosing showed that both ^{45}Ca and ^{89}Sr were absorbed faster by the controls and those pigs dosed 7 days after irradiation than by those pigs dosed 1 and 3 days post-irradiation. However, the radiochemical level of the plasma of the controls and the 7-day post-irradiated group dropped more rapidly than that of the other two groups. From 8 to 72 h after dosing, higher values were found for those pigs dosed at 1 and 3 days after irradiation. Average plasma values for each isotope fell into two distinct groups and are presented as pooled values (Fig. 1). These data show that both ^{45}Ca and ^{89}Sr absorption into plasma is delayed temporarily when animals are exposed to whole-body ionizing radiation. Perhaps there was an early reduction in rate of passage in these pigs as suggested for rats by Comar and Wasserman (1964). By 8 h, it appears that there was an increased permeability of the intestinal tract to provide higher plasma levels. These values at 24 h and later sampling periods are in agreement with those reported by Lowrey and Bell (1964) showing higher plasma levels for ^{45}Ca and ^{89}Sr for the irradiated animals.

Faecal excretion of ^{89}Sr was less in the pigs dosed 1 day after irradiation compared to the other groups (Table I). However, in contrast to previous

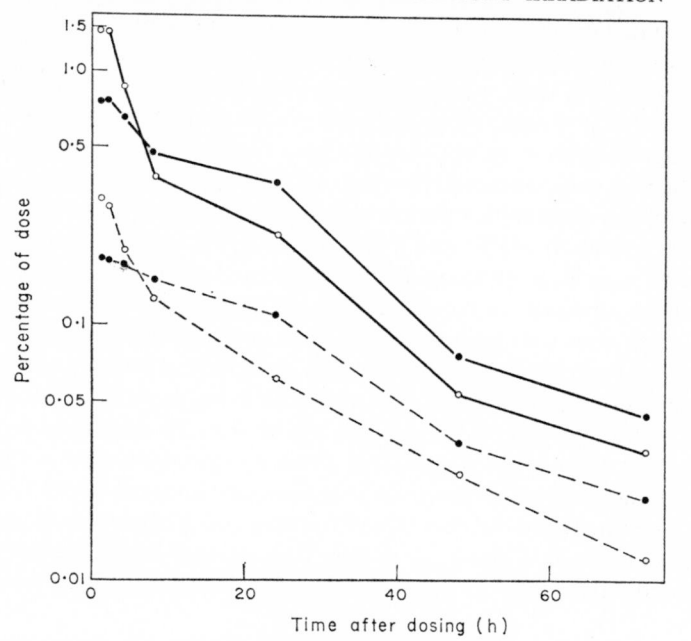

FIG. 1. Levels of ^{89}Sr and ^{45}Ca/litre of blood plasma of orally dosed pigs. The respective treatment groups were: A, controls; B, dosed 1 day after irradiation; C, 3 days; and D, 7 days. ●, B and C; ○, A and D. Continuous curves, ^{45}Ca broken curves, ^{89}Sr.

TABLE I

Excretion and retention of an oral dose of strontium-89 and calcium-45 given 1, 3 or 7 days post-irradiation (450 rads)

| | Excretion (percentage of dose) | | | |
	Controls (A)	1 Day (B)	3 Days (C)	7 Days (D)
Faeces				
^{89}Sr	73·29	69·02	74·94	74·39
^{45}Ca	28·65	29·11	25·70	24·94
Ratio ^{89}Sr/^{45}Ca	2·56	2·37	2·92	2·98
Urine				
^{89}Sr	2·73	3·13	4·36	3·44
^{45}Ca	0·74	0·48	0·78	1·22
Ratio ^{89}Sr/^{45}Ca	3·69	6·52	5·59	2·82
	Retention (percentage of dose)			
^{89}Sr	23·98	27·85	20·70	22·17
^{45}Ca	70·61	70·41	73·52	73·84
Ratio ^{89}Sr/^{45}Ca	0·34	0·40	0·28	0·30

data, no statistical difference in faecal excretion or retention of either ^{89}Sr or ^{45}Ca was found among the groups (P>0·05). The ^{89}Sr to ^{45}Ca ratio in the faeces was greater than in previous experiments, owing mainly to a smaller excretion of ^{45}Ca. These young pigs were growing rapidly as shown by the Ca and nitrogen retention values. Average gain for the collection period was 4·2 kg for the pigs, which accounts for the high nitrogen and Ca retention values. Treatments did not significantly affect feed consumption or rate of gain.

Urinary excretion of ^{89}Sr and ^{45}Ca was small and variable among animals. The ^{89}Sr/^{45}Ca ratios were again much greater than in the previously published experiments, especially in those dosed 1 and 3 days after irradiation. Animals dosed 7 days after irradiation excreted more urinary ^{45}Ca than at 1 day after irradiation (P<0·05). Data by Lowrey and Bell (1964) in the previous experiment showed higher ^{89}Sr/^{45}Ca ratios at 1 day after irradiation in comparison with the controls. The reverse was true by 21 days after irradiation (Lowrey and Bell, 1964). Data from these experiments show a temporary increase in the selective excretion of Sr in urine in comparison with Ca.

The ^{89}Sr content and the ^{89}Sr/^{45}Ca of the tibia and the femur averaged higher for those pigs dosed at 1 and 3 days after irradiation than for the controls and those dosed 7 days after irradiation (Table II). However, the only significant

TABLE II

Deposition of strontium-89 and calcium-45 in the tibia and femur of pigs dosed orally 1, 3 or 7 days post-irradiation (450 rads)

| | (Percentage of dose) | | | |
	Controls (A)	1 Day (B)	3 Days (C)	7 Days (D)
Femur				
^{89}Sr	0·74	0·99	0·92	0·75
^{45}Ca	2·06	2·26	2·02	1·84
Ratio ^{89}Sr/^{45}Ca	0·36	0·44	0·46	0·42
Tibia				
^{89}Sr	0·46	0·66	0·52	0·49
^{45}Ca	1·33	1·72	1·17	1·63
Ratio ^{89}Sr/^{45}Ca	0·34	0·38	0·44	0·30

differences (P<0·05) in ^{89}Sr was for the tibia between the 1-day and control group. Data reported by Mirzoyan (1964) showed a much more dramatic increase in bone retention of Sr in rats given u.v. irradiation and observed for 90 days. The ^{89}Sr/^{45}Ca ratios from 0·30 to 0·46 are within the range reported by McClellan (1964) for pigs of similar age.

There were no significant differences (P>0·05) among treatments in the nitrogen balance data (Table III). This is in contrast to the positive nitrogen balance in laboratory animals as reviewed by Kretchmar et al. (1965).

TABLE III

Nitrogen balance data (g) from pigs 1, 3 or 7
days post-irradiation (450 rads)

	Controls (A)	(Total for 6-day collection) 1 Day (B)	3 Days (C)	7 Days (D)
Total N intake	167	163	168	166
Total N faeces	56	59	53	57
Absorbed N	111	104	115	109
Percentage absorbed	66	64	68	66
Total N urine	66	65	68	68
Total N retained	45	39	47	41
Percentage retained	27	24	28	25
Percentage absorbed N retained	40	38	41	38

This manuscript is published with the permission of the Director of the University of Tennessee Agricultural Experiment Station, Knoxville.

REFERENCES

Bond, V. P., Fliedner, T. M., and Archambeau, J. O. (1965). "Mammalian Radiation Lethality: A Disturbance in Cellular Kinetics", pp. 231–275. Academic Press, New York.

Chambers, F. W., Jr., Biles, C. R., Bodenlos, L. J., and Dowling, J. H. (1964). *Radiat. Res.* **22**, 316–333.

Comar, C. L., and Wasserman, R. H. (eds) (1964) "Mineral Metabolism: An Advanced Treatise", vol. II, pp. 523–572. Academic Press, New York.

Kretchmar, A. L., McArthur, W. H., and Congdon, C. C. (1965). *J. Nutr.* **87**, 261–266.

Lowrey, R. S., and Bell, M. C. (1964). *Radiat .Res.* **23**, 580–593.

McClellan, R. O. (1964). *Nature, Lond.* **202**, 104–106.

Mirzoyan, A. A. (1964). Symposium on the Biological effects of Radioisotopes. Translated from *Atom. Energ.* **17**, 512–514.

Influence of Adsorption of Strontium in Food on the Strontium-85 Body Burden in Guinea Pigs

O. VAN DER BORGHT, R. KIRCHMANN
AND S. VAN PUYMBROECK

Department of Radiobiology, Belgian Nuclear Centre, Mol

SUMMARY

Solubility of ^{85}Sr in food material, as determined by extraction with water, sodium nitrate and acids, seems not the determining factor for ^{85}Sr assimilation in guinea pigs, and gives divergent results under varying conditions.

Presence of food in the stomach when given a liquid dose of ^{85}SrCl$_2$ can lower the assimilated ^{85}Sr about three times as compared with intubation on an empty stomach in fasted animals. Distribution of the absorbed Sr between intestine, femur, liver, kidney, muscle, eye and testes is compared with ^{226}Ra and ^{45}Ca content of some of these organs when radio-contamination with these isotopes occurred under similar conditions.

INTRODUCTION

The availability of Sr for intestinal absorption is one of the factors that will influence the body burden after ingestion of radiocontaminated food.

As the solubility of Sr varies in varying parts of contaminated plants (Myttenaere, 1964), it was decided to find out if these subtle differences in solubility could also influence the absorption of Sr in guinea pigs. A study was made of solubility of Sr in some constituents of food pellets and on the influence of some factors (such as the concentration of Ca in the extracting solvents and the presence of a bulk of non-radioactive food) on the results of the commonly used extraction techniques on which are based the interpretations of the more or less strong bond of the Sr in the food. Thereafter the dependence of the body burden in guinea pigs as depending on the presence of solid food in the stomach and on the Sr/Ca ratio in the administered solutions of ^{85}Sr was investigated. The distribution of the ^{85}Sr in the animals and the comparative behaviour of ^{45}Ca was also checked at this stage.

METHODS

Extraction techniques

A simplified version of the method described by Schilling (1960) was used and yielded essentially the same results as the more elaborate technique. After being dried at 70° C, the aerial parts and roots of *in vivo* contaminated pea plants were ground in a hammer mill (Culatti) and were extracted with or without addition of powdered carrots or standard cavia diet. The pea plants were grown in nutrient solution with a Ca/Sr ratio of 100/1, and they were harvested at an age of 20 days. Extraction was performed in 50 ml of solvent, for 1 to 5 g of plant material, or in 100 ml for 12·5 g of plant material with added standard diet. The extraction was carried out with continuous mixing, followed by centrifuging and transfer of the residue to the next solvent. Three samples of each solvent were analysed for radioactivity, each extraction was replicated three (exceptionally two) times. The residue was also assayed for ^{85}Sr after a complete extraction cycle. Solvents used were (1) water (90° C) for 45 min (extraction of water-soluble salts); (2) $NaNO_3$ (0·5 M) for 3 h (exchangeable Sr); (3) CH_3COOH (M) for 3 h (extraction of carbonates and phosphates); and (4) HCl (2 M) for 3 h (extraction of oxalates, pectates, etc.).

Radioactivity measurements

^{85}Sr was measured by using a NaI-well crystal with a three-channel pulse-height analyser. Voluminous samples (carcasses, dissected tissues) were measured by using a NaI-crystal with a 400-channel analyser. ^{45}Ca was counted by liquid scintillation, a correction for quenching being calculated from addition of known quantities of ^{45}Ca. Samples for ^{45}Ca assay were ashed at 500° C and dissolved in a minimum of hydrochloric acid. An aliquot of a dilution from this is added to the scintillation liquid.

Experiments with animals

Solutions were given by gastric intubation to 100–150 day old male albino animals kept without food for 3 days and without water for 15 h before intubation. Food pellets containing ^{85}Sr were eaten by the animals kept in the same conditions, but 3·5 g of powdered carrots were added to 1 g of powdered pea plants to make the pellets acceptable to the guinea pigs. Before, and for 10 days after the administration of radioactivity, the animals were fed a standard cavia diet, from the Hope Farms (Oost Knollendam, the Netherlands), which contains 0·84% Ca, 0·44% P and 13·2% fibres. On the 11th day, the animals were sacrificed, dissected and assayed for radioactivity.

RESULTS

The solubility of strontium under varying conditions

As indicated in Table I, the extraction of Sr from *in vivo* contaminated aerial parts of pea plants was compared with extraction of the same material after addition of 25 mg of Ca, 25 mg of Sr, powdered carrots and (12·5 g) of standard diet. The Ca was added to the powdered plant material as $CaCl_2.2H_2O$ and it represents about four times the quantity of Ca present in the extracted plant material. Finally, the *in vivo* contamination is compared with an *in vitro* contamination of powdered roots and carrots, resulting from the addition of 1 ml of $^{85}SrCl_2$ solution to the compressed pellet of the powdered materials. Table I gives the percentages of the total radioactivity, measured in aliquots of non-treated radioactive material recovered in each extracting solvent. Rather large quantities of ^{85}Sr are lost (-15% to $+10\%$), owing to the successive handlings, centrifuging and transfer of the material, but good reproducibility is generally obtained. Addition of Ca and Sr has a dramatic influence on the Sr, generally considered as soluble in water, and thus accessible for absorption; whereas the exchangeable Ca is not significantly influenced by this addition. Addition of more adsorbing material, such as powdered carrots or standard diet, will influence the fraction soluble in hydrochloric acid. More Sr will be retained in the residue after addition of a bulk of standard diet. No marked difference was observed between *in vivo* or *in vitro* contaminated food pellets, composed of powdered roots and carrots. To confirm the observation of the strong bond of Sr on the standard diet under physiological conditions, we compared the relative influence of a full and an empty stomach on the body burden after administration of $^{85}SrCl_2$ solution.

Assimilation of strontium-85 administered in various liquids and solids

A first group of seven animals received, by gastric intubation in an empty stomach, 1 ml of $^{85}SrCl_2$ solution containing about 3 μCi ^{85}Sr, 40 mg/l Ca and 500 mg/l Sr. A second group of five animals was given the same dose, but with 40 mg/l Ca and 1 mg/l Sr in the solution. A third group of seven animals was given the same solution as the latter, but the hungry animals were allowed to eat from the standard diet just before the intubation. Their stomach content was approximately 10 g of dry standard diet. A fourth group of two animals was added to indicate the assimilation of ^{85}Sr from *in vitro* contaminated food pellets made from 0·5 g of non-radioactive roots $+4·5$ g of carrots to which 1 ml of $^{85}SrCl_2$ solution was added.

Table II illustrates that the presence of the food in the stomach reduces body burden and the femur burden of ^{85}Sr by 2·7 times, whereas the other experimental factors such as Ca/Sr ratio in the intubated solution or presence of a moderate amount of plant material, do not influence the uptake of ^{85}Sr significantly.

TABLE I

Percentages of total strontium-85 recovered in extractions of plant material

Limits are the 95% fiducial limits ($=\bar{x}\pm t s_{\bar{x}}$)

	H_2O	$NaNO_3$ 0·5 M	CO_3COOH 1 M	HCl 2 M	Residual material	Lost
AP* 0·5 g without addition	27·3±0·5	18·9±0·2	18·2±0·4	22·6±0·4	0·3±0·6	−13
AP 0·5 g + powdered carrots 2·0 g	19·4±0·6	15·2±0·6	11·4±0·5	37·6±1·2	0·08±0·06	−16
AP 0·5 g + 25 mg Ca (added as $CaCl_2.2H_2O$)	52·2±2·3	13·3±1·6	7·6±0·4	12·3±0·2	0·02±0·02	−15
AP 0·5 g + 25 mg Sr (added as $SrCl_2.2H_2O$)	61·2±1·0	17·9±0·6	6·6±0·3	4·5±0·1	0·01±0·03	−10
AP 0·5 g + carrots 2·0 g + diet 12·5 g	24·2±1·4	18·0±1·2	24·4±8·9	26·5±7·4	6·1±7·4	−1
In vivo contaminated roots 0·2 g + carrots 2·3 g	10·7±0·2	15·1±2·6	8·8±2·7	43·6±2·4	6·5±6	−15
Non radioactive roots 0·5 g + carrots 4·5 g + $^{85}SrCl_2$ solution	11·3±1·5	16·0±1·7	12·2±2·5	64·7±6·5	6·7±3·2	+10

* AP = aerial parts of *in vivo* contaminated pea plants.

TABLE II

Percentage of the strontium-85 administered recovered after 11 days, as influenced by stomach content and calcium/strontium ratio

Limits $=95\%$ fiducial limits ($=\bar{x} \pm ts_{\bar{x}}$).

Mode of administration	% of dose recovered	
	As body-burden	In one femur
Intubation on empty stomach of		
1 ml ($H_2O + 3\ \mu Ci\ ^{85}Sr + 40\ \mu g\ Ca + 500\ \mu g\ Sr$)	31 ± 5	$0{\cdot}81 \pm 0{\cdot}25$
Intubation on empty stomach of 1 ml		
($H_2O + 3\ \mu Ci\ ^{85}Sr + 40\ \mu g\ Ca + 1\ \mu g\ Sr$)	30 ± 4	$0{\cdot}94 \pm 0{\cdot}26$
Food pellet (3 $\mu Ci\ ^{85}Sr + 0{\cdot}5$ g cold roots		
$+4{\cdot}5$ g carrots) eaten by the animals	$45\ (s_{\bar{x}}=5)$	$1\ (s_{\bar{x}}=0{\cdot}09)$
Intubation on filled stomach of 1 ml		
($H_2O + 3\ \mu Ci\ ^{85}Sr + 40\ \mu g\ Ca + 1\ \mu g\ Sr$)	$11{\cdot}5 \pm 3{\cdot}5$	$0{\cdot}32 \pm 0{\cdot}05$

Distribution of the strontium-85 in the animals after intubation on an empty stomach: comparative behaviour of calcium-45 and radium-226

In other experiments on ^{226}Ra metabolism (Van der Borght, 1966a) some animals were treated as for the ^{85}Sr intubation on an empty stomach. An

TABLE III

Distribution of strontium-85, calcium-45 and radium-226 in male adult caviae, about 11 days after gastric intubation

Statistical limits $=95\%$ confidence interval ($=\bar{x} \pm ts_{\bar{x}}$)

	^{85}Sr			^{45}Ca	^{226}Ra
	($\% \times 10^3$) of administered dose	$\mu Ci/g$ tissue per μCi dose	Weight of organ (g)	($\% \times 10^3$) of administered dose	($\% \times 10^3$) of administered dose
One femur	810 ± 250 ($=0{\cdot}81\%$)	36×10^{-4}	$2{\cdot}3 \pm 0{\cdot}2$	$1{,}500 \pm 300$	80
Gastro-intestinal					
tract	47 ± 5	10×10^{-6}	105 ± 34		
Liver	30 ± 20	6×10^{-6}	53 ± 3		$0{\cdot}6$
Kidneys	3 ± 1	6×10^{-6}	$5{\cdot}4 \pm 0{\cdot}4$		4
Thigh muscle					
(graciles)	$1 \pm 0{\cdot}2$	4×10^{-6}	$2{\cdot}8 \pm 0{\cdot}4$		
Eyes	$1 \pm 0{\cdot}2$	10×10^{-6}	$1{\cdot}1 \pm 0{\cdot}2$		40
Testes	$2{\cdot}6 \pm 0{\cdot}8$	5×10^{-6}	$4{\cdot}6 \pm 0{\cdot}3$		
Remainder of					
carcass	$30{,}000 \pm 5{,}000$ ($=30\%$)	$6{\cdot}2 \times 10^{-4}$	491 ± 34	$46{,}000 \pm 7{,}000$	$5{,}000$

intubation experiment was also run with ^{45}Ca. Table III gives the distribution of the ^{85}Sr in guinea pigs and gives comparative values for ^{45}Ca and ^{226}Ra.

The observed ratio (Comar *et al.*, 1956) $= \dfrac{\text{Sr/Ca animals}}{\text{Sr/Ca intubated}}$, between food and body burden or bone, has a mean value of $0 \cdot 61 \pm 0 \cdot 03$ in our experiments.

Although the Sr content of the eyes for unit tissue weight is about double that of other soft tissues, this is only 1/40th of the corresponding figure for ^{226}Ra, and when expressed as a percentage of the quantity absorbed as distinct from that administered, the Ra content of the eyes is some 250 times the Sr content.

DISCUSSION

Deductions relating the availability of strontium in food components and its assimilation by animals

The solubilities of ^{85}Sr from food components in different solvents, such as water, neutral salt solution and acids, do not seem to determine its availability for gastro-intestinal absorption. Moreover, *in vitro* and *in vivo* contaminated food pellets do not differ very much in their binding of Sr. This together with the marked influence on the extraction of Sr of the concentration of Ca in the extracting solvents and of the presence of a quantity of other non-radioactive material on which Sr can possibly adsorb, indicate that other factors predominate in determining, under physiological conditions, the availability of Sr for intestinal absorption.

The experiments thus concentrated on obtaining the greatest contrast in ^{85}Sr availability in food, by administering a $SrCl_2$ solution without any adsorbing material and comparing this with the results obtained when some 5 g or 10 g of adsorbing material were present together with the Sr in the stomach. Even the food pellets (5 g of powdered plant material) as adsorbing material were not able to reduce the absorption of Sr, and only the presence of a bulk of standard diet could reduce, by about a factor of three, the quantity of ^{85}Sr absorbed by the animals. If we compare the solubilities of Sr in water, in water + added diet and in water + added plant material, the figures are: 100%, water soluble; 24% H_2O soluble + 27% HCl soluble; and 11% H_2O soluble + 50% HCl soluble, respectively. The reduced solubility of diet-added Sr in HCl gives, very hypothetically, an explanation of the reduction in Sr absorption obtained. This also could indicate that the acidity of the stomach content could be the factor that determines the availability of Sr for absorption, rather than the quantity of water-soluble or exchangeable Sr present in the food. The similarity of the solubility of Sr in *in vitro* and *in vivo* contaminated food confirms results, such as those of Carr and Wall (1961), in rabbits. In rats, by using ^{89}Sr mixed with a commercial diet, Ichikawa and Enomoto (1963) obtained a reduction of a factor of $1 \cdot 4$ in the quantity of Sr absorbed when compared with the value obtained by using a milk diet, and Marcus and Lengemann (1962), again using rats, found that the quantity of ^{85}Sr absorbed when fed in a standard diet was lower by a factor of 2 than when given as an intubated solution. We obtain a reduction factor of 2·7 when comparing the ^{85}Sr absorbed from a solution intubated on an empty stomach and Sr intubated in an animal that has just had a meal.

The role in the retention of Sr of some constituents of the diet, such as the cellulose-like material, will be investigated for the reduction of the absorption of ^{85}Sr observed in these experiments could be related to the binding of Sr to polyuronates, as reported by Waldron-Edward et al. (see this book, p. 329) and Hesp and Ramsbottom (see this book, p. 314), and as was observed in piglets (van der Borght, 1966b).

The comparative behaviour of strontium, calcium and radium in guinea pigs

An OR of 0·61 was obtained when Ca and Sr are intubated on an empty stomach; this is close to the figure of 0·67 obtained by Marcus and Lengemann (1962) in rats, but is consistently higher than the values of 0·3 to 0·5 related by Ichikawa and Enomoto (1963) on rats and by Della Rosa et al. (1965) in beagles, and than the values of 0·4 to 0·25 for miniature pigs older than 6 weeks (McClellan, 1964) and of 0·26 in nearly adult swine (van der Borght, 1966b). The rather long period of deprivation of food (3 days) that was necessary to ensure the caviae ate the food pellets, could be responsible for this high OR, since the animals were possibly by then low in Ca and readily absorbed alkaline-earth ions.

The OR $\left(\dfrac{\text{Ra/Ca body}}{\text{Ra/Ca food}} \right)$ reached 0·17 and seems also to be somewhat high, although only a few of such OR values obtained under controlled conditions are obtainable for comparison.

Strontium is somewhat more concentrated in the eye as in other soft tissues, but does not reach the very high level of Ra in this organ (Hunt, 1966, van der Borght, 1966a), and the kidneys also play a more important part in the fixation of Ra than of Sr.

REFERENCES

van der Borght, O., et al. (1966a). In "Proceedings of the 1st International Congress of International Radiation Protection Association". Rome.

van der Borght, O., et al. (1966b). In "Proceedings of the International Symposium on Radioecological concentration Processes". Stockholm, pp. 589–593.

Carr, T. E. F., and Wall, J. (1961). Br. J. Nutr. **15**, 165–168.

Comar, C. L., Wasserman, R. H., and Nold, M. M. (1956). Proc. Soc. exp. Biol. Med. **92**, 859–863.

Della Rosa, R. J., Goldman, M., Andersen, A. C., Mays, C. W., and Stover, B. J. (1965). Nature, Lond. **205**, 197–198.

Hunt, V. R. (1966). In "Proceedings of the International Symposium on Radioecological concentration Processes". Stockholm.

Ichikawa, R., and Enomoto, Y. (1963). *Hlth. Phys.* **9**, 717–720.
McClellan, R. O. (1964). *Nature, Lond.* **202**, 104–106.
Marcus, C. S., and Lengemann, F. W. (1962). *J. Nutr.* **77**, 155–160.
Myttenaere, C. (1964). *Radiat. Bot.* **5**, 143–151.
Schilling, G. (1960). *Z. PflErnähr. Dung. Bodenk.* **91**, 212–224.

Discussion

VOLF: In one of our experiments we gave barium and sodium sulphates to fasting or to feeding rats shortly after ingestion of ^{85}Sr. The feeding controls retained significantly less ^{85}Sr than did the fasting animals (see Fig. A). Since treat-

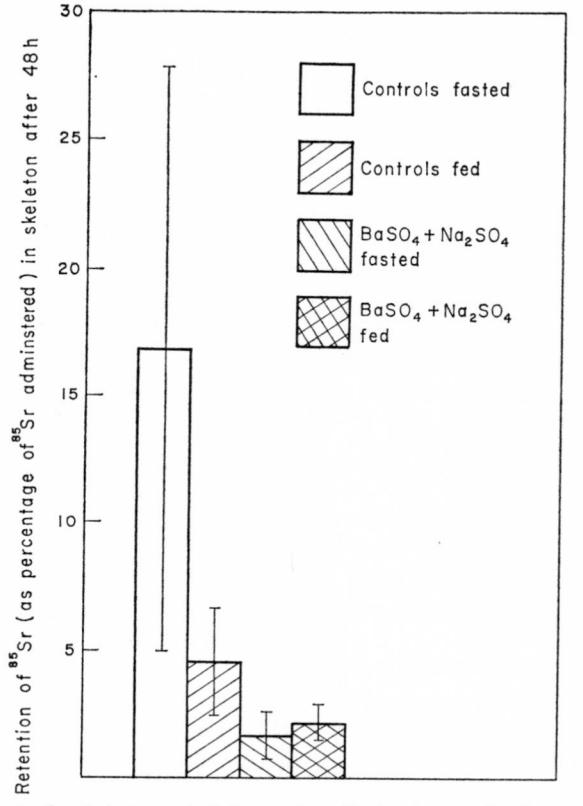

FIG. A. Effect of sulphates administered orally in doses of 0·8 mM, 10 min after oral administration of ^{85}Sr. (Details in *Acta Radiol. ther. phys. Biol.* **1966**, 4, 113.)

ment (with sulphate) reduced the retention of ^{85}Sr to approximately the same level both in fasting and in feeding animals, the relative effectiveness was greater in the fasting animals

Strontium-90 Transfer to Progeny via Placenta and Milk

L. N. BURYKINA

*Department of Public Health, U.S.S.R. Department of External Affairs,
Rachmanovsky Lane, 3, Moscow*

SUMMARY

Chronic exposure of dogs to ^{90}Sr at equilibrium with ^{90}Y caused a placental transfer to the whole litter of $1\cdot4\pm0\cdot3\%$ of the amount received within the period of pregnancy, or $4\cdot9\pm0\cdot8\%$ of the activity retained in the maternal skeleton. Bone concentrations in the progeny at birth were 17–46% of those in the maternal skeletons.

The amount transferred through the placenta to the whole litter 2·5–5 years after the cessation of mothers' exposure was 0·4–0·05% of that retained in the maternal skeleton or about 1–6% of the average placental transfer obtained at the time of chronic ^{90}Sr ingestion.

^{90}Sr intake with maternal milk in the conditions of chronic exposure of mothers exceeds the values of placental transfer by a factor of 10–17. The amount of radioactive Sr in the litter of one dog may be as high as 80% of the mother's body burden.

The kinetics of the metabolism of ^{90}Sr transferred to the progeny by the two routes specified can be well fitted into a two-compartment exponential model, the slowly metabolizing fraction having an effective half-life of about 2,000 days.

INTRODUCTION

The possibility of ^{90}Sr transfer from the mother to the offspring via placenta and milk is well established.

The quantitative nature of ^{90}Sr transfer from the mother to the progeny has been extensively studied on animals receiving single doses of the isotope. The literature contains many reports of the dependence of ^{90}Sr transfer via maternal placenta and milk on such factors as the time interval between ^{90}Sr ingestion and the onset of pregnancy and Ca content of the mothers' diet. However, relatively little is known quantitatively about ^{90}Sr transfer to progeny from mothers chronically receiving the isotope with diet (Rubanovskaya and Ushakova, 1957; Kurl'andskaya *et al.*, 1957; Downie *et al.*, 1959; Buldakov and Moskalev, 1960; Vorozheikina and Parfenov, 1962). The pertinent literature data are contradictory and present mainly the results of experiments with small animals whose mineral turnover is different from that of human subjects.

The purpose of this investigation was to study in dogs the quantitative nature of ^{90}Sr transfer from the mother to the offspring during and after chronic ingestion. Some characteristics of ^{90}Sr metabolism in the progeny both in early and late periods of life were also investigated.

MATERIALS AND METHODS

Seventeen dogs (ten females and seven males), aged 3–5 years and weighing 15–20 kg, were divided into three groups (three to eight dogs in each group) depending on the radioactive dose administered. An equilibrium mixture of ^{90}Sr and ^{90}Y (0.02×10^{-4}, 0.2×10^{-4} and 2.0×10^{-4} μCi/g/day) was given to the dogs with diet six times per week for 3–3.5 years.

^{90}Sr transfer was studied in litters produced by the treated animals at different times after the beginning of the experiment. Placental transfer of ^{90}Sr was estimated during the chronic ingestion of the isotope (19–41 months after the beginning of the experiment) and at different times after its cessation (3, 11, 29, 32, 50 and 58 months).

Fifty-eight of the newborn dogs were sacrificed as soon as they were born, and ^{90}Sr content of the bone tissue was determined.

The rest of the newborn dogs (43) were fed with maternal milk and sacrificed at the age of 9 and 13 days (2 dogs) and 1, 2, 3, 4, 6, 12, 35, 40 and 60 months (3–5 dogs each time).

The ^{90}Sr burden in the maternal skeleton was determined during and after the isotope ingestion both by radiometric analysis of amputated tail vertebrae and by direct counting of the activity of tooth samples (Keirim-Marcus et al., 1961).

RESULTS

The study of the kinetics of ^{90}Sr accumulation in dogs chronically receiving the isotope with diet has revealed the inverse relation between the value and rate of the skeletal ^{90}Sr deposition and time (Fig. 1). As can be seen from Fig. 1, about 14 months after the beginning of the experiment the skeletal ^{90}Sr accumulation reaches an equilibrium state, at which the radioactivity does not change significantly although the isotopic ingestion is continued. This effect was found for all the ^{90}Sr doses ingested.

Placental transfer

As a result of the placental ^{90}Sr transfer the skeletal burden in newborn pups was found to be 17–46% of that in the mother at the time of delivery (Table I). The total transfer to all of the offspring within the period of

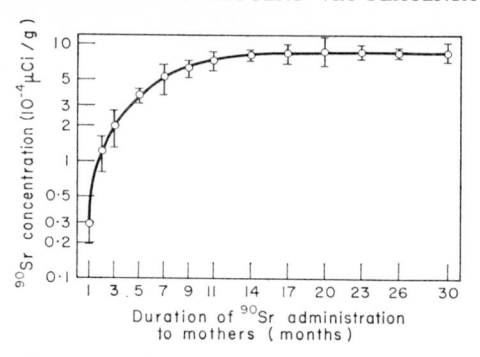

FIG. 1. Skeletal ^{90}Sr accumulation in dogs chronically receiving 0.02×10^{-4} μCi ^{90}Sr/g/day with diet.

pregnancy varied from 2·5 to 7·9% of the ^{90}Sr burden retained in the maternal skeleton, or was equal to $1.4 \pm 0.3\%$ of the activity ingested by the mother during pregnancy. As is clear from Table I, the absolute amount of ^{90}Sr transferred from the mother to the embryo is directly proportional to the radioactive dose administered to the mother, whereas the relative value of the placental transfer does not depend on the dose ingested.

The data obtained show that if the isotope is chronically administered to the mother and if the pregnancy occurs when the isotopic accumulation has reached the equilibrium state, the value of the placental ^{90}Sr transfer does not depend on the duration of the isotopic ingestion till delivery.

Thus, it was shown that the skeletal burdens of two new-born litters born 19 and 41 months after the beginning of the mothers' exposure to daily dietary doses of ^{90}Sr (0.2×10^{-4} μCi/g) were both 0·31% of the amounts received by the mothers within the period of pregnancy.

Studies on the nature of placental transfer of ^{90}Sr from the mother to the offspring after the cessation of the mothers' exposure have shown that the amount transferred was inversely related to the time interval between cessation of exposure and conception (Table II), the skeletal concentrations of ^{90}Sr in the new-born dogs relative to those in the mother decreasing sharply with time. The skeletal concentrations of ^{90}Sr in the pups born 29–32 months after the cessation of the mothers' exposure were only 5 to 7% of those in the animals born at the time of chronic ^{90}Sr intake. In the litters born 50–58 months after the mothers were denied ^{90}Sr-containing diet, the accumulation of the isotope per litter showed a 100 fold decrease compared to the placental transfer obtained during chronic ^{90}Sr intake.

A comparison of findings presented in Tables I and II suggests that the major part of the activity passing to the embryo through the placenta is contributed by ^{90}Sr contained in the mother's diet, or more precisely, ^{90}Sr

TABLE I

Placental transfer of strontium-90 to the progeny during chronic exposure of mothers to strontium-90

Mother No.	Weight (kg)	Daily dose of 90Sr (10^{-4} μCi/g)	90Sr concentration in maternal skeletal delivery at the time of delivery (10^{-4} μCi/g)	Overall 90Sr content in maternal skeleton treatment (μCi)	Time interval between delivery and onset of treatment (months)	Litter size (number of pups)	Number of pups subjected to radiometrical analysis	Average weight of new-born pups (g)	Mean skeletal weight of new-born pups* (10^{-4} g)	90Sr concentration in wet bone tissue (femur) of the progeny (10^{-4} μCi/g)	90Sr content in the skeleton of new-born dogs (μCi)	Pup-to-mother skeletal ratios of 90Sr concentrations per g weight (%)	Pup-to-mother ratios of 90Sr skeletal burdens (%)	Ratio of 90Sr skeletal burden in the whole litter to that in the mother (%)	Ratio of 90Sr skeletal burdens in the progeny to the total activity received by mothers during pregnancy (%) per pup	per litter
302	20·0	2·0	155·0	31·00	33	5	5	330	49·50	51·6±2·3	0·255±0·011	33·3	0·8	4·1	0·12	0·6
303	15·0	2·0	165·0	26·40	33	7	5	269	40·35	53·1±3·5	0·214±0·014	32·2	0·8	5·6	0·13	0·9
579	18·0	0·2	32·7	5·89	19	8	6	257	38·55	15·1±0·7	0·058±0·003	46·0	1·0	7·9	0·31	2·5
501	15·0	0·2	46·0	6·90	41	6	6	270	40·50	12·0±0·8	0·049±0·003	26·2	0·7	4·3	0·31	1·9
677	18·0	0·02	6·0	1·08	22+29	5	5	355	53·25	1·0±0·1	0·005±0·0001	17·0	0·5	2·5	0·26	1·3
											Mean	30·9 ±3·5	0·8 ±0·1	4·9 ±0·8	0·23 ±0·04	1·4 ±0·3

* The average skeleton weight of newborn pups in our experiment was 15% of the body weight.

TABLE II

Placental transfer of strontium-90 from the mother to the progeny in dogs after cessation of chronic exposure of the mother

Mother No.	Weight (kg)	Daily dose of ^{90}Sr $(10^{-4}\ \mu\text{Ci/g})$	^{90}Sr concentration in maternal skeleton at the time of delivery $(10^{-4}\ \mu\text{Ci/g})$	Overall ^{90}Sr content in maternal skeleton (μCi)	Time interval between delivery and the cessation of treatment (months)	Litter size (number of pups)	Number of pups subjected to radiometric analysis	Average weight of newborn pups (g)	Mean skeletal weight of newborn pups* (g)	^{90}Sr concentration in wet bone tissue (femur) of the progeny $(10^{-4}\ \mu\text{Ci/g})$	^{90}Sr content in the skeleton of new-born dogs (μCi)	Pup to mother ratios of ^{90}Sr skeletal concentrations per g weight (%)	Pup to mother ratios of ^{90}Sr in skeletal burdens (%)	Ratio of ^{90}Sr skeletal burden in the whole litter to that in the mother (%)	Ratio of placental-transfer levels at the end of ^{90}Sr chronic ingestion by the mother to those during ingestion (%)
667	20.0	0.2	36.0	7.20	3	6	3	323	48.45	8.86±0.91	0.043±0.004	24.70	0.60	3.57±0.38	73.0
418	14.0	2.0	130.0	18.20	11	5	5	389	58.35	9.00±1.15	0.053±0.007	6.93	0.29	1.45±0.19	30.0
579	19.0	0.2	25.3	4.81	29	8	8	246	36.90	0.89±0.10	0.003±0.0004	3.53	0.07	0.56±0.06	7.0
667	19.0	0.2	26.6	5.05	32	6	3	330	49.50	0.44±0.04	0.002±0.0002	1.65	0.04	0.26±0.02	5.0
421	20.8	2.0	109.2	22.71	50	4	4	290	43.50	0.59±0.10	0.003±0.0004	0.54	0.01	0.04±0.01	0.8
301	16.0	2.0	132.5	21.20	58	3	3	275	41.25	1.07±0.08	0.004±0.0003	0.81	0.02	0.06±0.005	1.4

* The average skeleton weight of newborn cubs in our experiment was 15% of the body weight.

TABLE III

Skeletal strontium-90 burdens and concentrations in young dogs
(Mothers had been fed 0·02 μCi ^{90}Sr/kg body weight daily)

Age	No. of pups investigated	Average weight (kg)	Skeletal weight of the pup (% ratio to the body weight)	Concentration the bone $(10^{-4}\ \mu\text{Ci/g})$	^{90}Sr burden in the maternal skeleton at the time of delivery (individual and average values) (μCi)	Skeletal burdens in the pups, calculated from femoral values per pup (μCi)	% of the amount retained in maternal skeleton per pup	per litter (5 pups)
New-born	6	0·26	15·0	15·05±0·70	5·9	0·06±0·003	1·0	4·9
9 days	1	0·55	15·0	23·80	5·9	0·20	3·4	17·0
13 ,,	1	0·71	15·0	29·00	6·0	0·31	5·2	25·8
1 month	5	0·96	15·0	65·42±1·57	6·0	0·94±0·02	15·7	78·3
2 ,,	5	2·54	13·0	31·81±1·90	6·4	1·05±0·06	16·4	82·0
3 ,,	5	4·48	10·0	18·80±1·04	6·2	0·84±0·05	13·6	67·8
4 ,,	5	6·82	10·0	7·14±0·28	6·6	0·49±0·02	7·4	37·1
6 ,,	5	9·22	10·0	2·54±0·12	6·6	0·23±0·01	3·9	19·5
12 ,,	5	11·32	10·0	2·38±0·14	5·9	0·27±0·02	4·4	21·8
35 ,,	4	14·45	10·0	1·78±0·19	6·2	0·26±0·03	3·9	19·7
40 ,,	3	12·60	10·0	1·70±0·12	6·6	0·21±0·02	3·3	16·4
60 ,,	4	13·00	10·0	1·29±0·10	6·0	0·17±0·01	2·8	14·2

contained in the mother's circulating blood; this ^{90}Sr level is determined by gastro-intestinal resorption and by metabolic processes elsewhere in the body.

^{90}Sr transferred from the mother to the offspring showed a non-uniform distribution in the growing skeleton. The non-uniformity factor of ^{90}Sr distribution in different skeleton compartments of new-born dogs (expressed as the ratio of maximum to minimum concentrations) was shown by our earlier experiments to be about 5 (Burykina and Parfenov, 1964), whereas in the work of Ekman and Åberg (1958) a factor of about 2 was obtained for new-born goats born to mothers repeatedly exposed to ^{90}Sr during pregnancy (for 19 successive days up to the time of delivery).

Transfer via milk

Maternal milk is another route of ^{90}Sr intake by the progeny. ^{90}Sr intake with maternal milk at the time of mothers' chronic exposure gave rise to a 10–17 fold increase in the body burdens of the progeny. The highest concentration thus obtained was four times the skeletal concentration at birth and twice the concentration in the maternal skeleton (Table III). The maximum skeletal burdens of ^{90}Sr in the progeny within the period of lactation were 16·4 and 82% cent of the amount retained in the maternal organism for one pup and for the whole litter, respectively.

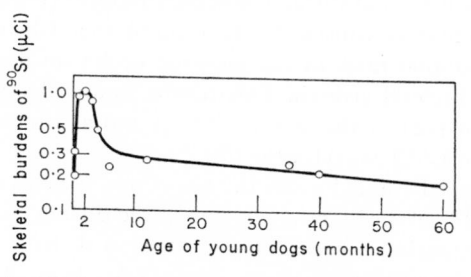

FIG. 2. Changes in ^{90}Sr body burdens in pups from birth up to 5 years of age.

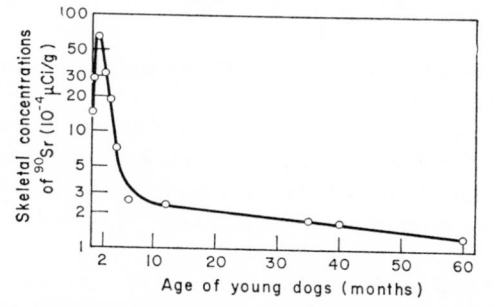

FIG. 3. Skeletal ^{90}Sr concentrations (calculated from femoral data) in pups.

K

Retention and excretion

In Figs. 2 and 3 the metabolism of ^{90}Sr transferred via placenta or milk is traced in young dogs for relatively long periods of time (up to 5 years). Analysis of the curves shows that 70 % of the skeletal ^{90}Sr is excreted with an effective half-life of 23 days, whereas the excretion of the remaining 30 % is exceedingly slow, the effective half-life totalling 2,160 days. [90]Sr metabolism in the progeny can be expressed by the following equation:

$$Q_t = 0.73e^{-0.925(t-2.6)} + 0.31e^{-0.00963(t-2.6)} \quad t \geqslant 2.6 \text{ months,}$$

where Q_t is the amount of ^{90}Sr in μCi in the skeletons of young dogs aged t months.

Owing to low excretion rates, dogs aged 1 year retained 25.6 % of ^{90}Sr transferred from the maternal organism, and 5 years later they still had about 16 % of that value.

This is in keeping with the findings of other workers (Kulikova, 1960; Parfenov, 1963) who showed that young rats can retain 10–30 % of maternal ^{90}Sr for periods of up to 1 year.

In evaluating the dynamics of changes observed in skeletal concentrations of ^{90}Sr in young dogs, it should be pointed out that the decrease in ^{90}Sr concentration is rather rapid within the first six months (Fig. 3).

The rapid decrease is caused by two simultaneously occurring factors namely, high metabolic rates in the growing organism and the dilution of activity owing to skeletal growth. This results in an effective half-life of 23 days for the major part of the activity (96 %) and a slow excretion rate (an effective half-life of 1,650 days) for the rest.

In young dogs aged 1 and 5 years the respective skeletal concentrations of ^{90}Sr were 3.6 and 2 % of the highest value observed during lactation.

The following equation expresses the changes in skeletal concentrations of ^{90}Sr in the progeny as a function of age:

$$q_t = 6.2e^{-0.925(t-1.2)} + 0.27e^{-0.0126(t-1.2)} \quad t \geqslant 1.2 \text{ months,}$$

where q_t is the concentration of radioactivity (in μCi/kg) in the skeletons of young dogs aged t months.

By comparing ^{90}Sr excretion rates in young dogs with those in their parents, it was found that the slowly excreted fractions had similar effective half-lives. An effective half-life of about 1,800 days has been obtained by me (Burykina, 1966) after the cessation of administration of ^{90}Sr with diet in daily doses of 0.2×10^{-4} and 2×10^{-4} μCi/g for 3–3.5 successive years.

The data obtained in the present study on the transfer of ^{90}Sr from the mother to the offspring should allow a more reliable estimate of the biological significance of this isotope, with special reference to its effect on the progeny. The above equations describing ^{90}Sr metabolism in young animals can be used for calculating the doses absorbed in their skeletons.

REFERENCES

Buldakov, L. A., and Moskalev, Yu. I. (1960). *Bull. exp. Biol. Med. U.S.S.R.* **10**, 111–113.

Burykina, L. N. (1966). *In* "Distribution and Biological Effects of Radioisotopes", p. 66. Atomizdat, Moscow.

Burykina, L. N., and Parfenov, Yu. D. (1964). *In* "Distribution, Biological Effects and the Stimulation of Excretion of Radioisotopes", p. 124. Medgiz, Moscow.

Downie, E. D., *et al.* (1959). *Br. J. Cancer,* **13**, 3, 408–423.

Ekman, L., and Åberg, B. (1958). *Proc. 2nd Int. Conf. peaceful Uses of atom. Energy,* Report No. 175.

Keirim-Marcus, I. B., *et al.* (1961). *Medskaya Radiol.* **6**, 51–55.

Kulikova, V. G. (1960). *Dokl. Akad. Nauk SSSR,* **131**, 1433–1436.

Kurl'andskaya, E. B., Beloborodova, N. L., and Baranova, E. F. (1957). *In* "Data on the Toxicology of Radioactive Substances", vol. I, p. 16. Medgiz, Moscow.

Parfenov, Yu. D. (1963). *In* 'The Effect of Radioactive Substances on the Reproductive Function and on the Progeny", p. 193. Medgiz, Moscow.

Rubanovskaya, A. A., and Ushakova, V. F. (1957). *In* "Data on the Toxicology of Radioactive Substances", vol. I, p. 23. Medgiz, Moscow.

Vorozheikina, T. B., and Parfenov, Yu. D. (1962). *Bull. exp. Biol. Med. U.S.S.R.* **8**, 96–100.

Effect of Dietary Strontium on Reproductive Performance of the Laying Hen

F. R. MRAZ, P. L. WRIGHT AND T. M. FERGUSON*

Agricultural Research Laboratory of the University of Tennessee, Oak Ridge, Tennessee†

SUMMARY

Thirty yearling SCW Leghorn hens were fed a diet containing 2·8% Ca plus 0·0, 0·5, 1·0, 2·0 or 4·0% Sr for 4 weeks before receiving ^{45}Ca and ^{89}Sr orally. Egg production, strength of egg shell, and hatchability were reduced on the 2 and 4% Sr diets. The deposition of both ^{89}Sr and ^{45}Ca in the egg shell and the tissues increased as the level of Sr in the diet increased.

INTRODUCTION

Investigators have been interested in Sr for some time because of its chemical similarity to Ca. This chemical similarity often causes Sr, under normal physiological conditions, to follow metabolic routes similar to those of Ca. Interest in Sr increased following the discovery that radioactive Sr from nuclear fission could present a considerable hazard to health. Engfeldt *et al.* (1962) reported that 1·6% Sr in a low (0·001%) Ca diet increased the severity of rickets, thus demonstrating an antagonism between Sr and Ca at low dietary intakes of Ca. Mraz (1961) fed a diet containing 1·09% Sr and 1·50% Ca to chicks and found that, although growth rate was reduced, no rickets was apparent even though the ratio of stable Sr to Ca in the bone approached 0·2. Grant *et al.* (1960) stated that when a ration containing 0·2% Sr was fed to hens, the concentration of Sr in the ash of the egg shells increased by a factor of 40–50 within 3 days with no further increases after that time. Oshima and Nozaki (1961) and Edwards and Mraz (1961) reported that the shell of the first egg after dosing contained the greatest amount of the ^{89}Sr administered to a laying hen. Monroe *et al.* (1961) concurred with Edwards and Mraz (1961) that radiocalcium was preferentially used over radiostrontium. Little work on the prolonged feeding of high levels of stable Sr

* Research Participant, ORINS, on leave from the Department of Poultry Science, Texas A & M University, College Station.

† Operated by the Tennessee Agricultural Experiment Station for the U.S. Atomic Energy Commission under Contract AT–40–1–GEN–242.

and its effect on reproductive performance has been conducted with laying hens and such information is essential if the metabolic behavior of Sr is to be understood.

EXPERIMENTAL METHODS

Thirty Single-Comb White Leghorn yearling hens were housed in individual cages with individual feeders and waterers and fed a commercial all-mash laying ration during the pre-experimental period. They were randomly divided into five groups of six hens each and fed a corn–soybean oil meal-type diet (Table I) containing 2·8 % Ca and the following levels of Sr in the carbonate form: 0·0, 0·5, 1·0, 2·0 or 4·0 %; sand replaced strontium carbonate to

TABLE I

Composition of basal diet*

	Percentage
Ground yellow corn	66·33
Soybean meal, solvent 44%	25·00
Calcium carbonate	6·00
Dicalcium phosphate	2·00
Sodium chloride	0·470
Potassium iodide	0·000291
Manganese sulphate monohydrate	0·0238
Zinc carbonate	0·00634
Vitamin A acetate (10,000 I.U./g)	0·110
Vitamin D ((30,000 I.U./g)	0·00441
α-Tocopheryl acetate (110 I.U./g)	0·0100
Menadione	0·000220
Riboflavin	0·000551
Calcium pantothenate	0·00132
Niacin	0·00176
Vitamin B_{12} (0·1% trituration)	0·00110
Choline chloride (25%)	0·0361

* Sand, strontium carbonate, or combinations of the two comprised 6·72 lb of every 100 lb of diet.

maintain isocaloric and isonitrogenous diets. The hens were artificially inseminated at 5-day intervals before and during the experiment. All sound eggs collected during a 4-week period were incubated for 22 days, whereupon all eggs or chicks were examined for abnormalities. Time of death for each dead embryo was estimated. After 4 weeks on the experimental diets, the hens were administered, *per os*, 200 μCi. [45]Ca (0·82 mg Ca) and 100 μCi [89]Sr (carrier free) in the chloride form. The hens were sacrificed 48 h later and the radionuclide contents of the whole blood, superficial pectoralis muscle, liver,

TABLE II

Egg production and incidence of cracked eggs from hens fed various levels of strontium

Figures are percentages

| Dietary Sr | Pre-experimental eggs* | | Weeks on experimental diets | | | | | | | |
| | | | 1 | | 2 | | 3 | | 4 | |
	Produced	Cracked	Produced	Cracked	Produced	Cracked	Produced	Cracked	Produced	Cracked
0·0†	..	..	51	6	63	9	63	5	63	5
0·5	57	0	67	0	64	0	57	4	50	0
1·0	63	5	55	13	43	0	43	11	52	0
2·0	50	7	50	10	38	19	29	42	19	0
4·0	67	0	67	29	14	67	2	100	5	50

* Pre-experimental data on hens before dietary Sr was administered.
† Results from five hens.

kidney and eggs laid after dosing were determined by using the techniques described by Edwards and Mraz (1961). Other groups of hens were fed the 0, 0·5, 1·0 or 2·0% Sr diets for 100 days and their egg shells removed by dissolution by using the ethanol–hydrochloric acid technique of Itoh and Hatano (1964) to produce five artificial shell strata. Ca and Sr were determined on an atomic-emission spectrophotometer.

RESULTS AND DISCUSSION

Egg production (Table II) declined quite rapidly after the first week of feeding the 2 and 4% Sr diets, with the decline being more severe for hens fed the 4% Sr diet. Sr also tended to reduce shell strength as evidenced by the higher percentages of cracked eggs. The eggs laid during the first week the hens were fed the Sr diets had a rough chalky shell. In succeeding weeks the appearance of the egg shells returned to normal suggesting an adaption by the bird to the dietary Sr.

Hens fed 1% or more Sr lost more weight than those fed lower levels (Table III). Egg weight and fertility, however, were not affected by the dietary Sr. Hatchability was reduced considerably by the 2 and 4% Sr diets as a result of higher embryonic mortality during the third week of incubation. This mortality coincided with the time when the embryo would be drawing rather heavily upon the shell for Ca (Nozaki *et al.*, 1954) and would, therefore, be metabolizing the Sr present with the shell Ca.

TABLE III

Reproductive performance of hens fed varying levels of strontium

Sr (%)	Hen weight loss (g)	Average egg weight (g)	No. of eggs*	Fertility (%)	Embryonic mortality (week) 1st (%)	Embryonic mortality (week) 2nd (%)	Embryonic mortality (week) 3rd (%)	Hatch of fertile eggs (%)
0·0†	150	59	79	81	13	3	22	62
0·5	137	61	99	97	8	0	31	61
1·0	403	58	76	97	11	0	38	51
2·0	429	62	47	81	8	3	71	18
4·0	498	51	23	96	23	0	41	36

* Cracked eggs not included in number of eggs.
† Results from five hens.

That this high rate of late embryonic mortality stabilized in the second week of Sr feeding would be expected, since Grant *et al.* (1960) found no further increase in Sr in the shell after 3 days of stable Sr feeding. None of the chicks that hatched showed any abnormalities attributable to Sr fed to their dams.

Levels of both ^{45}Ca and ^{89}Sr from an oral dose increased in shell of the eggs laid the first day after dosing, as the Sr content of the diet increased (Table IV). However, the ratio of ^{89}Sr/^{45}Ca also increased, indicating that the ^{89}Sr content in the egg shell increased more rapidly than the ^{45}Ca content. The contents of

TABLE IV

Deposition of calcium-45 and strontium-89 in the shell of eggs laid within 24 h post-dosing

On the second day 4, 2, 1 and 3 eggs were obtained from the 0, 0·5, 1·0 and 2·0% Sr diets, respectively.

Sr %	No. of eggs	^{45}Ca (% dose/egg)	^{89}Sr (% dose/egg)	^{89}Sr/^{45}Ca
0·0	2	50	24	0·48
0·5	4	56	30	0·54
1·0	5	59	40	0·68
2·0	1	64	40	0·63
4·0	0	..	..	..

the eggs laid on the first day after dosing and the entire eggs laid on the second day were not included because of the wide variations encountered, which were dependent upon the stage of oviposition at the time of dosing and whether an egg was laid the first day. This variation is further explained by Edwards and Mraz (1961). The relation between dietary Sr content and increased ^{89}Sr and ^{45}Ca deposition in the egg shell was also observed in tissue retention of ^{89}Sr and ^{45}Ca (Table V). The whole blood, muscle, liver and kidney data suggest that more ^{45}Ca and ^{89}Sr were absorbed as the Sr content

TABLE V

Calcium-45 and strontium-89 contents of tissues from hens fed varying levels of strontium

Dietary Sr (%)	Hen weight (kg)	Whole blood ^{45}Ca	^{89}Sr	Pectoral muscle ^{45}Ca	^{89}Sr	Liver ^{45}Ca	^{89}Sr	Kidney ^{45}Ca	^{89}Sr	Tibia ^{45}Ca	^{89}Sr
0·0	2·1	0·52	0·42	0·41	0·04	0·28	0·16	0·91	0·49	180	120
0·5	1·8	0·84	0·53	0·37	0·06	0·25	0·17	1·20	0·69	150	90
1·0	1·6	1·39	0·86	0·49	0·10	0·40	0·20	1·50	1·06	180	110
2·0	1·8	1·50	0·86	0·86	0·16	0·58	0·29	2·46	0·99	250	130
4·0	1·5	2·61	0·96	1·49	0·21	1·68	0·41	7·10	1·89	220	200

Radionuclide content of tissue (% dose $\times 10^3$/g)

of the diet increased, but the tibia retention data were quite variable. The number of eggs produced per group was more important than dietary-Sr intake in altering radionuclide retention by tissues.

This was evident in rather high standard deviations (approaching the mean) observed in some tissues, which reflected to a large degree the amount of ^{45}Ca (as high as 65%) and ^{89}Sr (as high as 50%) deposited in the egg shell. Hens not laying the first day would tend to deposit most of this activity. The extremely high retention of ^{45}Ca and ^{89}Sr in tissues of hens fed the 4% Sr diets could be due to this lack of ^{45}Ca and ^{89}Sr excretion in egg shell. No consistent relationship between body weight and radiocalcium or radiostrontium concentration was observed, probably because it was masked by the variations due to the excretion of these radionuclides through egg shell.

The feeding of Sr and Ca in ratios of 0·18, 0·36 or 0·71 resulted in Sr/Ca ratios of 0·11, 0·16 or 0·30, respectively, in the shells of eggs laid 4 or more days after initiation of Sr feeding. In every instance the inner layer of egg shell contained less Sr than the outer layer (Table VI).

TABLE VI

Strontium-to-calcium ratios in 5 layers of egg shells from hens fed diets varying in strontium

Dietary Sr	No. of eggs	Shell layers				
		1 (outside) Sr/Ca	2 Sr/Ca	3 Sr/Ca	4 Sr/Ca	5 (inside) Sr/Ca
0·0	30	0·002	0·002	0·002	0·001	0·001
0·5	36	0·126	0·113	0·108	0·107	0·105
1·0	51	0·180	0·167	0·160	0·160	0·153
2·0	56	0·341	0·337	0·308	0·276	0·251

Thus it appeared that an embryo would have a lower Sr/Ca ratio available to it for bone formation than would be predicted from the Sr/Ca ratio of entire shell. The greater Sr/Ca ratio of the outer layer would contribute greatly to the susceptibility of the eggs to breakage, since the calcite lattice of the outer egg shell, which contributes to its strength, is distorted by the larger Sr ion.

The hen has the ability to adapt to diets containing as much as 1% Sr without adversely affecting egg production or egg-shell strength and with only a slight increase in late embryonic mortality. None of the levels of Sr fed in addition to 2·8% Ca reduced ^{89}Sr uptake, concurring with the observation of Mraz (1961) that the gut could adapt to increasing levels of dietary Sr, regardless of the level of dietary Ca, and permit more Sr to pass across the gut wall. Stable Sr, therefore, could not only be ineffective in reducing the uptake of the Sr radionuclides, but might result in an increase in their uptake by hens.

This manuscript is published with the permission of the Director of the University of Tennessee Agricultural Experiment Station, Knoxville.

REFERENCES

Edwards, H. M., Jr., and Mraz, F. R. (1961). *Poult. Sci.* **40**, 493–503.
Engfeldt, B., Hjertquist, S. O., and Lagergren, C. (1962). *Acta Soc. Med. upsal.* **67**, 239–257.
Grant, C. L., Ringrose, R. C., and Downer, R. (1960). *J. biol. Chem.* **235**, 2157–2159.
Itoh, H., and Hatano, T. (1964). *Poult. Sci.* **43**, 77–80.
Monroe, R. A., Wasserman, R. H., and Comar, C. L. (1961). *Am. J. Physiol.* **200**, 535–538.
Mraz, F. R. (1961). *Poult. Sci.* **40**, 958–965.
Nozaki, H., Horii S., and Takei, Y. (1954). *Bull. natn. Inst. agric. Sci.(Tokyo) Ser. G*, No. 9, 89–95.
Oshima, M., and Nozaki, H. (1961). *Bull. natn. Inst. agric. Sci. (Tokyo), Ser. G*, No. 20, 225–236.

Urinary/Fecal Excretion Ratios of Strontium and Calcium in the Rat

R. C. THOMPSON AND P. L. HACKETT

Biology Department, Battelle Memorial Institute,
Pacific Northwest Laboratory, Richland, Washington

SUMMARY

Following the intraperitoneal injection of ^{45}Ca and ^{85}Sr to mature rats on a high Ca diet, the ratio of urinary/fecal excretion of both radionuclides decreased with time after injection. The decrease after 47 days was to halve the initial values for ^{45}Ca, and to one-sixth the initial values for ^{85}Sr. This effect was shown to be due to an increase in excretory clearance via the intestine.

INTRODUCTION

The observations described in this report, while they may be unique to the animal species and particular experimental conditions employed, are nevertheless of some general interest because they were unexpected and because they remain unexplained. They may offer a lead to the understanding of more fundamental problems.

Several years ago we studied the urinary and fecal excretion of ^{90}Sr and ^{45}Ca in the rat, following a single injection, as a function of age and dietary Ca level. The results were in general agreement with conclusions previously reached from retention studies involving serial sacrifice and femur analysis (Thompson and Palmer, 1960; Palmer and Thompson, 1964). One peculiarity in the data, however, attracted our attention. The ratio of radionuclide excretion in the urine as compared to that in the feces decreased with time following injection. This change was more evident for Sr than for Ca, was more evident in the adult than in the growing rat, and was most marked on a high Ca diet. The experiment reported here was designed to check carefully these observations under conditions where a maximum effect might be expected.

EXPERIMENTAL METHOD

Twenty-eight, 10-month-old, female, Sprague–Dawley rats were placed on a diet containing 2 % Ca and 0·5 % P (Thompson and Palmer, 1960). After 5 days on this diet, half of the animals (Group I) were injected intraperitoneally with ^{85}Sr (30 μCi) and ^{45}Ca (60 μCi). The remaining animals (Group II) received an injection of saline. The Group II animals were injected 40 days later with the radionuclides, and Group I animals were sham-injected. All animals were sacrificed 7 days after the second injection, and plasma samples were obtained. The plan of the experiment is summarized in Fig. 1. Throughout the experiment, animals were caged individually, food consumption, urine volume, and feces weights were recorded daily and, after the radionuclide injection, ^{85}Sr and ^{45}Ca were determined in daily urine and feces collections from each animal. ^{45}Ca was determined by β-proportional counting, corrected for self-absorption and ^{85}Sr activity. ^{85}Sr was measured directly in a well-type γ-spectrometer.

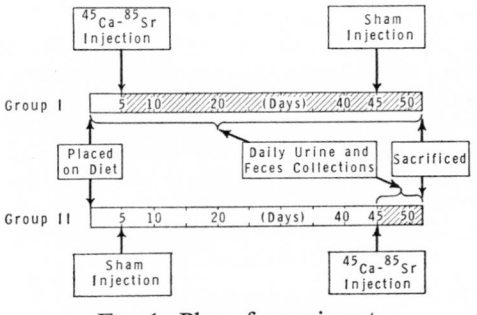

FIG. 1. Plan of experiment.

RESULTS AND DISCUSSION

Animal weights did not change greatly during the experimental period. The average initial weight was 380 g and the average weight at sacrifice was 393 g. Food consumption, urine volume, and feces weights were quite constant.

Excretion curves for ^{45}Ca in urine and feces of the Group I animals are shown in Fig. 2(a). Each point is an average of separate determinations on fourteen animals, and the data are quite coherent. Fecal excretion exceeded urinary excretion at all times, but the spread between these two routes clearly increased with time after injection. Most striking is the increase in fecal excretion which occurred after about the 35th day. This increase is statistically significant.

Even more surprising are the ^{85}Sr excretion curves shown in Fig. 2(b). Again there was a statistically significant and sustained increase in fecal excretion beyond about the 25th day after injection. By the end of the experiment fecal excretion of ^{85}Sr substantially exceeded urinary excretion.

These data are displayed in a somewhat different manner in Fig. 3(a), in which a graph of the ratio of urinary to fecal excretion is plotted as a function of time after injection. Over the 47-day period there is an approximately sixfold decrease in this ratio for ^{85}Sr, and an approximately twofold decrease for

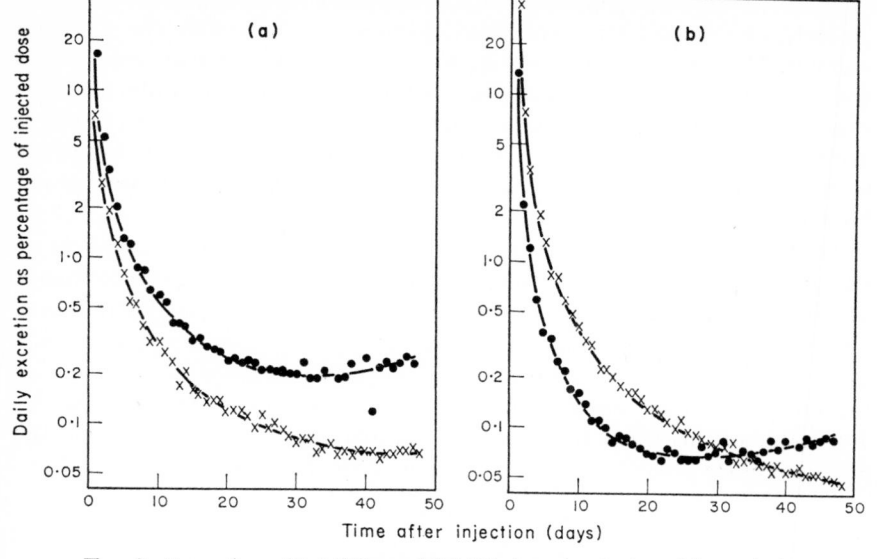

FIG. 2. Excretion of (a) ^{45}Ca and (b) ^{85}Sr in urine (×) and feces (●).

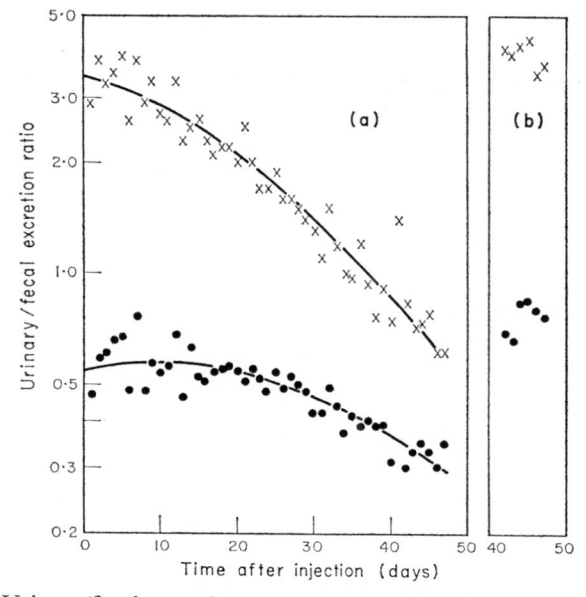

FIG. 3. (a) Urinary/fecal excretion ratios for ^{45}Ca (●) and ^{85}Sr (×) for Group I animals.

(b) Urinary/fecal excretion ratios for Group II animals who received radionuclide injections 40 days after Group I animals: ● , ^{45}Ca; × , ^{85}Sr.

^{45}Ca. In Fig. 3(b) are shown the urinary/fecal excretion ratios for the Group II animals who received radionuclide injections 40 days after the Group I animals. These ratios correspond quite closely between the two groups when compared at comparable times after injection, but are very different when compared at the time of measurement. This indicates that the change in urinary/fecal excretion ratio is a function of residence time of the radionuclide in the animal and not related to age, time on diet, or other factors common to both groups.

The plasma clearance data summarized in Table I sheds additional light on the situation. For both ^{45}Ca and ^{85}Sr, renal clearance values do not change appreciably between 7 and 47 days after injection. Clearance via the intestine, however, increases approximately twofold for ^{45}Ca, and approximately fivefold for ^{85}Sr. It seems evident, therefore, that the anomalous behavior of the urinary/fecal excretion ratio must be attributed to changes in the denominator of this fraction. Both the changing intestinal clearance, and the peculiar upturn to the fecal excretion curves, point in this direction.

TABLE I

Clearance data

	After 7 days deposition (Group II)	After 47 days deposition (Group I)
Calcium-45		
Circulating in plasma*	0.51 ± 0.11	0.084 ± 0.022
Excreted/day in urine*	1.1 ± 0.4	0.17 ± 0.07
Excreted/day in feces*	2.2 ± 0.4	0.74 ± 0.28
Renal clearance†	2.3 ± 0.9	2.2 ± 1.1
Intestinal clearance†	4.3 ± 1.1	9.1 ± 3.4
Strontium-85		
Circulating in plasma*	0.20 ± 0.04	0.018 ± 0.007
Excreted/day in urine*	3.2 ± 0.5	0.22 ± 0.05
Excreted/day in feces*	1.3 ± 0.4	0.49 ± 0.19
Renal clearance†	15 ± 3	13 ± 6
Intestinal clearance†	6.6 ± 1.2	30 ± 19

* Expressed as % of current body burden $\pm$s.d.
† Expressed as plasma volumes/day $\pm$s.d.

We have confirmed these observations in three completely separate experiments, and somewhat similar observations have been reported by Fujita and Iwamoto (1965). One of two possible kinds of explanation would seem to be required. Either (1) the form of Ca or Sr released to the blood from bone changes with time following deposition to one more readily excreted via the intestine (or less readily re-absorbed from the intestine), or (2) the mechanism of intestinal excretion is altered as a result of the deposition of [85]Sr and [45]Ca in bone. The latter alternative seems the less likely. It would involve the assumption of radiation damage, and although the radionuclide levels used might conceivably have caused significant damage to the bone, it is difficult to see how such damage could affect the processes of intestinal excretion of Ca and Sr.

The possibility of a changing form of Ca and Sr in the blood seems more attractive. The source of blood [45]Ca and [85]Sr clearly changes with time following injection. Early after injection it is composed principally of Ca and Sr in equilibrium with freely exchangeable sites on bone surfaces. With the passage of time, however, it is, to an increasing extent, derived from firm binding sites in bone that are unearthed by irreversible bone-resorption processes. It is not inconceivable that Ca and Sr released from such stable binding sites might exist in a different chemical form, more subject to excretion via the intestinal route.

It is interesting to note that, after 47 days' deposition, the total amounts of [85]Sr and [45]Ca excreted per day, when expressed as fractions of the currently retained body burden, are almost identical (Table I). This fact is well illustrated in Fig. 4, in which a graph of the ratio of [85]Sr/[45]Ca in urine, feces, and

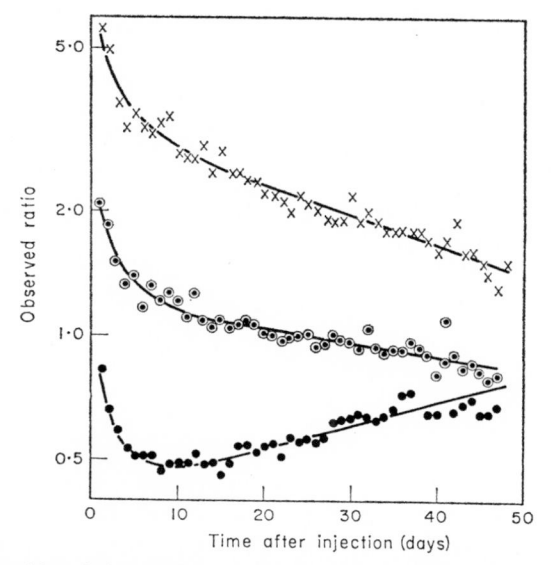

FIG. 4. Relationship of observed ratio and time: ×, urine; ⊙, urine + feces; ●, feces.

total excreta $\left(OR = \dfrac{{}^{85}Sr_{excreta}/{}^{85}Sr_{bone}}{{}^{45}Ca_{excreta}/{}^{45}Ca_{bone}}\right)$ is plotted as a function of time after injection. Beyond about the 10th day after injection the total excretion of ${}^{45}Ca$ and ${}^{85}Sr$, expressed as fractions of retained body burden, are very nearly the same, as indicated by an OR approximating to unity. This is what one might expect if the release of these radionuclides from bone was the result of an irreversible bone-resorption process, and if re-utilization was held to a minimum by a high dietary Ca intake. The fact that the ORs for urine and feces both tend to a value of unity is also interesting, and no doubt pertinent to an understanding of the anomalous excretory behavior observed in this experiment.

This paper is based on work performed under U.S. Atomic Energy Commission Contract AT(45–1)–1830.

REFERENCES

Fujita, M., and Iwamoto, J. (1965). *Hlth Phys.* **11**, 271–281.
Palmer, R. F., and Thompson, R. C. (1964). *Am. J. Physiol.* **207**, 561–566.
Thompson, R. C., and Palmer, R. F. (1960). *Am. J. Physiol.* **199**, 94–102.

Radiostrontium, Stable Strontium, Stable Calcium and Phosphorus in Sperm DNA

BERTIL ÅBERG AND MARGOT GILLNER

Department of Clinical Biochemistry,
Royal Veterinary College, Stockholm

SUMMARY

In spermatozoa from rams the DNA has been analysed as well as the stable Ca and Sr content. The numbers of atoms of radiostrontium (^{89}Sr and ^{85}Sr) taken up by the spermatozoon has been determined. The effects of the ionizing radiation from radiostrontium on the sperm DNA are discussed.

RESULTS

We have recently published some preliminary results showing that in ram sperm cells the ratio Sr/Ca is 1/118 (range 1/45–1/239) (Åberg and Gillner, 1966). It was further shown that the intravenous administration of ^{89}SrCl$_2$ to rams results in a spermatozoal uptake of the radionuclide. The intravenous administration of ^{32}P as phosphate also results in a spermatozoal uptake of the nuclide as already shown by Dawson (1958). Fig. 1 shows an example of the concentration curves achieved in rams. Obviously there is a difference in behaviour between ^{32}P and ^{89}Sr, inasmuch as the latter first is rapidly taken up and then decreases until it finally is taken up again. It has to be borne in mind, however, that the normal life time for a spermatozoon from formation until it is ejaculated is 40–50 days in rams (Dawson, 1958). Out of that time, some 10–14 days are spent in the epididymis (Dawson, 1958; Crabo, 1965) where there is a considerable exchange of ions between the spermatozoon and the epididymal plasma (Crabo, 1965).

CONCLUSION

We concluded among other things from our first results (Åberg and Gillner, 1966) that the first ^{89}Sr peak is due to a diffusion of Sr in the epididymis whereas the second concentration peak indicates binding of Sr to the contents of the spermatozoon.

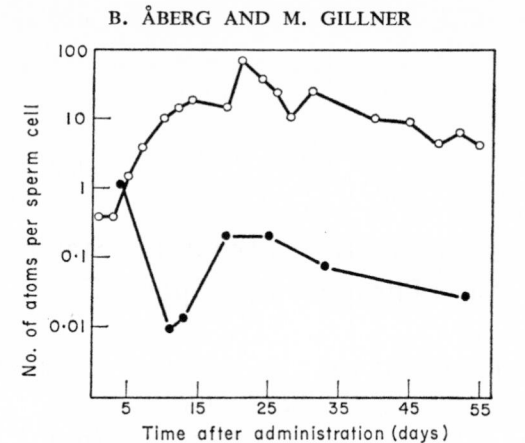

FIG. 1. Relationship between time and concentration of radionuclide in sperm cells of rams who had intravenously received 10 mCi ^{85}Sr and 5·86 mCi ^{32}P: ○, ^{32}P; ●, ^{89}Sr.

It seems very probable that this binding takes place, at least partly, in the ionic binding of protein to DNA via divalent cations, such as Ca^{+2} or Sr^{+2}. It cannot yet be excluded, however, that the Sr is bound to some other structures, and especially it cannot be ruled out that it is bound to that 0·02 % of the sperm cell nucleic acids which is RNA. The presence of enough $^{89}Sr^{+2}$ in the spermatozoon kills the cell (Åberg and Gillner, 1966), and the necessary concentration is roughly 10^{-1} atoms of ^{89}Sr per spermatozoon on the 50th day after the administration of the radionuclide (30mCi $^{89}SrCl_2$ intravenously). Morphologically, the sperm cells seem normal when the ^{89}Sr concentration has declined to some 10^{-3} atoms per spermatozoon, which occurs about 60 days after the administration of a single dose of the nuclide.

It seems of interest in connection with this study to try to establish the pool of stable Sr in sperm cells into which the radiostrontium enters. We have determined stable Sr with the Perkin–Elmer model 303 atomic-absorption flame spectrophotometer. The detection limit on scale 10 for Sr was 0·01 mg/l (presence of down to 0·005 mg/l can be shown). On scale 2 our detection limit for Ca was 0·05 mg/l.

In rams the mean Ca concentration is $2·4 \times 10^8$ atoms per spermatozoon, whereas the Sr concentration is 2×10^6. In bulls the corresponding values are, for Ca, $3·1 \times 10^8$ and, for Sr, $2·2 \times 10^6$. In the blood serum from the bulls, the mean Ca content was 4·8 mequiv./l and the mean strontium content 0·014mequiv./l.

It is interesting that the presence of as little as 1 atom of ^{89}Sr among some 10^6 atoms of stable Sr can cause so very profound pathological changes in the sperm cells. We do not yet find the time ripe to make any guesses as to what it means that the ratio Sr/Ca seems to change from some 1/500 in bull serum to some 1/100 in the spermatozoon. We think however that for the time being we should concentrate on the turnover of radiostrontium in the pool of stable Sr.

REFERENCES

Åberg, B., and Gillner, M. (1966). *Acta physiol. scand.* **66,** 106–114.
Crabo, B. (1965). *Acta vet. scand. Suppl.* 5, **6,** 1–94.
Dawson, R. M. C. (1958). *Biochem. J.* **68,** 512–519.

Removal of Radiostrontium from the Mammalian Body

A. CATSCH

Lehrstuhl für Strahlenbiologie, Technische Hochschule Karlsruhe, and Institut für Strahlenbiologie, Kernforschungszentrum, Karlsruhe

INTRODUCTION

It is a somewhat thankless task to survey experimental studies of the modification of radiostrontium metabolism, as most of them yielded either negative or only marginal results. Anticipating one of the main conclusions of this review we can say that the efficacy of a few particular measures as well as the failure of many attempts to influence skeletal retention of radiostrontium share—paradoxical as this statement might appear—a common reason. This is, the close resemblance between Sr and Ca with regard to their chemical and biological features. The metabolic behaviour of Sr simulating to a certain extent that of Ca, is hereby, of course, subject to the influence of numerous physiological factors, that are difficult to control, and thus, the inconsistencies, which are so frequently encountered in experiments with radiostrontium, are by no means surprising.

In order to permit consideration of the vast amount of experimental data, it is proposed to restrict this discussion to two main topics. Firstly, "modification of metabolism" will be defined as enhancement of radiostrontium excretion from the mammalian body. Such a definition is, admittedly, one-sided, but justified if one is aiming primarily at the elaboration of therapeutical or preventive measures in cases of radiostrontium incorporation. Secondly, emphasis will be placed on experimental designs that are concerned with the metabolic fate of a single dose of radiostrontium or with the diminution of an already existing skeletal burden.

ISOTOPIC DILUTION

If we are dealing with carrier-free radiostrontium and take into account the mode of action by which it is retained by bone tissue, isotopic dilution appears, at least at first glance, as the most straightforward approach in attempting the removal of radiostrontium. In fact, it has been demonstrated by numerous investigators (Copp and Greenberg, 1947; MacDonald, 1956; Rubanovskaya and Ushakova, 1957; Catsch and Melchinger, 1959 a; Kawin,

1959; Carlqvist and Nelson, 1960; Nelson *et al.*, 1963) that simultaneous administration of stable carrier does lead to an elevated excretion of radio-strontium. The dependence of efficacy upon dosage and mode of administration is shown in Fig. 1. Quite a few peculiarities are obvious. Following

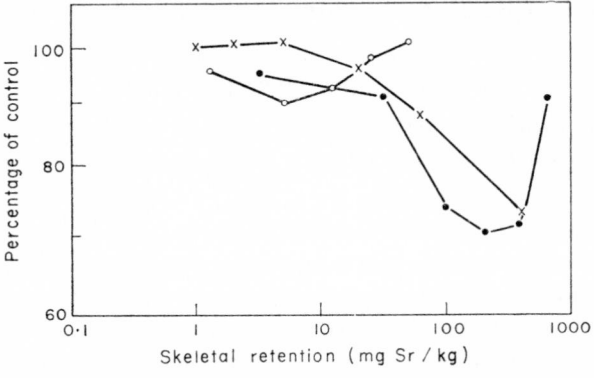

FIG. 1. Skeletal retention of isotopically diluted radiostrontium in the rat, expressed as percentage of control, i.e. carrier-free radiostrontium. Dose administered intravenously, ○ (Catsch and Melchinger, 1959a); intraperitoneally, ● (Catsch and Melchinger, 1959a); orally, × (Harrison *et al.*, 1957).

intraperitoneal injection the increase of carrier dosage up to the near-toxic range leads, paradoxically enough, to an enhanced skeletal retention of radiostrontium, whereas there is no significant effect from intravenous carrier doses. The failure of intravenous, as well as of high intraperitoneal carrier doses, to increase radiostrontium excretion, may tentatively be explained by the assumption that a threshold value of the ion product $Sr^{2+}.HPO_4^{2-}$ is exceeded and that the formation of new apatite crystals proceeds at an elevated rate. As to the oral administration, the figures in the graph are based on the absorbed dose; the actual radiostrontium content of the skeleton, however, remains virtually constant over the whole dose range (Harrison *et al.*, 1957; Wasserman *et al.*, 1957; Kawin, 1959). Consequently, a twofold effect has to be assumed: high carrier doses seem to bring about an elevated absorption from the intestine as well as an enhanced urinary excretion of the absorbed fraction.

The therapeutic applicability and value of isotopic dilution are both severely limited because of the extreme time dependence. As a rule, a significant reduction of skeletal retention is obtained only if stable Sr is given simultaneously or within a few hours before or after the administration of radiostrontium. If treatment is initiated later, all investigators (Copp and Greenberg, 1947; Rubanovskaya and Ushakova, 1957; Catsch and Melchinger, 1959a) failed to achieve a statistically significant diminution of the body burden, even if the supplementation with stable Sr was maintained for a fairly long time.

Taking for granted that hetero-ionic exchange of Sr against Ca from the crystal surface is one of the basic (and by definition reversible) mechanisms by which radiostrontium is retained by bone, the aforementioned marked dependence of carrier effectiveness upon time may appear somewhat conflicting. It is true that radiostrontium, after being deposited on the crystal surface, is gradually withdrawn from equilibrium and becomes fixed irreversibly. This is thought to be due to intracrystalline exchange, recrystallization, and accretion of new bone. It is doubtful, however, whether these relatively slow processes can account for the rapid loss of carrier effectiveness in its entirety. On the other hand, we should not overlook a less sophisticated explanation; namely, the fact that the Sr ions in the fluid phase are not only in exchange with the radiostrontium located on the crystal surface, but to a much larger extent with Ca ions, because of their exceedingly high concentration.

So far, we have discussed results obtained with small rodents. In humans, however, according to Spencer (1963), approximately 10% of the radiostrontium body burden could be mobilized by daily infusions of strontium gluconate from the 15th till the 17th day following the injection of ^{85}Sr. It is questionable, however, whether in this case a genuine mobilization of radiostrontium from the bone is involved, for it is known that in humans the radiostrontium concentration of soft tissues is maintained at a relatively high level over a longer period than in rodents (Schulert et al., 1959). The studies of Cohn et al. (1962) give clear evidence that the long-term turnover of radiostrontium is not affected by supplementation of stable Sr.

The possibility, however, that the response of different mammalian species to a particular treatment may differ, should not be denied a priori. We have seen, for instance, that in small rodents isotopic dilution virtually does not interfere with the absorption rate of radiostrontium from the gut. This also holds for goats and dairy cows. A surprisingly high carrier effect, i.e. a diminution of the absorption by approximately 80% has, however, been observed in young pigs by Bartley and Reber (1961). Unfortunately, the experimental conditions are not strictly comparable, as a Sr-rich diet was fed for 4 weeks before the oral administration of radiostrontium.

At this juncture, a different approach should be mentioned briefly. Since the solubility of strontium phosphates is distinctly higher than that of calcium phosphates, one may assume for physico-chemical reasons that the attractive forces to remove ions from solutions during crystal growth are less pronounced for a more soluble crystal, and, hence, that a smaller fraction of radiostrontium should be retained by bone previously enriched with stable Sr. This conjecture has been substantiated (Cohn et al., 1961; Kriegel et al., 1963). Pre-feeding of a high Sr diet reduced markedly the retention of a single intraperitoneal dose of radiostrontium in rats. No sound explanation can yet be given why in a similar experimental design in mice (except that radiostrontium was given orally) pre-feeding of stable Sr was followed by a significantly higher retention of radiostrontium (Müller, 1963).

DILUTION OF STRONTIUM WITH CALCIUM CARRIER

Taking into account the similarity of the metabolic pathways of Sr and Ca, it is sensible to use a Ca carrier in order to reduce the skeletal retention of radiostrontium. The actual experimental data (MacDonald, 1956; Catsch and Melchinger, 1959a; Kawin, 1959; Carlqvist and Nelson, 1960; Nelson et al., 1963; Ogawa et al., 1961b, 1964), however, are at variance. No effect at all, or a slightly reduced as well as an enhanced deposition of parenterally administered radiostrontium have all been reported. An explanation of these discrepancies may be the different doses of Ca used by these authors, since Catsch and Melchinger (1959a) have shown that the dependence of the efficacy upon the dosage of Ca bears exactly the same characteristic already encountered in the case of isotopic carrier.

One should expect that disturbances of Ca homeostasis and measures leading to a positive Ca balance or a mobilization of skeletal Ca, might be reflected, to some extent, by a similar response of Sr. In keeping with this, it has been demonstrated (Thompson and Palmer, 1960; Cohn et al., 1961; Harrison, 1964; Palmer and Thompson, 1964) that pre-feeding of a high Ca diet distinctly lowers the retention of a single parenteral dose and of chronically ingested radiostrontium. Particularly instructive and pertinent to this point are the observations of Spencer et al. (1961b) in humans. The data, presented in Fig. 2, show an inverse relationship between retention of radiostrontium and the urinary Ca level. But we also see that a high intake of Ca does not lead necessarily and per se to a reduced retention of radiostrontium,

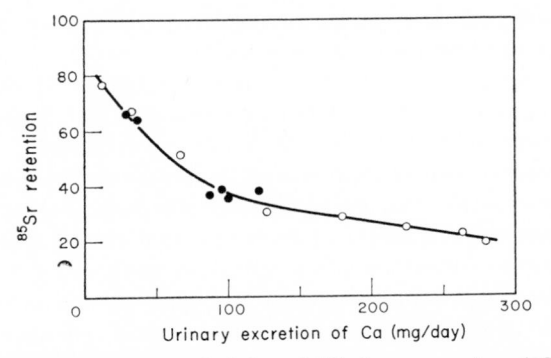

FIG. 2. Retention of intravenously injected ^{85}Sr in man on low (150 mg), ●, and high (1,550 mg), ○, Ca intake. Experimental results of Spencer et al. (1961b).

but only if it is accompanied by an elevated excretion of Ca. Obviously, the physiological condition of the Ca metabolism is the relevant and decisive factor.

Reference must also be made to the studies of Thompson and Palmer (1960); the results are presented in Fig. 3. The retention of radiostrontium in rats can be approximated satisfactorily by a power function. The graph shows that the coefficient of the function, i.e., the initial retention, depends upon the Ca level of the diet, whereas the exponent of the power function, representing the speed of elimination, remains virtually constant. This finding, that the elimination of radiostrontium, once it is fixed in the bone, is unaffected by supplementation of Ca, is at variance with observations in humans (Spencer *et al.*, 1965). Delayed oral or parenteral administration of Ca was shown to enhance the urinary excretion of radiostrontium at least within the first two weeks. The explanation, proffered in discussing the efficacy of stable strontium, applies to this case.

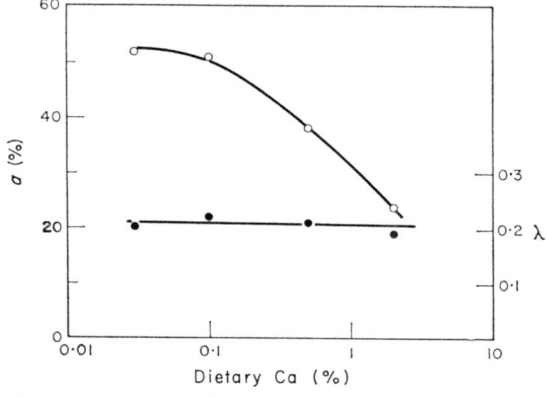

FIG. 3. Retention and elimination of intraperitoneally injected ^{90}Sr in rats maintained on diets with varying content of calcium. Experimental data of Thompson and Palmer (1960). Power function is given by $R_t = at^{-\lambda}$, where t is the time in days. $\bigcirc$, curve for $a\,(\%)$; $\bullet$, curve for λ.

LOW PHOSPHORUS DIET

A profound influence on the turnover of bone mineral is exerted by diets deficient in phosphate, the main consequences being hypophosphataemia, hypercalcaemia and a high renal clearance of Ca. The, now classical, studies of Copp *et al.* (1951) and Ray *et al.* (1956) showed that a low phosphorus regimen prevents the skeletal retention of radiostrontium as well as enhances its elimination from the bone. Meanwhile these findings have been substantiated and supplemented (Ito *et al.*, 1958b; Putten, 1962; Kostial *et al.*, 1963; Kriegel *et al.*, 1963). In order to exert an influence upon the turnover of radiostrontium, a reduction of the phosphorus intake in itself is not sufficient, but rather a high dietary Ca/P ratio is an essential prerequisite. Further, it has been ascertained that diets with a slightly lowered content of phosphorus are effective without impairing the bone structure. Even variations of the phosphorus concentration within the physiological range are reflected in correspondingly changed excretion rates of radiostrontium (Fig. 4). From the

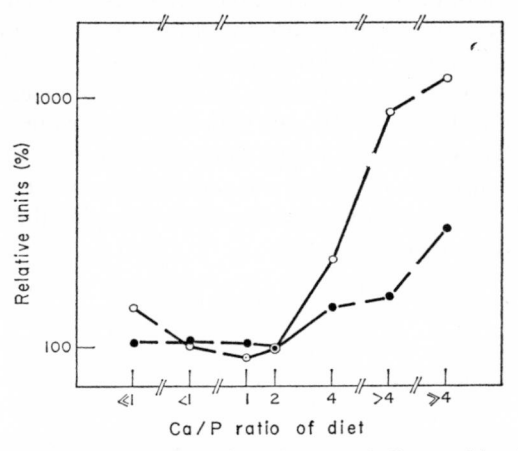

FIG. 4. Excretion of intravenously injected ^{89}Sr as influenced by the dietary Ca/P ratio The excretion in the control (Ca/P = 2) was arbitrarily set at 100%. Experimental data of Clark *et al.* (1965). ○, urinary excretion; ●, total excretion.

practical point of view, it is noteworthy that the effect of a low phosphorus intake is relatively independent of the time when treatment is initiated, and that even a transient restriction of phosphorus intake is sufficient to bring about a significant diminution of the body burden. Referring to the last point, the data of Uchiyama (1958) and Uchiyama and Ukita (1960) are most impressive: oral administration, for 6 days, of aluminium citrate or related compounds, which render the phosphates in the intestine insoluble, reduced the skeletal radiostrontium by 50%.

So far no data are available about the efficacy of low phosphorus diets in man. There are, of course, legitimate doubts about the practicability of such a regimen since a restriction of phosphorus intake can only be achieved by simultaneous reduction of dietary protein. One should, further, keep in mind the difference in bone metabolism between man and small rodents, and it is questionable that a similar effect will be obtained in man. In spite of these objections, appropriate studies in humans are worthwhile and desirable, since in experimental animals phosphorus depletion proved to be the one and only measure by which the long-term turnover of radiostrontium can be influenced to a higher extent.

PHYSIOLOGICAL APPROACH

Among numerous calciuric agents that have been tested and, as a rule, proved to be ineffective, only a few are worth dealing with in more detail. For parathyroid hormone, the experimental data (Copp and Greenberg, 1947; Bacon *et al.*, 1956; Gorodetzkiy *et al.*, 1958; Catsch and Melchinger, 1959b; Schmid, 1960; Della Rosa *et al.*, 1961; Schmid and Zipf, 1961; Ogawa *et al.*, 1961a) are ambiguous. No effect at all and an elevated as well as diminished retention of radiostrontium have all been reported. These discrepancies may be partly due to the fact that the effect to be measured is only marginal. It is also known that the efficacy of commercial hormone preparations in mobilizing Ca from bone may be subject to a pronounced variability. Finally, experiments with ^{45}Ca give evidence that the parathyroid exerts an influence on the Ca from sites that are not in close contact with the easily exchangeable Ca pool (Elliot and Talmage, 1958; Jeffay and Bayne, 1964). According to studies of Spreng (1967) in our laboratory, radiostrontium seems to respond in a similar manner. In this study the parathyroid hormone was given over 10 days, the treatment starting at different times following the injection of ^{85}Sr. As can be seen from Fig. 5, a slight, but statistically significant, diminution of the body burden was achieved only when treatment was initiated on the 24th or 47th day. Massive oral doses of vitamin A, which are known to accelerate the processes of bone resorption, were effective over the whole period. A mode of action, different from that of the parathyroid hormone, is thus indicated.

Repeated doses of ammonium chloride are followed by a metabolic acidosis and a demineralization of bone. Its efficacy with regard to radiostrontium is

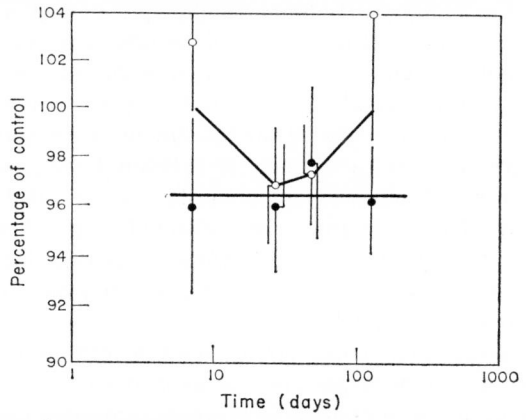

FIG. 5. Efficacy of parathyroid hormone, ○, 10 times 1 USP i.u./g/day, and vitamin A, ●, 10 times 200 i.u./g/day, in removing ^{85}Sr from the skeleton of rats. The treatment was initiated at different time intervals after the intraperitoneal injection of ^{85}Sr. Fiducial limits for $P = 0.05$. Experimental data of Spreng (1967).

equivocal. Whereas in humans a considerable increase of the urinary excretion has been reported (Spencer *et al.*, 1958, 1965), at least when given within the first two weeks, studies with mice (Putten, 1962) and dogs (Della Rosa *et al.*, 1961) yielded negative results.

As early as 1909 it was shown that magnesium is a potent calciuric agent (Mendel and Benedict, 1909). This also applies to radiostrontium (Catsch and Melchinger, 1959a; Nelson *et al.*, 1963; Clark *et al.*, 1964; Annenkov, 1965). Magnesium salts, provided administration is early, are nearly twice as effective as equimolar amounts of isotopic carrier, but unfortunately only in high and near-toxic doses. As to the delayed treatment, several authors (Rubanovskaya and Ushakova, 1957; Catsch and Melchinger, 1959a; Annenkov, 1965) failed in demonstrating a substantial effect. On the other hand, studies by Clark *et al.* (1962, 1964, 1965) gave clear evidence that supplementation of magnesium increases the urinary excretion of radiostrontium by a factor of about 1·5.

The major fraction of alkaline–earth metal ions (comprising more than 90%) filtrated by the kidney is subject to tubular re-absorption. Obviously, attempts aiming at the interruption of this physiological "recycling" of radiostrontium by blocking the processes of re-absorption are promising. Many diuretics, including diamox and mercurial compounds, have been tested, although with negative results (MacDonald *et al.*, 1957; Catsch, 1961b; Ogawa *et al.*, 1964; Zander-Principati and Kusma, 1964a, b). Daily administration of sodium salicylate (its mode of action is thought to be a tubular block) enhances the urinary excretion of radiostrontium by a factor that during the first days following the injection of radiostrontium equals 3·5, but gradually drops to lower figures of the order of 1·5 (Smith and Bates, 1965). The efficacy of a slow infusion of sodium sulphate in dogs has been likewise imputed to an impaired tubular re-absorption (Walser *et al.*, 1961).

Pyridoxal, pyridoxine, and pyridoxic acid exhibit a pronounced saliuretic action. Interestingly enough, the urinary excretion of Sr is affected to a higher degree than that of Ca, and administration daily for 14 days of pyridoxal to rats, pre-fed with stable Sr, reduced the Sr content of the spongiosa by 15% and of the compacta by 10% (Schmid, 1965; Schmid and Zipf, 1965; Zipf *et al.*, 1965). The mode of action is not yet fully understood. It has still to be proven that carrier-free radiostrontium is affected by pyridoxal to the same degree. In view of the well known differences in pyridoxine metabolism in various animal species, it is also questionable whether these results are valid for other mammals and human beings.

The same approach, i.e., an interference with the physiological recycling of radiostrontium, can in principle also be applied to its excretion into the intestine. Since this excretory pathway, however, is only of minor importance and since a small fraction of the excreted amount is re-absorbed, one ought not to expect a very marked effect, and, in fact, experiments pursuing this line have had virtually negative results (Volf, 1960a, b, 1961, 1963). It would be beyond the scope of this review to enumerate the vast number of compounds that have been tested for their ability to prevent the enteral absorption of radiostrontium. It may suffice to mention here that various sulphates, in particular barium sulphate (Volf, 1960a, b, 1964a, b, 1965; Volf and Roth, 1965), as well as calcium phosphate (Bruce, 1963b; Volf, 1960a, b, 1965) and sodium alginate (Paul et al., 1964; Waldron-Edward et al., 1965) have to be considered as antidotes of choice.

It is somewhat surprising that hitherto no one has taken the trouble to repeat the experiments of Setälä (1962) and Setälä et al. (1964), who claimed to have achieved a drastic diminution of the radiostrontium body burden by repeated doses of pilocarpine, an agent that promotes the secretion of saliva and stimulates the excretory activity of the intestinal mucosa.

CHELATING AGENTS

Let us now turn to the action of those compounds that have been shown to be very effective with most of the potentially hazardous radionuclides, but have disappointed when applied to radiostrontium, namely, the chelating agents. This unfavourable state of affairs has been reviewed and discussed by several authors (Schubert, 1958; Heller and Catsch, 1959; Heller, 1963; Catsch, 1961a, 1964; Rubin, 1963; Smith and Bates, 1965) so that only relatively brief comment is required here. The chelation of alkaline-earth metal ions obeys the rule that the stability constant is an inverse function of the ionic radius. Consequently, in the presence of a high and (owing to its homeostatic control) virtually constant physiological Ca concentration the fraction of chelated Sr will be negligible. In keeping with this, the efficacy of most ligands so far tested has been found to be nought or poor. The maximum effect which can be achieved under optimal conditions, i.e., the simultaneous administration of high doses of ligands with relatively small differences in their affinity towards Ca and Sr ions, has been a reduction of skeletal retention by 15 to 25% only. The following should be mentioned as effective chelators: citrate (Catsch, 1957; Akiya and Uchiyama, 1958; Catsch and Melchinger, 1959b; Ito et al., 1958a, b; Carlqvist and Nelson, 1960; Nelson et al., 1963; Ogawa et al., 1964), tricarballylate (Ito et al., 1958a, c), sodium thiosulphate (Catsch and Melchinger, 1959a, b; Ogawa et al., 1964), low-molecular-weight metaphosphates (Spencer et al., 1958; Catsch and Melchinger, 1959b; Semenov and Tregubenko, 1960), 2,2'-bis[di(carboxymethyl)amino]diethyl sulphide (Catsch and Melchinger, 1959b), and 2,2'-bis[dicarboxymethyl]-amino]diethyl ether (BADE) (Catsch and Melchinger, 1959b). It may be added that BADE has been tested with positive results in humans, although its administration was started 14 days following the injection of [85]Sr (Spencer et al., 1962). It failed, however, to influence the urinary excretion in a case of accidental internal contamination that occurred several months before treatment was initiated (G. Möhrle, personal communication).

The comprehensive compilation of stability constants by Sillén and Martell (1964) shows that there are quite a few ligands, whose affinity towards alkaline earth metals exhibits an anomalous behaviour; likewise ligands have been reported to bind Sr equally strongly or even better than Ca. Taking into account, however, that the chelation properties of these particular compounds are, as a rule, extremely weak and that at a physiological pH the competition of hydroxonium ions may become more relevant than that of Ca, gross effects are unlikely and, as a matter of fact, have not been observed in experimental animals. However, the possibility of developing ligands that (owing to an appropriate number of donor atoms and a favourable spatial configuration) may bind Sr preferentially, cannot be excluded. All attempts either to find or to design and synthetize such ligands have failed so far (Catsch and Melchinger, 1959b; Kroll and Gordon, 1960; Catsch, 1964).

For the sake of completeness, quite a few of the more recent studies should be quoted explicitly. The stability constants of the Sr and Ca chelates of 1,2-diaminopropanetetra-acetic acid are (according to Smith, 1959) $10^{10.7}$ and $10^{10.4}$, respectively. In spite of this favourable ratio, this ligand proved to be ineffective in rats (Catsch and Melchinger, 1959b), so that this negative result throws serious doubts on the validity of the aforementioned figures. This has been confirmed by Grimes et al. (1963); these authors determined the stability of the Ca chelate as $10^{11.5}$ and of the Sr chelate as $10^{9.6}$.

It was claimed by Ogawa et al. (1961b, 1964) and Richards et al. (1961) that tetracycline lowers the skeletal retention of radiostrontium and this result was ascribed to the chelating properties of this compound. The actual affinity of tetracycline for Ca, however, is distinctly higher than for Sr (Maxwell et al., 1963). Further, other investigators (Nakatsuka et al., 1959; Smith, 1963; MacDonald et al., 1964) failed to confirm the earlier positive results. In our own studies (Catsch, 1963) tetracycline was found to be effective only if it was administered for several days before the injection of radiostrontium. It may be assumed tentatively that insoluble Ca chelates are formed on the surface of the apatite crystals and the exchange capacity of bone is lowered.

The positive results of Lindenbaum and Fried (1958) with rhodizonates, which precipitate Sr in the presence of Ca, were not confirmed by other authors (Kriegel and Melchinger, 1959; Dooronbekov et al., 1960; Volf, 1960a, b; Nelson et al., 1963). A drastic reduction of radiostrontium deposition in bone by more than 60% has been found by Kriegel and Melchinger (1959) for Diamond Fast Blue. Bruce (1963a) was unable to substantiate this finding, and pointed out that the positive results are most likely due to methodic artifacts, caused by the low solubility of the dye.

It is known that, for ion exchangers and under certain conditions the binding affinity for Ca and Sr can be reversed. However, zirconium citrate

(Schubert and Wallace, 1950; Cohn and Gong, 1953; Catsch and Melchinger, 1959b; Volf, 1960a, b; Zander-Principati and Kusma, 1964a) and condensed polyphosphates (Catsch, 1957; Catsch and Melchinger, 1959b; Tregubenko and Semenov, 1960), which *in vivo* presumably act as soluble ion exchangers, show an effectiveness no greater than that of the aforementioned ligands. We should also add that the therapeutic index of both compounds is relatively small. At this point, the data of Zander-Principati and Kusma (1960, 1964b) should be mentioned in order to stress a generally important point. These authors demonstrated that even a relatively small diminution of radiostrontium deposition in bone, achieved by early treatment with zirconium citrate, was followed by a correspondingly lessened incidence of osteosarcomas.

Finally, I want to present data of a screening test we have just performed in the very last months. The choice of compounds was based on the idea that ligands possessing a cyclic or heterocyclic configuration might be able to proffer a favourable spatial grouping of donor atoms. As can be seen from Table I, some of the compounds tested proved indeed to be effective, unfortunately, in the wrong direction, i.e., they potentiate the deposition of radiostrontium. So far we are not able to suggest a sound explanation for these results, and it is not yet clear whether chelation is involved at all or whether the effect has to be ascribed to an interference with excretory functions. It is noteworthy that the efficacy of other ligands, such as kojic acid and pyridine-2,6-dicarboxylic acid, is not inferior to that of BADE. A distinctly higher skeletal retention of radiostrontium, by the way, was also obtained by L-ascorbic and D-isoascorbic acids (Fig. 6).

Interesting as the search for ligands with a preferential affinity for Sr might be from the theoretical point of view, doubts may be expressed in principle about the practical value of this approach. Experience with other metal ions has shown clearly that really effective chelating agents must have a conditional stability constant (defined in this context as the ratio K_{ML}^{M}/K_{CaL}^{Ca}) of the order

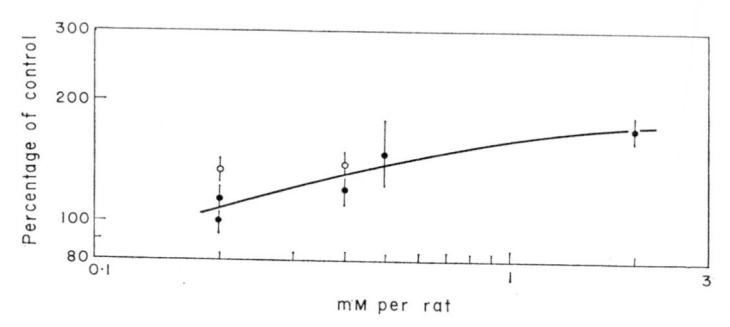

FIG. 6. Retention of intraperitoneally injected [85]Sr by the skeleton of rats as influenced by varying doses of L-ascorbic, ●, and D-isoascorbic, ○, acids, which were administered intraperitoneally immediately after [85]Sr. Fiducial limits for $P = 0.05$.

L

TABLE 1

Effect of different compounds on the retention of strontium-85 by the skeleton of rats (% of control)

The compounds were injected intraperitoneally within a few minutes after the intraperitoneal administration of ^{85}Sr. Six animals per group. The figures in the brackets indicate the statistical significance (P values)

β-Truxinic acid 100 μM: 99% (0·8) 400 μM: 153% ($<10^{-5}$)	Pyridine-2-carboxylic acid 100 μM: 109% (0·4) 200 μM: 123% (0·05)
α-Truxillic acid 100 μM: 78% (10^{-4}) 400 μM: 90% (0·3)	Pyridine-2,3-dicarboxylic acid 100 μM: 90% (0·1) 200 μM: 99% (0·9)
Cyclobutane-1,1-dicarboxylic acid 100 μM: 112% (0·2) 200 μM: 179% (10^{-3})	Pyridine-2,6-dicarboxylic acid 100 μM: 75% ($5 . 10^{-3}$) 200 μM: 68% ($<10^{-4}$)
Pyrrole-2,5-dicarboxylic acid 100 μM: 108% (0·1) 200 μM: 148% (10^{-2})	Pyrazine-2,3-dicarboxylic acid 100 μM: 80% ($<10^{-4}$) 400 μM: 79% (10^{-3})
Pyrazole-3,5-dicarboxylic acid 100 μM: 81% (0·07) 200 μM: 112% (0·2)	Piperazine-2,6-dicarboxylic acid 50 μM: 113% (0·04) 200 μM: 132% ($<10^{-4}$)
Kojic acid 200 μM: 73% ($<10^{-4}$)	Diethylmalonic acid 200 μM: 100% (0·9)
Chelidonic acid 50 μM: 91% (0·02) 200 μM: 90% (0·02)	BADE 100 μM: 74% (10^{-4})

of at least 10^4. Owing to the high Ca concentration in the hydration shell of apatite, even higher conditional stabilities are needed to mobilize radiostrontium fixed in the bone. It is extremely unlikely that further research will succeed in meeting this rigorous demand for efficacy with Sr.

FUTURE POSSIBILITIES

In summary we can say that although there are quite a few positive results and promising starting points, techniques so far investigated cannot be considered as real remedies for radiostrontium poisoning. One can go even further and say that future studies will prove unavailing as long as they are aimed solely at finding the one and only pharmaceutical or procedure. Future work ought to follow a suggestion of Loutit (1963) and main emphasis should be placed on studies of the additivity of factors with different sites or modes of action. This task is, of course, by no means a new one, and some pertinent experimental results are already available.

For instance, the effectiveness of demineralizing agents, such as the parathyroid hormone, might be masked and attenuated by the physiological recycling of mobilized radiostrontium. If this should be so one might expect that the combined use of compounds, though ineffective in mobilizing radiostrontium from bone *per se*, could act as a "trap" in the extracellular phase, and thus potentiate the efficacy of the parathyroid hormone. In keeping with this assumption positive results have been reported for the combined treatment by parathyroid hormone and citrate (Elliot and Talmage, 1958; Takeuchi *et al.*, 1964), EDTA (Graul *et al.*, 1958), condensed polyphosphates (Catsch and Melchinger, 1959b), pilocarpine (Setälä *et al.*, 1964) and sodium trimetaphosphate (Catsch and Melchinger, 1959b), respectively. An almost complete additivity has also been obtained in humans with the administration of calcium gluconate and ammonium chloride (Spencer *et al.*, 1961a), whereas supplementation with the latter failed in potentiating the efficacy of a phosphorus deficient regimen (Putten, 1962). On the other hand, Rogacheva (1961) observed a higher effect of phosphorus restriction, if sodium trimetaphosphate was administered in addition. A striking example of the synergistic action of two different principles is provided by the simultaneous use of chelation and isotopic dilution. Although each of the factors alone prevents the retention of radiostrontium by approximately 25% only, their combination, i.e., the administration of Sr–BADE, lowers it by 50% (Catsch, 1961b, 1962). Finally, brief mention can be made of the results of an experiment, just performed in our laboratory, concerned with the additivity of three different factors, vitamin A, pilocarpine and Ca–BADE. As can be seen from Table II, none of these treatments, pilocarpine included, was found to be effective.

TABLE II

Effectiveness of different treatments in removing strontium-85 from the skeleton. Daily administration from 10th till 19th day
Six rats per group

No.	Treatment	Percentage of control
1	Vitamin A (200 I.U./g p.o.)	100
2	Na_2 Ca-BADE (250 μM i.p.)	98
3	Pilocarpine (0·1 mg/g s.c.)	99
4	1 + 2	101
5	1 + 3	99
6	2 + 3	100
7	1 + 2 + 3	100

In designing and evaluating multifactorial experiments, one ought to be very cautious. Studies of this kind may be extremely susceptible to the order in which the different agents are administered, which in turn may lead to a failure to detect additive effects or even to the production of an antagonistic action. The fact that these investigations must necessarily be extremely laborious and tedious, should not deter us from the only approach by which more satisfying results can be achieved.

REFERENCES

Akiya, S., and Uchiyama, M. (1958). *Seikagaku*, **28**, 154–158.
Annenkov, B. N. (1965). *Radiobiologiya*, **5**, 620–621.
Bacon, J. A., Patrick, H., and Hansard, S. L. (1956). *Proc. Soc. exp. Biol. Med.* **93**, 349–351.
Bartley, J. C., and Reber, E. F. (1961). *J. Nutr.* **75**, 1754–1762.
Bruce, R. S. (1963a). "Diagnosis and Treatment of Radioactive Poisoning", p. 162. International Atomic Energy Agency, Vienna.
Bruce, R. S. (1963b). *Nature, Lond.* **199**, 1107–1108.
Carlqvist, B., and Nelson, A. (1960). *Acta radiol.* **54**, 305–315.
Catsch, A. (1957). *Naturwissenschaften*, **44**, 94.
Catsch, A. (1961a). *Fedn. Proc. Fedn. Am. Soc. exp. Biol.* **20,** suppl. **10**, 206–219.
Catsch, A. (1961b). *Int. J. Radiat. Biol.* **4**, 75–83.
Catsch, A. (1962). *Atomkernenergie*, **7**, 65–70.
Catsch, A. (1963). *Nature, Lond.* **197**, 302.
Catsch, A. (1964). "Radioactive Metal Mobilization in Medicine". Charles C. Thomas, Springfield, Illinois.
Catsch, A., and Melchinger, H. (1959a). *Strahlentherapie*, **108**, 63–72.
Catsch, A., and Melchinger, H. (1959b). *Strahlentherapie*, **109**, 561–572.
Clark, I., and Smith, M. R. (1962). *Proc. Soc. exp. Biol. Med.* **109**, 135–139.
Clark, I., Rivera-Cordero, F., and Cohn, S. H. (1965). *Proc. Soc. exp. Biol. Med.* **118**, 879–882.

Clark, I., Gusmano, E. A., Nevins, R., and Cohn, S. H. (1964). *Proc. Soc. exp. Biol. Med.* **116**, 984–987.

Cohn, S. H., and Gong, J. K. (1953). *Proc. Soc. exp. Biol. Med.* **83**, 550–553.

Cohn, S. H., Nobel, S., and Sobel, A. E. (1961). *Radiat. Res.* **15**, 59–69.

Cohn, S. H., Spencer, H., Samachson, J., Feldstein, A., and Gusmano, E. A. (1962). *Proc. Soc. exp. Biol. Med.* **110**, 526–528.

Copp, D. H., and Greenberg, D. M. (1947). MDDC–1001, 11–26.

Copp, D. H., Hamilton, J. G., Jones, D. C., Thompson, D. M., and Cramer, C. (1965). Trans. 3rd Conf. Metabol. Interrel., pp. 226–252.

Della Rosa, R. J., Smith, F. A., and Stannard, J. N. (1961). *Int. J. Radiat. Biol.* **3**, 557–578.

Dooronbekov, Zh., Kasatkin, Yu., and Fedorov, N. A. (1960). *Med. Radiol.* **5**, 876–879.

Elliot, J. R., and Talmage, R. V. (1958). *Endocrinology*, **62**, 709–716.

Gorodetzkiy, A. A., Sivachenko, T. P., Khomutovskiy, O. A., and Ryabova, E. Z. (1958). "Deystviye Ioniziruyshchikh Izlucheniy na Zhivotnyi Organism", pp. 31–37. Medgiz, Moscow.

Graul, E. H., Hundeshagen, H., and Schömer, W. (1958). *Strahlentherapie*, **106**, 391–396.

Grimes, J. H., Huggard, A. J., and Wilford, S. P. (1963). *J. inorg. nucl. Chem.* **25**, 1225–1238.

Harrison, G. E. (1964). *Int. J. Radiat. Biol.* **8**, 177–186.

Harrison, G. E., Jones, H. G., and Sutton, A. (1957). *Br. J. Pharmac. Chemother.* **12**, 336–339.

Heller, H.-J. (1963). "Diagnosis and Treatment of Radioactive Poisoning", pp. 347–373. International Atomic Energy Agency, Vienna.

Heller, H.-J., and Catsch, A. (1959). *Strahlentherapie*, **109**, 464–483.

Ito, Y., Tsurufuji, S., Shikita, M., and Ishibashi, S. (1958a). *Chem. pharm. Bull Tokyo*, **6**, 287–290.

Ito, Y., Tsurufuji, S., Shikita, M., and Matsushima, Y. (1958b). *J. pharm. Soc. Japan*, **78**, 76–82.

Ito, Y., Tsurufuyi, S., Ishidate, M., Tamura, Z., and Takita, H. (1958c). *Chem. pharm. Bull. Tokyo*, **6**, 34–38.

Jeffay, H., and Bayne, H. R. (1964). *Am. J. Physiol.* **206**, 415–418.

Kawin, B. (1959). *Experientia*, **15**, 313–314.

Kostial, K., Lutkić, A., Gruden, S., Vojvodić, S., and Harrison, G. E. (1963). *Int. J. Radiat. Biol.* **6**, 431–439.

Kriegel, H., and Melchinger, H. (1959). *Atompraxis*, **5**, 425–430.

Kriegel, H., Kollmer, W. E., and Weber, E. (1963). *Int. J. Radiat. Biol.* **7**, 289–299.

Kroll, H., and Gordon, M. (1960). *Ann. N. Y. Acad. Sci.* **88**, 341–352.

Lindenbaum, A., and Fried, J. F. (1958). ANL–6093, 97–99.

Loutit, J. F. (1963). "Diagnosis and Treatment of Radioactive Poisoning", p. 167. International Atomic Energy Agency, Vienna.

MacDonald, N. S. (1956). ANL–5584, 83–90.

MacDonald, N. S., Ibsen, K. H., and Urist, M. R. (1964). *Proc. Soc. exp. Biol. Med.* **115**, 1125–1128.

MacDonald, N. S., Noyes, P., and Lorick, P. C. (1957). *Am. J. Physiol.* **188**, 131–134.

Maxwell, D. C., Smith, P. J. A., and Wilford, S. P. (1963). *Nature, Lond.* **198**, 577–578.

Mendel, L. B., and Benedict, S. R. (1909). *Am. J. Physiol.* **25**, 1–5.

Müller, W. A. (1963). *Atompraxis*, **9**, 408–411.

Nakatsuka, M., Arantani, H., Tokoki, K., Tanaka, Y., and Ishii, K. (1959). *Folia pharmac. jap.* **55**, 144–146.

Nelson, A., Rönnbäck, C., and Rosén, L. (1963). *Acta radiol. ther. Phys. Med.* **1**, 129–139.

Ogawa, E., Fukuda, R., Suzuki, S., and Shibata, K. (1961a). *Gunma J. med. Sci.* **10**, 109–116.

Ogawa, E., Fukuda, R., Suzuki, S., and Shibata, K. (1961b). *Gunma J. med. Sci.* **10**, 117–120.

Ogawa, E., Suzuki, S., Fuji, S., Honma, T., and Tsuzuki, H. (1964). *Gunma J. med. Sci.* **13**, 214–220.

Palmer, R. F., and Thompson, R. C. (1964). *Am. J. Physiol.* **207**, 561–566.

Paul, T. M., Waldron-Edward, D., and Skoryna, S. C. (1964). *Canad. med. Ass. J.* **91**, 553–557.

van Putten, L. M. (1962). *Int. J. Radiat. Biol.* **5**, 471–476, 477–484.

Ray, R. D., Stedman, D. E., and Wolff, N. K. (1956). *J. Bone J. Surg.* **38A**, 637–654.

Richards, Y., Lowenstein, J. M., Philips, J. W., and Armitage, C. (1961). *Proc. Soc. exp. Biol. Med.* **107**, 550–551.

Rogacheva, S. A. (1961). *In* "Raspredeleniye, Biologicheskoye Deystviye u Migratsiya Radioaktivnykh Isotopov", pp. 123–135. Medgiz, Moscow.

Rubanovskaya, A. A., and Ushakova, V. F. (1957). *Mater. Toks. Radioakt. Vesh.* **1**, 197–202.

Rubin, M. (1963). *In* "Transfer of Calcium and Strontium across Biological Membranes" (Wasserman, R. H., ed.), pp. 25–46. Academic Press, New York.

Schmid, A. (1960). *Arch. exp. Path. Pharmak.* **240**, 35.

Schmid, A. (1965). *Arzneimittel Forsch.* **15**, 28–30.

Schmid, A., and Zipf, K. (1961). *Biochem. Z.* **333**, 529–533.

Schmid, A., and Zipf, K. (1965). *Arch. exp. Path. Pharmak.* **250**, 282.

Schubert, J. (1958). *Atompraxis*, **4**, 393–395.

Schubert, J., and Wallace, H. (1950). *J. biol. Chem.* **183**, 157–166.

Schulert, A. R., Peets, E., Laszlo, D., Spencer, H., Charles, M. L., and Samachson, J. (1959). *Int. J. appl. Radiat. Isotopes*, **4**, 144–153.

Semenov, D. I., and Tregubenko, I. P. (1960). *Trudÿ Inst. Biol., Sverdlovsk*, **12**, 20–56.

Setälä, K. (1962). *Naturwissenschaften*, **49**, 302–303.

Setälä, K., Lindroos, B., and Kuikka, A. O. (1964). *Strahlentherapie*, **123**, 545–562.

Sillén, L. G., and Martell, A. E. (1964). "Stability Constants of Metal-Ion Complexes." The Chemical Society, London.

Smith, H. (1963). *Int. J. Radiat. Biol.* **6**, 197–198.

Smith, H., and Bates, T. H. (1965). *Nature, Lond.* **207**, 799–801.

Smith, R. L. (1959). "The Sequestration of Metals." Chapman and Hall, London.

Spencer, H. (1963). "Diagnosis and Treatment of Radioactive Poisoning", pp. 145–155. International Atomic Energy Agency, Vienna.

Spencer, H., Feldstein, A., and Samachson, J. (1961a). *Proc. Soc. exp. Biol. Med.* **108**, 308–312.

Spencer, H., Feldstein, A., and Samachson, J. (1962). *J. Lab. clin. Med.* **59**, 445–455.

Spencer, H., Li, M., and Samachson, J. (1961b). *J. clin. Invest.* **40**, 1339–1345.

Spencer, H., Samachson, J., Hardy, E. P., and Rivera, J. (1965). *Radiat. Res.* **25**, 695–705.

Spencer, H., Samachson, J., Kabakow, B., and Laszlo, D. (1958). *Clin. Sci.* **17**, 291–301.

Spreng, P. (1967). *Nature, Lond.* In press.

Takeuchi, T., Seki, M., Takagi, C., Enomoto, Y., Matsuo, Y., Kantake, N., Mashimo, T., Ishimochi, R., and Yoneyama, T. (1964). *Acta path. jap.* **14**, 405–412.

Thompson, R. C., and Palmer, R. F. (1960). *Am. J. Physiol.* **199**, 94–102.

Tregubenko, I. P., and Semenov, D. I. (1960). *Trudÿ Inst. Biol. Sverdlovsk*, **12**, 5–33.

Uchiyama, M. (1958). *J. pharm. Soc. Japan*, **78**, 255–257.

Uchiyama, M., and Ukita, T. (1960). *Chem. pharm. Bull., Tokyo*, **8**, 384–388.

Volf, V. (1960a). *Physiologia bohemoslov.* **9**, 423–427.

Volf, V. (1960b). *Physiologia bohemoslov.* **9**, 428–434.

Volf, V. (1961). *Physics Med. Biol.* **6**, 278–294.

Volf, V. (1963). "Diagnosis and Treatment of Radioactive Poisoning", pp. 131–141. International Atomic Energy Agency, Vienna.

Volf, V. (1964a). *Experientia*, **20**, 626–628.

Volf, V. (1964b). *Int. J. Radiat. Biol.* **8**, 509–511.

Volf, V. (1965). *Experientia*, **21**, 571–572.

Volf, V., and Roth, Z. (1965). *Acta radiol. ther. Phys. Biol.* **3**, 216–228.

Waldron-Edward, D., Paul, T. M., and Skoryna, S. C. (1965). *Nature, Lond.* **205**, 1117–1118.

Walser, M., Payne, J. W., and Browder, A. A. (1961). *J. clin. Invest.* **40**, 234–242.

Wasserman, R. H., Comar, C. L., and Papadopoulou, D. (1957). *Science*, **126**, 1180–1182.

Zander-Principati, G. E., and Kusma, J. F. (1960). *Radiat. Res.* **13**, 489–495.

Zander-Principati, G. E., and Kusma, J. F. (1964a). *Hlth. Phys.* **10**, 473–477.

Zander-Principati, G. E., and Kusma, J. F. (1964b). *Int. J. Radiat. Biol.* **8**, 427–437.

Zipf, K., Schenkel, R., and Schmid, A. (1965). *Arch. exp. Path. Pharmak.* **251**, 190–192.

Discussion

QUESTION: Could it be that osteoporosis was induced by underfeeding of the rats? Do you think that the pyridoxal or the vitamin A treatment might work through this mechanism?

ANSWER: Vitamin A is known to induce demineralization of bone. The action of pyridoxal is not known; the metabolism of this drug is, of course, quite different in dogs and human beings on the one hand and in small rodents on the other.

H. SPENCER: You mentioned various substances that are effective in experimental animals and suggested that they be tried in human subjects. Our experience leads to the following conclusions or speculations: (*a*) so far, we have found administration of stable Sr more effective (for the removal of ^{90}Sr) than either stable Ca or Mg; (*b*) removal of Sr in saliva is not very promising in man. Studies with ^{45}Ca and ^{85}Sr indicate that the concentration of Sr in saliva is rather low—about 0·1 of the level in urine—and that it is not practicable to stimulate the secretion of saliva sufficiently to give useful results in this regard; (*c*) we have tried sodium citrate in six subjects, at different dosage levels, but without effect.

High Strontium Diet and Radiostrontium Retention

DANUTA DEPCZYK, TOMISLAW DOMANSKI
AND JULIAN LINIECKI

Institute of Occupational Medicine, Lodz, Poland

SUMMARY

High Sr diet reduces the bodily retention of [85]Sr by about the same proportion in the rat and in man. Several possible explanations of this observation are discussed.

INTRODUCTION

It is extremely difficult—if not impossible—to change the retention of radiostrontium when it becomes fixed in the skeleton. Therefore, measures that are capable of preventing or reducing this fixation are of intrinsic interest.

Cohn *et al.* (1961) described a fivefold reduction of radiostrontium retention by young weanling rats, when they were fed a diet containing high amounts of inactive Sr for a few days preceding a single parenteral administration of [85]Sr.

A further investigation has been made of the effect, in rats, of a high Sr diet on the retention of a radioactive dose of Sr including the period on the high Sr diet and the degree of Sr supplementation of the diet, the route of administration of radioactive dose, the P content of the diet and the age of the animals.

In one age group [133]Ba was used as well as [85]Sr to compare the effect of the high Sr diet on the retention of the two tracers.

A preliminary experiment has also been made on an adult male to compare the effect of this therapy in rats and man.

METHODS

Rat experiments

A total of 144 inbred albino rats of both sexes were used. The animals were 2·5 months old at the time of [85]Sr administration; those given ^{90}Sr were 1·5, 7·5 and 13·5 months old. [133]Ba was injected at 7·5 months of age only. [85]Sr and ^{90}Sr were carrier-free, the latter in equilibrium with ^{90}Y and the activity administered was 2 and 5 μCi, respectively. About 3 μCi of high specific activity [133]Ba was used. All nuclides were injected into the tail vein in 0·5 ml of the acetate buffer solution at pH 4. [85]Sr was also given as a single dose with food to animals previously fasted over 12 h.

The basic diet was prepared of milled barley (26%), whole wheat flour (24%), whole rye flour (24%), casein (13%), margarine (3%) and beet sugar (10%). This diet was supplemented with 5,000 units of vitamin D_2/kg and contained 1·3 and 1·4 mM % of Ca and P respectively.

Experimental diets were prepared by adding calcium chloride, strontium chloride and monosodium phosphate to the concentrations given in Table I.

TABLE I

Concentrations of phosphorus, calcium and strontium in different diets

Diet	mM/100 g of food		
	P	Ca	Sr
Basic "BP"	16	12·5	..
Basic low P "Bp"	3	12·5	..
High Ca "CaP"	16	30	..
High Sr "SrP"	16	1	20
High Sr–low P "Srp"	3	1	20

For technical reasons, inactive Sr was not determined in the basic diet. From the natural content of minerals in foodstuffs, the Sr/Ca ratio was probably of the order of 10^{-3} (ARCRL, 1959; Alexander and Nusbaum, 1959; Bryant and Loutit, 1961).

Food and water were given *ad libitum*. All control rats were fed the basic diet "BP" for 15 days before administration of the radioelements. 2·5-month-old rats were given basic diet for 5 days and afterwards the high Sr diet "SrP" for varying periods. The 1·5-, 7·5- and 13·5-month-old experimental rats were fed diets "SrP", "SrP" and "CaP" for 10 days preceding the injection of ^{90}Sr or [133]Ba. Immediately after administration of these isotopes all groups except one (see below) were again given the basic diet "BP".

Collection of excreta was performed in 24-h periods and every day the metabolic cages were carefully rinsed with solutions of both NaEDTA and acidified strontium chloride. [85]Sr was counted conventionally in a fixed geometry. Whole-body activity was measured in a fixed geometry with a NaI(Tl) scintillation detector coupled to a single-channel γ-ray spectrometer. The counts from [85]Sr and [133]Ba were recorded in photo-peaks only; bremsstrahlung from [90]Sr + [90]Y was measured in the photon energy interval from 36–184 keV. Fractional biological retention (R_b) was obtained by expressing the activity retained at a given time (t), as a percentage of the initial count-rate measured 0·5 h after administration of nuclide, correction in all cases being made for radioactive decay. R_b was described quantitatively as a sum of three exponentials of a general form:

$$R_b = A_1 e^{-\lambda_1 t} + A_2 e^{-\lambda_2 t} + A_3 e^{-\lambda_3 t}$$

assuming that retention between the 50th and 100th day can be described as a single exponential defined by the rate constant, λ_3. In the equation, $A_i =$ fractional coefficients of exponential terms and $\lambda_i =$ rate constants for three respective fractions of the tracer excreted at different rates.

Parameters of the third term were calculated by the least-squares method, assuming equal statistical weight of results obtained between 50 and 100 days since administration. The short-term exponentials were then estimated graphically.

After sacrifice of the rats (at 100 days) the contents of ^{85}Sr and ^{90}Sr and ^{90}Y in ashed bones were determined by conventional methods (see Table II).

TABLE II

Concentration of strontium-90 + yttrium-90 in femora of rats injected with the isotope

The groups were sacrificed on the 100th day after the injection with the exception of the 1·5-month-old animals given "BP", "SrP", "Srp" and "CaP", which for technical reasons were killed 14 days earlier.
The systematic difference introduced by this fact is almost negligible.

Age at injection (months)	Percentage of administered dose ±standard error of the mean				
	"BP"	"Bp"	"SrP"	"Srp"	"CaP"
1·5	2·41 ±0·08	1·79 ±0·03	0·33 ±0·03	0·32 ±0·04	2·26 ±0·08
7·5	0·76 ±0·08	0·47 ±0·11	0·24 ±0·02	0·14 ±0·01	0·69 ±0·09
13·5	0·44 ±0·06	0·34 ±0·02	0·23 ±0·11	0·25 ±0·02	0·49 ±0·1

Human experiment

A single experiment was performed on a young healthy adult male, J.L., 35 years old. During an initial period of 98 days, normal food was supplemented daily with 240 mg of calcium as gluconate and 180 mg of phosphorus as sodium monophosphate. At the end of this period a complete balance of Ca and P was determined over an interval of 5 days by analytical methods described elsewhere (Liniecki, 1966). On the 63rd day of this period, 0·3 μCi of ^{85}Sr in neutral physiological saline was injected intravenously and the retention was followed for 44 days by means of whole-body counting as described previously (Liniecki et al., 1965; Liniecki, 1966). Calcium gluconate was then replaced by strontium lactate in the amount corresponding to 2·0 g

of Sr ion daily for a further 28 days. On the 15th day of the high Sr diet
0·6 μCi of ^{85}Sr was administered and the retention was followed for another
59 days by whole-body counting. Retention was obtained after correction for
radioactive decay and has been expressed as a sum of power function and a
single exponential:

$$R_b = At^{-b} + ae^{-\lambda t}$$

To correct the second series of measurements for ^{85}Sr retained from the first
injection, the body contents of ^{85}Sr from the latter was calculated by extrapolation on the assumption that the power function obtained between 19th
and 44th day held over the extended period.

RESULTS

Rats

The relationship between the duration of feeding the rats the "SrP" diet and
the reduction of radiostrontium retention in 2·5-month-old rats is presented in
Fig. 1. The most effective reduction of retention is observed after 5 days,
further feeding producing practically no additional effect.

The biological retention of a single dose of strontium in 2·5-month-old rats

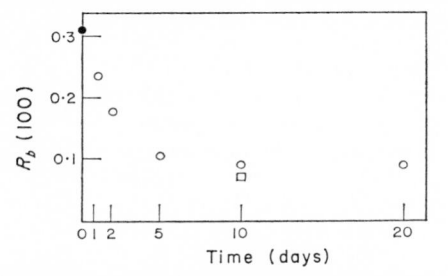

FIG. 1. Reduction of radiostrontium retention in 2·5-month-old rats against the
time of feeding the "SrP" diet. ○, feeding only before injection of radiostrontium;
□, feeding 10 days before and 10 days after injection of isotope; and ●, control.

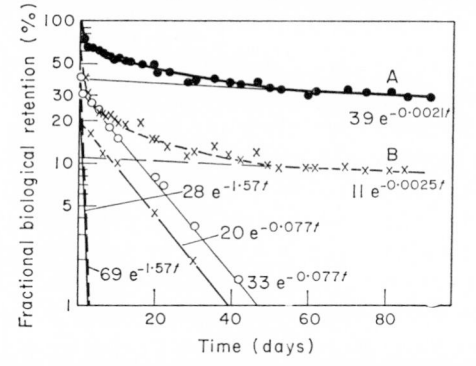

FIG. 2. ^{85}Sr retention in control A, rats and animals fed "SrP" diet, B, for 10 days,
corrected for radioactive decay. Age of rats, 2·5 months.

kept on the control diet ("BP") and high Sr diet ("SrP") for 10 days preceding injection of ^{85}Sr is presented in Fig. 2.

It seems that the difference between the retention of ^{85}Sr in the two groups consists only of changes in the fractional coefficients A_i. In the rats fed "SrP" diet the slowly excreted third fraction is reduced, hence A_3 (control)$>A_3$ (experimental).

The cumulative excretion of ^{85}Sr over 30 days in the control and experimental group was 46 and 78% of the injected dose, respectively. This corresponds to a retention of 54 and 22% of the dose as opposed to 40 and 12% when determined from whole-body counting. This discrepancy in the total balance is most probably due to unavoidable cumulative errors in collection of excreta over a prolonged period.

Over the first 30 days, the ratio of the cumulative excretion of ^{85}Sr in the urine to that in the faeces changed only slightly and was almost identical in the control and the "SrP" diet (Table III). It seems, therefore, that the high Sr diet, although reducing the retention of radiostrontium did not affect the ratio of urinary to faecal elimination of the radioactive marker.

TABLE III

The ratio of cumulative excretion of strontium-85 in urine to that in faeces

Time (days)	Cumulative excretion of ^{85}Sr in urine / Cumulative excretion of ^{85}Sr in faeces	
	Control group	Experimental group
1	1·31	1·31
2	1·21	1·20
3	1·30	1·18
4	1·19	1·16
5	1·17	1·16
6	1·16	1·15
7	1·17	1·15
8	1·17	1·15
9	1·17	1·15
10	1·18	1·15
15	1·15	1·16
20	1·12	1·15
25	1·10	1·15
30	1·09	1·15

The level of ^{85}Sr in the femora and tibiae of the control and high Sr groups measured over the period of 10 days after the administration of the isotope is presented in Fig. 3. Peaks in ^{85}Sr activity were observed at a few hours after injection, disappearing rapidly before the 12th hour.

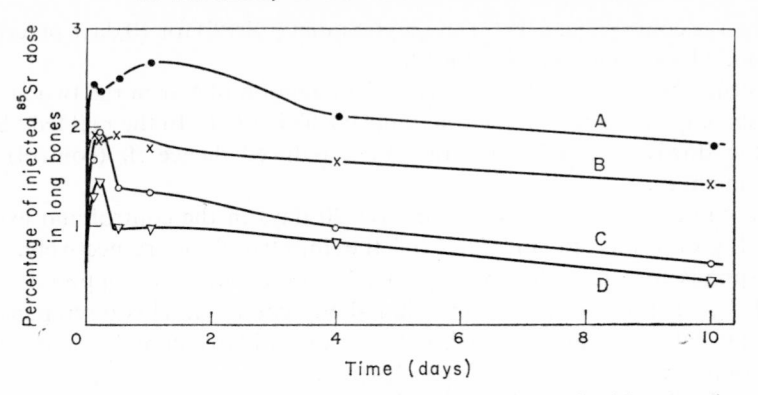

FIG. 3. ^{85}Sr retention in femora and tibiae of 2·5-month-old rats: A and B, controls; C and D, rats fed the "SrB" diet for 10 days.

These maxima may reflect temporary presence of the tracer in the rapidly exchangeable fraction of bone Ca (Bauer *et al.*, 1961).

The activity remaining in the skeleton after 12 h is mainly due to the deposition of Ca and inactive Sr in processes of accretion. Ten days after injection of the Sr tracer, the ratio of ^{85}Sr in the bones of control rats to that in bones of rats fed the high Sr diet "SrP" was about 3, which is in agreement with the ratio of the respective whole-body retentions.

The retention of ^{85}Sr for rats given the tracer orally in dietary pellets for 10 days is shown in Fig. 4. The reduction in retention is almost exactly the same as when the isotope was given parenterally, i.e., the ratio of retention in control to experimental animals is about 3·5. This ratio of the body retention

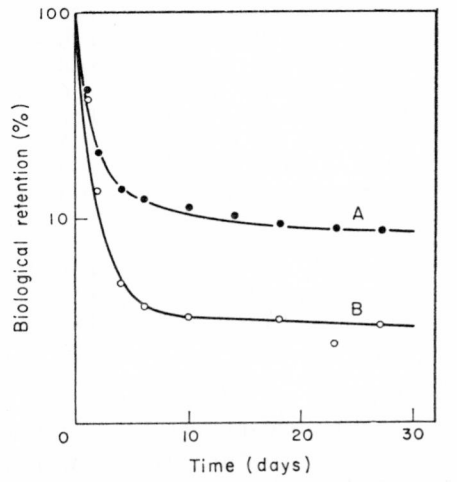

FIG. 4. Influence of the "SrP" diet upon retention of ^{85}Sr administered with the food: A, retention in controls; B, retention in animals fed "SrB" diet.

of the tracer when given with food to that when given by intravenous injection represents the fraction of the element absorbed from the gastro-intestinal tract into the blood stream, provided the unabsorbed tracer has left the gut. This fraction, measured 10 days after administration of ^{85}Sr, is shown to be independent of the type of diet given to animals ("SrP", "BP"). In both cases the absorbed fraction of the tracer amounted to about 20%.

Effect of age

It is striking that the "SrP" diet reduced the retention of radiostrontium to the same level in rats of all except the youngest age group (Fig. 5) and the well known age effect upon the retention almost completely disappeared. It seems, therefore, that a single function may be used to describe the fractional biological retention in all rats fed the "SrP" diet, with the exception of the youngest animals (1·5 months old), for which the long-term excretion was slightly slower than in the remaining groups.

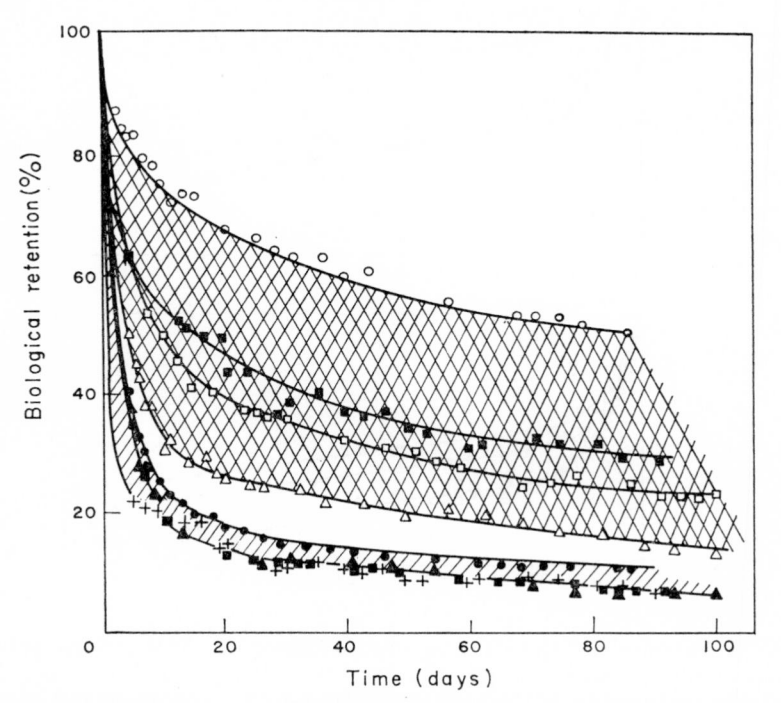

Fig. 5. Disappearance of the age dependence of radiostrontium retention in animals fed "SrP" diet for 10 days: ○, ×, □ and △, retention in 1·5-, 2·5-, 7·5- and 13·5-month-old control rats, respectively; ●, +, ▧ and ▲, retention in experimental rats of corresponding ages to control rats. The tracer for 1·5-, 7·5- and 13·5-month-old rats was ^{90}Sr and ^{90}Y; for 2·5-month-old rats, the tracer was ^{85}Sr.

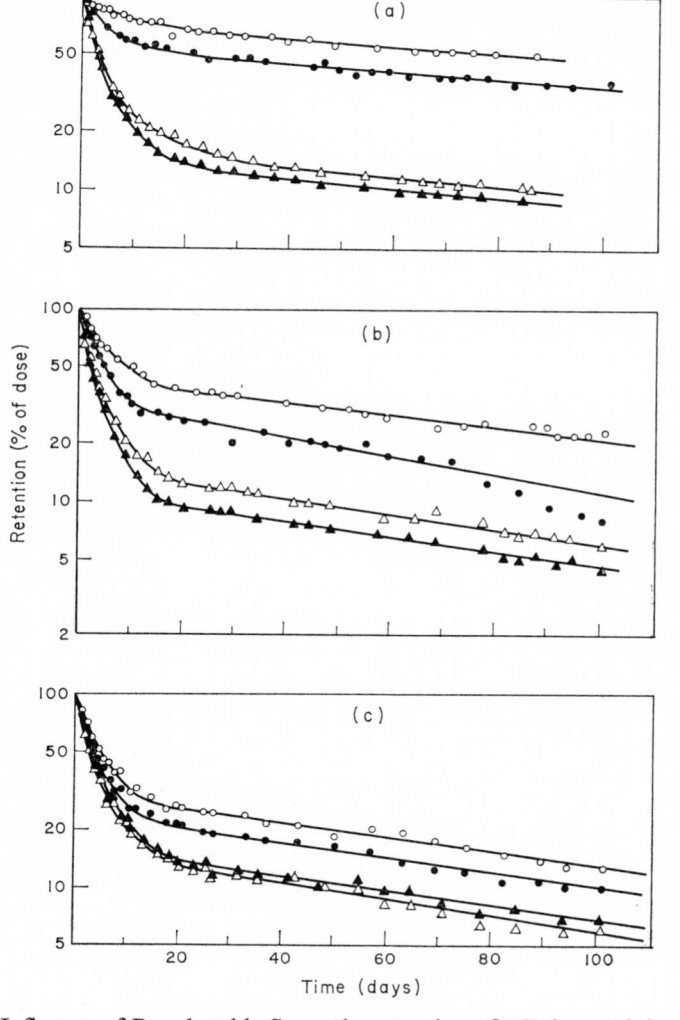

Fig. 6. Influence of P and stable Sr on the retention of ^{90}Sr in rats injected with isotope at different ages: ○, "BP" diet; ●, "Bp" diet; △, "SrP" diet; ▲, "Srp" diet. (a) 1·5-month-old rats; (b) 7·5-month-old rats; (c) 13·5-month-old rats,

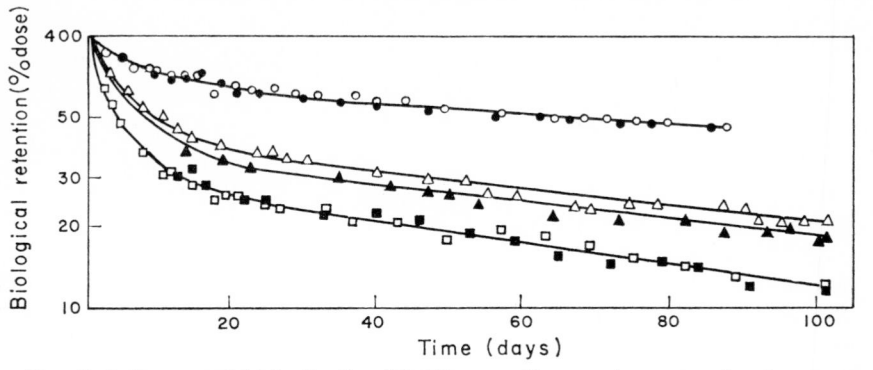

FIG. 7. Influence of high Ca diet "CaP" on radiostrontium retention in rats of different ages: ○ and ●, 1·5-month-old control and experimental rats, respectively; △ and ▲, 7·5-month-old control and experimental rats, respectively; □ and ■, 13·5-month-old control and experimental rats, respectively.

Influence of phosphorus and calcium level in the diet

The effect of the low phosphorus diet on body retention of the tracer is presented in Fig. 6. When the P content was reduced from 16 to 3 mM %, the retention of radiostrontium was reduced in all age groups to about the same degree, i.e. by some 20–30%.

In Fig. 7 the retention of ^{90}Sr in rats of three age groups (1·5, 7·5 and 13·5 months) is shown when they were fed the "CaP" diet containing 30 mM % of Ca as opposed to that of 12·5 in the basic diet (BP). There is no statistically significant difference between any corresponding control and experimental group.

Figure 8 shows the effect of "SrP" diet on radiobarium retention. On the high Sr diet the reduction of the retained ^{133}Ba was about 60%, which is less pronounced than that observed with radiostrontium in rats of corresponding age (7·5 months).

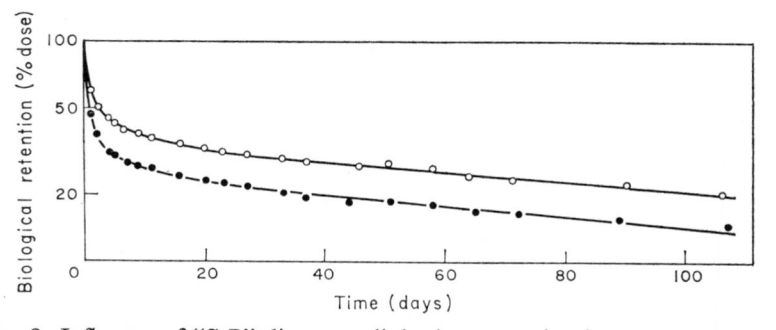

FIG. 8. Influence of "SrP" diet on radiobarium retention in 7·5-month-old-rats: ○, control rats; ●, experimental rats.

Man

The balance of calcium and phosphorus in subject J. L. is presented in Table IV. It may be assumed that metabolic equilibrium with respect to these elements existed at the time of investigation. The levels of Ca and P in serum amounted to 10·3 and 8·7 mg %, respectively, which are within the range regarded as normal. In the course of the strontiuml actate feeding, no abnormal symptoms were noted.

The fractional retention of a single intravenous administration of ^{85}Sr is demonstrated in Fig. 9, for the control and strontium lactate feeding period. A few days after administration of the tracer, differences between body retentions on the two diets became apparent. After the 20th day, the retention lines on the two diets are almost parallel and the resulting reduction of Sr retention was then almost twofold.

TABLE IV

Calcium and phosphorus balance in subject J.L.

Day	Intake Ca (mg)	Intake P (mg)	Excretion (urine + faeces) Ca (mg)	Excretion (urine + faeces) P (mg)
1st	986	764	728	819
2nd	601	1,094	1,108	1,240
3rd	501	804	695	802
4th	1,180	1,180	757	952
5th	1,223	1,156	1,197	1,206
Total in 5 days	4,491	5,008	4,485	5,019

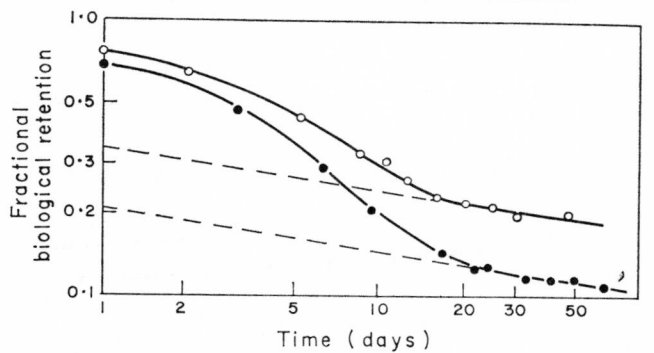

FIG. 9. High Sr diet and biological retention of a single administration of ^{85}Sr in subject J.L.: ○, control retention $[R_b(t) = 0.34t^{-0.16} + 0.46e^{-0.22t}(t \geqslant 1)]$; ● retention after the feeding of a strontium lactate $[R_b(t) = 0.21t^{-0.16} + 0.48e^{-0.21t}]$.

DISCUSSION

Results of the present experiments confirm quantitatively the influence of a high Sr diet on radiostrontium retention observed earlier by Cohn et al. (1961). It is interesting to note that in this experiment, the reduction of ^{85}Sr retention in man is approximately the same as that observed in adult rats. This points to the possibility that the investigated phenomenon may not be limited to small rodents.

Cohn and his co-workers have postulated a physio-chemical mechanism which they assumed was responsible for reduced radiostrontium retention after feeding the high Sr diet to rats. According to their hypothesis the observed effect resulted from a diminished pool of exchangeable Ca and reduced accretion rate of bone mineral. The change in these parameters according to Cohn et al. (1961) was due to differences in the solubility of tertiary calcium and strontium phosphates leading, on thermodynamic grounds, to a lower incorporation of Sr ions from solution by the crystal phase, when calcium phosphate was replaced by the strontium salt. However, this hypothesis does not explain the present findings, where under the influence of high Sr diet the retention was reduced to the same level irrespective of the age at the time of injection of the tracer.

It seems that all facts observed so far could be explained on the basis of metabolic principles of alkaline-earth transport between blood, or rapidly exchangeable mineral pool, and bone as postulated by Marshall (1960) (see Fig. 10), if, in the fractional rate of removal of Ca and Sr from the exchangeable pool, bone formation decreases with age and long-term ionic exchange and excretion are age independent.

The observed effect of the high Sr diet on radiostrontium retention in rats of different ages could be explained if it is assumed that the high concentration of Sr in the body fluids blocked processes of new bone formation, leaving the physico-chemical processes of long-term ionic exchange unaffected or altered to a similar degree in all age groups investigated. The following facts may be advocated to support this hypothis. Müller (1962) showed that feeding mice a high Sr diet (Ca/Sr = 0·5–1·0) leads to a rapid saturation of the bones with Sr in 12 to 13 days and further accumulation of this element is relatively small. At the same time stunting of growth indicated that a possible blocking of true bone formation was taking place.

According to Rowland and Marshall (1959) the activity of a single administration of an alkaline-earth tracer is partitioned in the skeleton—when the rapidly exchangeable fraction has been already eliminated—approximately equally between "hot spots" (true bone formation) and diffuse labelling (the results of long-term ionic exchange). Inhibition of true bone formation or remodelling in the adult should result in twofold reduction of Sr retention, as observed in the human experiment described above.

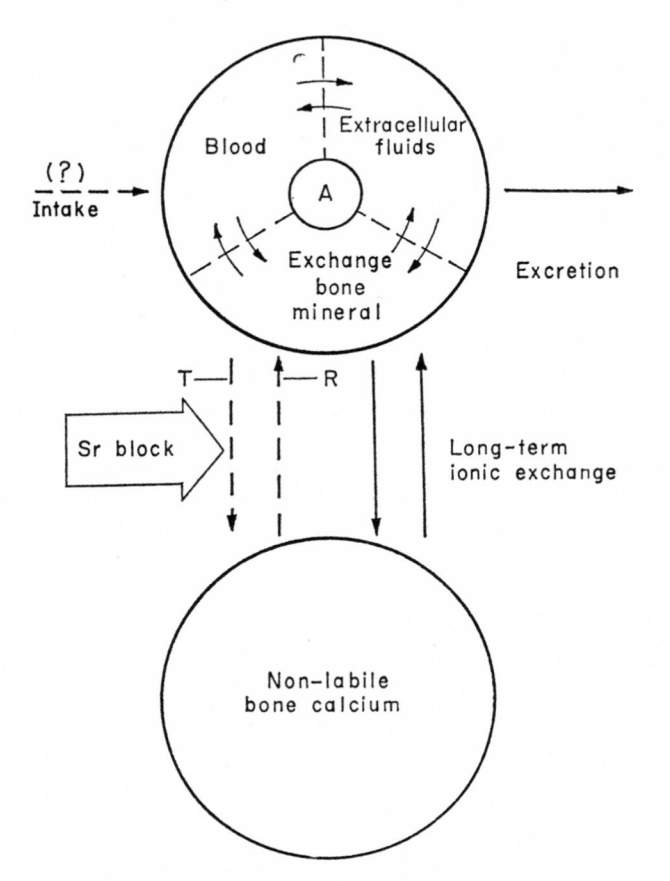

FIG. 10. Hypothetical influence of high Sr diet on the metabolic processes trans-
ferring alkaline-earth elements from rapidly exchangeable mineral into "fixed"
phase.
Scheme based on the concept of Marshall (1960). T, true bone formation; R, bone
retention; ------, fractional rates (absolute amount of Ca taken into or excreted
from A per unit of time/total amount of Ca in A) which depend on biological age
over 2 months; ————, fractional rates, independend of age over 2 months.

It seems that changes in serum clearance of Sr by the kidneys and intestines
when the "SrP" diet was given do not play an important role in the accelera-
tion of radiostrontium excretion from the body. Such a conclusion is supported
by the fact that partition of the tracer between faeces and urine of rats remained unaltered. The alternative conclusion implies that both excretory
processes changed similarly, which seems improbable in view of the different
physiological mechanisms involved (Cramer, 1963).
It seems noteworthy that a high Sr diet may be used to modify the retention

of other alkaline earths. Moreover, if the hypothesis is true, it might be possible to separate by experimental dietary procedure the diffuse labelling of bone from formation of "hot spots". The hypotheses presented, however, need direct confirmation by means of autoradiographic techniques and additional kinetic studies. These are under way in this laboratory.

The conclusion reached by Cohn and his co-workers (1961) on the influence of high Ca diet on Sr retention does not appear to be correct. In their experiments, elevation of Ca in the diet was accomppanied by a simultaneous reduction of P concentration and the latter procedure *per se* seems responsible for the observed effect on retention. No such postulated Ca effect was observed in the present experiments.

The practical applications of a high Sr diet—if any—seem to be limited to predictable high exposure to radiostrontium of short duration, and in this connection the following features of the procedure appear advantageous: rapid manifestation of metabolic effects; effectiveness with different routes of entry of radiostrontium into the body; and the low toxicity of inactive Sr in short-term administration. Chronic application seems precluded by findings of Wasserman and Comar (1960), Tessari and Spina (1961) and Müller (1962), which clearly point to deleterious effects of large amounts of strontium on skeletal tissue, especially in young, growing mammals.

The assistance of Mrs W. Karniewicz in the whole-body counting of the human subject is highly appreciated.

REFERENCES

Agricultural Research Council Radiobiological Laboratory (1959). ARCRL. Report No. 1. H.M. Stationery Office, London.

Alexander, G. V., and Nusbaum, R. (1959). *J. biol. Chem.* **234**, 418–421.

Bauer, G. C. H., Carlsson, A., and Lindquist, B. (1961). *In* "Mineral Metabolism" (Comar, C. L., and Bonner, F., eds.), vol. IB, pp. 609–688. Academic Press, New York.

Bryant, F. J., and Loutit, J. F. (1961). U.K. AERE Report 3718. H.M. Stationery Office, London.

Cohn, S. H., Nobel, S., and Sobel, A. E. (1961). *Radiat. Res.* **15**, 59–69.

Cramer, C. (1963). *In* "The Transfer of Calcium and Strontium across Biological Membranes" (Wasserman, R. H., ed.), pp. 75–84. Academic Press, New York.

Liniecki, J. (1966). Ośrodek Informacji Pełnomocnika Rządu d/s Wyrkozystania Energii Jadrowej, Warsawa.

Liniecki, J., Karniewicz, W., and Kosterkiewicz, A. (1965). *Nukleonika*, **10**, 35–49.

Marshall, J. H. (1960). *In* "Bone as a Tissue" (Rodahl, K., *et al.*, eds.), pp. 144–155. McGraw-Hill, New York.

Müller, W. A. (1962). *Naturwissenschaften*, **49**, 38–39.

Rowland, R. E., and Marshall, J. H. (1959). *Radiat. Res.* **11**, 299–313.

Tessari, L., and Spina, G. M. (1961). *Sperimentale*, **111**, 27–35.

Wasserman, R. H., and Comar, C. L. (1960). *Proc. Soc. exp. Biol. Med.* **103**, 124–129.

Discussion

QUESTION: Did you make any allowance for changes in counting efficiency owing to the increase in weight of the animals during the experiment?

ANSWER: This change may have occurred in respect of the younger animals, which were still growing; the other groups kept a steady weight.

Ingested or Injected Strontium as Influenced by Oral Treatment Shortly Before or After Exposure

V. VOLF

Institute of Radiation Hygiene, Prague, Czechoslovakia

SUMMARY

This paper reviews experiments with [85]Sr on rats in three subject areas:

(*a*) Attempts to minimize the radiostrontium absorption from the gastro-intestinal tract by sulphates, phosphates or carbonates. Several mixtures proved most effective provided treatment started almost immediately after Sr ingestion.

(*b*) Effect of these orally administered substances on the excretion, via the gut, of injected radiostrontium. Phosphates increased the faecal excretion of Sr and lowered its urinary elimination, so that the total excretion and retention remained unchanged. Shortly after exposure, sulphates increased the amount of Sr in the small intestine, but later mainly promoted its urinary excretion.

(*c*) The effect of orally administered absorbable chlorides or sulphates on urinary excretion of injected radio-strontium. The excretion pattern and the retention of radio-strontium were influenced when the substances were given shortly before radiostrontium and in sufficient concentrations.

ORAL TREATMENT FOLLOWING INGESTION OF RADIOSTRONTIUM

The effects of various substances are summarized in Fig. 1. For comparison the average skeletal retention [85]Sr in the treated groups is expressed as percentage of control values. All the tested substances were administered in single equimolar doses (regardless of their solubility) to fasting male adult rats, 10 min after oral contamination with [85]Sr. The animals were sacrificed 48 h later. At least five rats were used for each point.

When 0·8 mM of various sulphates were administered, the skeletal [85]Sr content decreased by 40–60% except with strontium sulphate, which was least effective (Volf and Roth, 1965). A significantly greater effect was produced by a mixture of barium sulphate with magnesium or sodium sulphates (Volf and

Roth, 1965), and especially when calcium-activated barium sulphate was used. The treated animals retained only 24, 22 and 7% of the control values, respectively.

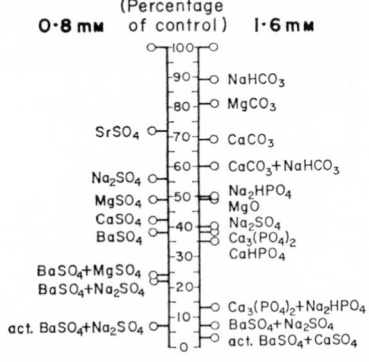

FIG. 1. Average femoral ^{85}Sr retention 48 h after ^{85}Sr administration. At least five rats per point.

When the tested substances were administered in greater amounts, it was noted that magnesium oxide, phosphates and sodium sulphate decreased the skeletal retention of ^{85}Sr by 50–60%, but carbonates were ineffective (Volf, 1965). Rats treated by mixtures of calcium and sodium phosphates (Volf, 1965), of barium and sodium sulphates (Volf and Roth, 1965) and by calcium-activated barium sulphate in saturated calcium sulphate solution (Volf, 1964) retained 13, 7 and 3% of the control values, respectively.

Although the effect produced in these conditions is considerable, in practice the efficacy would depend on several factors.

One of them is delay in the start of treatment. Thus, when a mixture of calcium-activated barium sulphate and sodium sulphate was given orally immediately after ^{85}Sr ingestion, its skeletal retention decreased by more than 99% in comparison with controls. This extremely high effectiveness decreased rapidly with time, so that 80 min after ^{85}Sr ingestion, by which time the bulk of it was already removed from stomach, treatment proved ineffective (Volf, 1964).

This clearly indicates the limitations of first aid when the methods described are used, although the decrease of effectiveness will be probably slower in man than in rats owing to a slower removal of ingesta from the stomach.

ATTEMPTS TO INFLUENCE EXCRETION OF RADIOSTRONTIUM VIA THE INTESTINE

The above conclusions were drawn from ^{85}Sr retention data only. Further metabolic studies should explain whether the tested substances merely prevent intestinal absorption of radiostrontium or also influence its movement from blood to the gut.

Mixtures of phosphates or sulphates were administered orally to fasted rats, 80 min before the injection of ^{85}Sr. Further, pilocarpine was injected subcutaneously immediately and 2 h after the injection of ^{85}Sr. Finally, by the same schedule, oral phosphates or sulphates were combined with pilocarpine or atropine injections.

Figure 2 shows the average intestinal distribution of ^{85}Sr 160 min after its intravenous administration. Immediately after sacrifice, the abdomen was incised and ligations made at the oesophageal and pyloric orifices of the stomach, along the small intestine, ileocaecal juncture and anal orifice. The intestinal tract was removed, the segments separated and measured with a scintillation counter. No attempt was made to isolate ^{85}Sr in the gut wall from the intestinal contents.

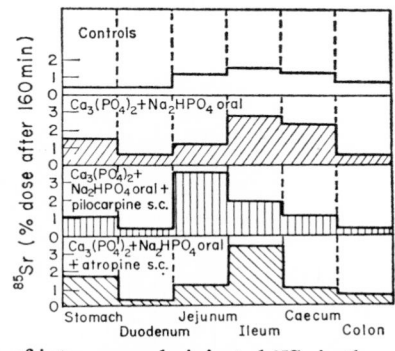

FIG. 2. Distribution of intravenously injected ^{85}Sr in the gastro-intestinal tract as affected by oral treatment with phosphates (80 min before ^{85}Sr) and pilocarpine/atropine injections (immediately, and 120 min after ^{85}Sr). Each column represents the mean of six rats.

In all groups treated with calcium and sodium phosphates, the percentage of ^{85}Sr recovered in the stomach exceeded the control values several times. In the group with phosphates and pilocarpine there was also a significant increase in jejunal ^{85}Sr.

Figure 3 presents results obtained under identical experimental conditions, but instead of phosphates a mixture of calcium-activated barium sulphate with calcium sulphate was administered, and to one group, pilocarpine only was injected. A significant, but slight, increase of ^{85}Sr in the stomach was observed after sulphates and in all treated groups ^{85}Sr in the small intestine was substantially elevated.

As can be seen in Fig. 4, the average ^{85}Sr content of the whole gastro-intestinal tract was significantly increased in rats to which phosphates and/or pilocarpine or sulphates with pilocarpine were administered. However, skeletal content of ^{85}Sr was reduced only after phosphates with pilocarpine.

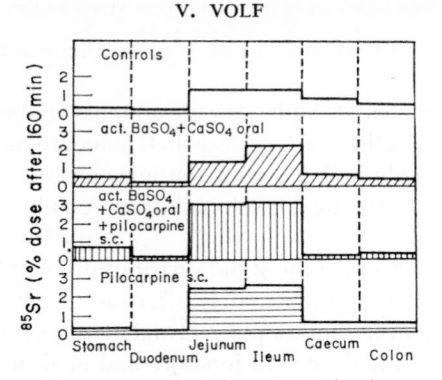

FIG. 3. Distribution of intravenously injected ^{85}Sr in the gastro-intestinal tract as affected by oral treatment with sulphates (80 min before ^{85}Sr) and/or pilocarpine injections (immediately, and 120 min after ^{85}Sr). Each column represents the mean of six rats.

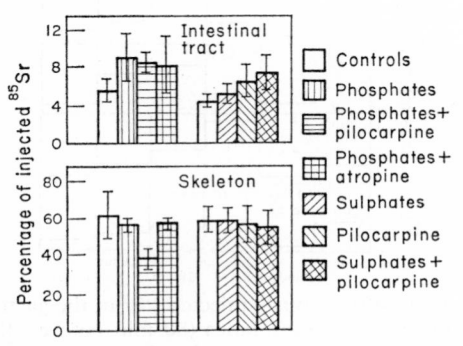

FIG. 4. Retention of intravenously injected ^{85}Sr as affected by treatment with oral phosphates/sulphates and injected pilocarpine/atropine (for schedule see Figs. 2 and 3). Arithmetic mean ± standard error of the mean multiplied by t value for 95% confidence level. Skeleton, content of ^{85}Sr in 1 femur × 20.

In a further study, the latter combination was equally used, rats were placed in metabolic cages and sacrificed 48 h after intraperitoneal injection of ^{85}Sr. No change was observed in the total excretion of ^{85}Sr (Table I). Although in the faeces of treated animals there was twice as much ^{85}Sr, this was compensated by the decreased urinary elimination. Since less Ca was excreted in the urine of treated rats, the ratio $^{85}Sr/Ca$ was not significantly altered.

In similar series, sulphates were administered orally and pilocarpine injected subcutaneously in consecutive doses. Although the elimination of ^{85}Sr in the faeces was somewhat higher than in the controls, treatment resulted mainly in substantial elevation of ^{85}Sr in urine. This was attributed to the effect of sulphate ions absorbed from the intestine.

TABLE I

Excretion of intraperitoneally injected strontium-85 as affected by oral phosphates and injected pilocarpine

^{85}Sr values are expressed as percentage of dose 24 h after its administration. Arithmetic means ± standard error of the mean multiplied by t value for 95% confidence level

Group	No. of rats	^{85}Sr (% of dose)	Urine Ca (mg)	Urine P (mg)	$^{85}Sr/Ca$ (%/mg)	Faeces ^{85}Sr (% of dose)	Total excretion ^{85}Sr (% of dose)
Control	6	14·1 ±4·3	10·4 ±2·9	46·3 ±17·3	1·4 ±0·3	4·7 ±1·0	18·8 ±4·2
Phosphates + pilocarpine	5	6·5* ±2·6	5·6* ±1·3	26·5* ±3·7	1·1 ±0·2	10·3* ±1·2	16·8 ±2·7

* Significantly different from control rats. For dosage schedule see Fig. 2.

ATTEMPTS TO ENHANCE URINARY EXCRETION OF INJECTED RADIOSTRONTIUM BY ORAL TREATMENT

Fasted male rats received orally single doses (0·8 mM) of sodium or magnesium sulphates or magnesium chloride. Each substance was administered in three various concentrations, 10 min before the intraperitoneal injection of ^{85}Sr. The average skeletal retention 48 h later is shown in Table II.

TABLE II

Effect of various salts administered orally in doses of 0·8 mM 10 min before the intraperitoneal injection of strontium-85

Values are expressed as arithmetic means of six rats ±standard error of the mean multiplied by t value for 95% confidence level. Values are percentage of dose in 1 femur times 20, 48 h after ^{85}Sr injection

	Na$_2$SO$_4$	MgSO$_4$	MgCl$_2$
Controls	53·7 ±4·2	59·7 ±2·4	46·2 ±7·1
0·1 M	53·3 ±7·6	55·8 ±5·0	42·6 ±8·4
0·2 M	51·6 ±6·9	49·8 ±4·2	39·3* ±2·5
0·8 M	42·1* ±9·5	49·7* ±5·4	34·1* ±4·6

* Significantly different from control rats.

After treatment, there was a significant reduction in retained ^{85}Sr with all three substances tested, provided the concentrations were sufficient.

The experiment was repeated with the highest concentrations of the above substances (Fig. 5). Urinary ^{85}Sr increased significantly in all the treated groups. Although after magnesium chloride or sulphate faecal ^{85}Sr was significantly lowered, the total excretion remained higher than in controls, and this caused significant decrease in its retention.

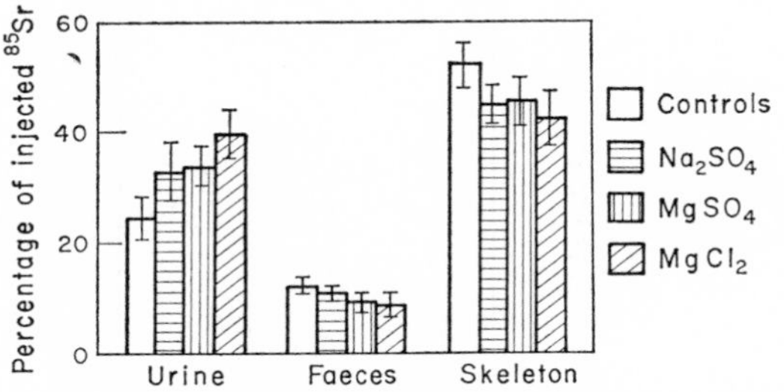

FIG. 5. Effect of various salts administered orally in doses of 0·8 mM 10 min before the intraperitoneal injection of ^{85}Sr. Each column is a mean from five to ten rats, 48 h after ^{85}Sr administration. For further details see Fig. 4.

TABLE III

Urine constituents as affected by various salts administered orally in doses of 0·8 mM 10 min before the intraperitoneal injection of strontium-85

Values represent urinary excretion during the first 24 h after ^{85}Sr administration

Substance tested	No. of rats	^{85}Sr (% of dose)	Ca (mg)	Mg (mg)	Ca + Mg (mg)	^{85}Sr/Ca (%/mg)	^{85}Sr/(Ca + Mg) (%/mg)	SO_4 (mg)
Controls	9	18·5 ± 3·0	6·3 ± 1·3	1·4 ± 0·4	7·8 ± 1·3	3·0 ± 0·5	2·4 ± 0·3	47·5 ± 11·7
Na₂SO₄	5	26·8* ± 5·5	4·6* ± 0·9	1·3 ± 1·4	5·9* ± 1·1	6·2* ± 1·6	4·6* ± 1·4	97·4* ± 7·9
MgSO₄	10	28·5* ± 2·7	9·2* ± 1·9	4·7* ± 0·9	13·9* ± 1·7	3·2 ± 0·4	2·1 ± 0·3	83·6* ± 9·2
MgCl₂	5	34·9* ± 3·6	7·6 ± 2·0	8·7* ± 2·4	16·1* ± 3·1	4·7* ± 1·2	2·2 ± 0·5	49·9 ± 9·5

* Significantly different from control rats.

Table III presents data on the urinary excretion of several stable ions during the first 24 h of the experiment. With magnesium sulphate the Ca and Mg excretions were elevated, whereas magnesium chloride promoted mainly the elimination of Mg. With both substances, however, the sum of urinary Ca and Mg was equal. In both groups, with sulphates more of the ion appeared in the urine. The ratio $^{85}Sr/Ca$ was significantly higher with sodium sulphate and magnesium chloride, whereas the ratio $^{85}Sr/(Ca + Mg)$ was elevated in the former group only. This seems to indicate that ^{85}Sr was excreted in a rather constant ratio to the sum of alkaline-earth ions, except after sodium sulphate administration, where it might have been eliminated also in another form than in the remaining groups, perhaps electrostatically associated with sulphate (Walser *et al.*, 1961).

In another series of rats, strontium and calcium chlorides were administered orally (0·8 mM in a single dose), 10 min before the intraperitoneal injection of ^{85}Sr. The results are illustrated in Fig. 6. Calcium chloride probably formed an intestinal reservoir of Ca enhancing "endogenous faecal Sr", and after strontium chloride administration, there was also an increase in alkaline-earth elimination via the kidney and a corresponding rise in urinary ^{85}Sr. This resulted in reduced skeletal retention of ^{85}Sr (by about 30 %) in the latter group.

Under similar conditions, the observed effect of oral magnesium and sodium sulphates and of calcium, magnesium and strontium chlorides was comparable to that reported by other investigators after injection of these salts. (Catsch and Melchinger, 1959; Kawin, 1959; Ogawa *et al.*, 1962; Nelson *et al.*, 1963; Annenkov, 1965).

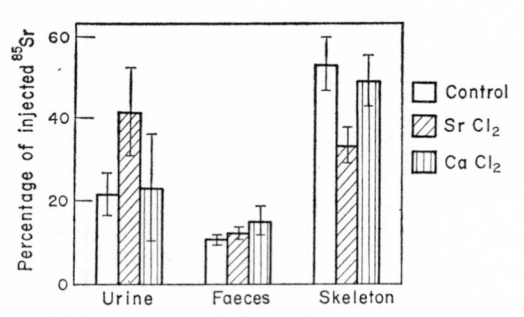

FIG. 6. Effect of strontium and calcium chlorides administered orally in doses of 0·8 mM 10 min before the intraperitoneal injection of ^{85}Sr. Each column is the mean of five rats, 48 h after ^{85}Sr administration. For further details see Fig. 4.

CONCLUSIONS

The data show that oral sulphates and phosphates effectively reduce the retention of ingested radiostrontium in rats when administered in time. Besides the known mixture of calcium and sodium phosphates highly effective combinations of barium and sodium sulphates and of calcium-activated barium sulphate with calcium sulphate were used.

These oral mixtures alone or in combination with pilocarpine also enhanced the excretion of injected radiostrontium via the intestine shortly after exposure.

Phosphates merely changed the pattern of excretion, increasing the faecal and reducing the urinary radiostrontium. Sulphates, which were partly absorbed, mainly promoted the elimination of radiostrontium in urine.

The latter effect was most pronounced when readily absorbable sulphates and alkaline-earth chlorides where administered orally before the injection of radiostrontium, except for calcium chloride, which enhanced faecal radiostrontium only. Excepting the sodium sulphate group higher urinary excretion of radiostrontium was accompanied by increased alkaline-earth elimination.

REFERENCES

Annenkov, B. N. (1965). *Radiobiologiia*, **5**, 620–621.
Catsch, A., and Melchinger, H. (1959). *Strahlentherapie*, **109**, 561–572.
Kawin, B. (1959). *Experientia*, **15**, 313–314.
Nelson, A., Rönnbäck, C., and Rosén, L. (1963). *Acta. radiol. ther. Phys. Biol.* **1**, 129–139.
Ogawa, E., Suzuki, S., and Fukuda, R. (1962). *Gunma J. med. Sci*, **11**, 205–213.
Volf, V. (1964). *Int. J. Radiat. Biol.* **8**, 509–511.
Volf, V. (1965). *Experientia*, **21**, 571.
Volf, V., and Roth, Z. (1965). *Acta. radiol. ther. Phys. Biol.* **3**, 216–228.
Walser, M., Payne, J. W., and Browder, A. A. (1961). *J. clin. Invest.* **40**, 234–242.

Effect of a Low Phosphorus Diet on Strontium-90 Toxicity in Mice

L. M. VAN PUTTEN

Radiobiological Institute TNO, Rijswijk Z.H., The Netherlands

SUMMARY
Treatment with a low P diet increases the life span and reduces the incidence of bone tumours in mice injected with ^{90}Sr at a level of $0 \cdot 3 \mu Ci/g$ body weight.

INTRODUCTION

An earlier paper (van Putten, 1962b) discussed the effect of a P-deficient diet on the bone-tumour frequency and the life span of mice injected with ^{90}Sr ($1 \mu Ci/g$ body weight). It was shown that the diet treatment, which started 2 days after injection of the nuclide, slowly reduced the body burden of the treated mice to about 50% of that of the control mice.

For the evaluation of the toxic effects, use was made of 2 parameters: bone-tumour frequency and survival time of the mice. By interpolation from these parameters, the toxic effect can be expressed in dose equivalents. These indicate that dose which, upon injection without further treatment, would give rise to similar bone-tumour frequency and life-span shortening. Thus the mice that received $1 \mu Ci$ ^{90}Sr/g body weight had their body burden reduced by the diet to that expected in untreated mice after injection with $0 \cdot 5 \mu Ci$ ^{90}Sr/g but the toxicity was equivalent to that seen after an injection of $0 \cdot 7 \mu Ci$ ^{90}Sr/g without treatment. The explanation for this incomplete reduction of toxicity by the slow removal of the Sr is not completely clear; since the treatment reduces the body burden slowly, the initial high dose-rate irradiation of the bone is hardly affected, but the cumulative radiation dose received over a year was reduced to about 60%.

It could not be established whether the small discrepancy between this reduction of the radiation dose to 60% and the parallel reduction of the toxicity to 70% should be ascribed to the inaccuracies of the method by which these figures were obtained.

To confirm the earlier findings the experiment has now been repeated with a lower ^{90}Sr dose.

M

EXPERIMENTAL METHODS

Methods were similar to those used in the previous experiments (van Putten, 1962b) except for the ⁹⁰Sr doses and the mouse strain employed. Female (C57BL/Rij × CBA/Rij)F₁ hybrids were used in five groups: after ⁹⁰Sr injection (0·3 μCi/g), group A mice remained untreated, and group B mice received the low P diet. Similarly, of the control mice, group D remained untreated, but group E received the diet in order to permit evaluation of possible effects of the diet itself on the life span of the animals. With the aim of facilitating interpolation of toxicity results, another group (C) was included. This group received 0·15 μCi/g and no further treatment. Follow-up of the mice consisted of periodic determinations of the ⁹⁰Sr body burden; the mice were observed daily for mortality, and at autopsy a radiograph of the skeleton was made to determine the number of bone tumours.

RESULTS

The ⁹⁰Sr retention of the mice is represented in Fig. 1. As a result of the low P diet the mean body burden of the group B mice decreased to 52% of the group A average. The mean body burden of group C mice was 45% of group A. The numbers of bone tumours and the survival times of the mice are presented in Table I. The dose equivalent of the toxic effects in group B mice interpolated on the basis of the results for groups A, C and D and earlier studies (van Putten and de Vries, 1962) is approximately 0·2 μCi ⁹⁰Sr/g or 67% of the group A dose. The relative radiation doses received by group A and B mice were estimated by integration of the retention curves. Over the

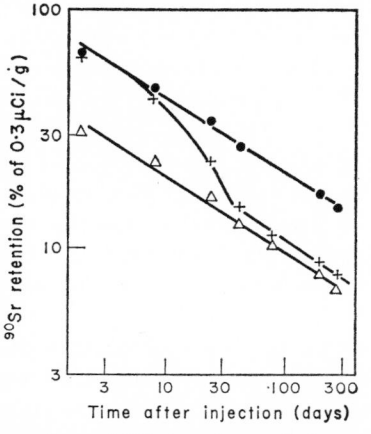

FIG. 1. The average retention of ⁹⁰Sr in mice as influenced by the low P diet in comparison with the untreated control groups: ●, 0·3 μCi/g ⁹⁰Sr, standard diet; +, 0·3 μCi/g ⁹⁰Sr, low P diet; △, 0·15 μCi/g ⁹⁰Sr standard diet.

TABLE I

Effect of treatment with phosphorus-deficient diet on bone tumour frequency and survival time in mice injected with strontium-90

	^{90}Sr injected*	P-deficient diet†	No. of mice	No. of bone tumours at autopsy	Average survival time (days)
A	0·3 μCi/g	No	50	39	534
B	0·3 μCi/g	Yes	50	18	618
C	0·15 μCi/g	No	50	3	653
D	0‡	No	25	0	782
E	0	Yes	25	0	789

* ^{90}Sr was injected intraperitoneally in 0·2 ml saline.

† From day 2 till 44 after injection, the mice received a diet containing 1 % Ca and less than 0·02 % P (van Putten, 1962a).

‡ Injected with saline.

relevant time interval, 450 days after injection, group B mice received 60 % of the radiation dose of group A animals. These and earlier results are summarized in Table II. The slow action of the diet is responsible for the relatively small reduction in radiation dose, and it is tempting to ascribe the even smaller reduction in toxicity to the fact that it is especially the late (low dose-rate) radiation that is reduced, with a relative excess of the early (high dose-rate) radiation.

TABLE II

Effects of treatment with phosphorus-deficient diet

^{90}Sr dose injected	Percentages of corresponding values for control animals on normal diet		
	Body burden	Radiation dose	Toxicity
1·0 μCi/g	50	60	70
0·3 μCi/g	52	60	67

DISCUSSION

It seems that a delayed treatment of radiostrontium intoxication is less effective not only because a smaller fraction of the deposited nuclide will eventually be eliminated, but also because even if a late treatment is highly effective in enhancing elimination of the deposited radiostrontium it does not produce an equivalent decrease of the toxic effects. It should be stressed that in longer-lived animal species, the relative contribution of the late radiation dose becomes increasingly important, and it seems therefore likely that delayed removal of radioactive Sr will be more effective in reducing toxic effects in those species.

Another observation in the present experiment seems to be of interest; there is no difference in survival between the treated and the untreated control mice. Neither the mineral loss from the skeleton nor the hazards of urogenital tract complications owing to the hypercalciuria have affected survival. It is doubtful, however, whether a much more severe treatment of this type could be tolerated. In rats, application of a similar diet severely jeopardized the health of the animals (Ray *et al.*, 1956).

All the successful methods that have been described to enhance the late excretion of radiostrontium also increase the elimination of Ca, and the same holds true for many of the less effective methods. Two mechanisms are involved: an increase of the exchange of bone-mineral ions with the extracellular fluid; and an increase of the kidney clearance of Ca, preventing return of a large fraction of the mobilized divalent ions into the bone.

Parathyroid hormone (Bacon *et al.*, 1956) and citrate (Ito and Tsurufuji, 1958) probably act by both mechanisms. The effect of ammonium chloride (Spencer and Samachson, 1961) and salicylate (Smith and Bates, 1965) is primarily on the kidney and only secondarily on the bone. Ca salts increase the kidney Ca clearance, but do not affect the bone-mineral exchange ratio. A P-deficient diet (van Putten, 1962a) increases the bone-mineral exchange ratio, and, if in addition, the Ca content in the diet is high, an enhanced absorption from the intestinal tract leads to a high Ca clearance by the kidney.

Since all these treatments have common mechanisms it should be possible to evaluate the limits of effectiveness of this general approach. Before comparing the results of different treatments, the circumstances under which they have been obtained must be taken into account, since these conditions may have a marked influence on the effectiveness of each method.

(1) As has already been referred to, the effectiveness of the treatment is dependent on the time interval between the deposition of radioactive Sr and the initiation of the treatment.

(2) The choice of the experimental animal determines the latitude left for treatment. If the animal is a fast-growing rat on a low Ca diet, large amounts of parenteral Ca salts enhance the elimination of the nuclide far more than in an adult mouse on a high Ca diet.

(3) The tolerance for treatment seems also to be dependent on the species used. The P-deficient diet in rats seriously jeopardized the health of the animals, but in mice no deleterious effects were observed. However, daily parenteral injection of an effective dose of salicylate is tolerated by rats (Smith and Bates, 1965) but not (in our experience) by mice.

With these variables in mind it seems possible to outline a maximally effective treatment. The most marked enhancement of bone-mineral exchange is obtained by administration of citrate and by a P-deficient diet. The former is more effective as an early acute treatment since it has no lag period as is the case with the diet; the latter is the most effective mobilizing agent at a later time after deposition of radioactive Sr if only by the fact that it can be applied continuously for 6 weeks. Increase of the Ca clearance by the kidney at the cost of return of Ca into bone is also maximally stimulated by the low P–high Ca diet, which causes a marked polyuria as a consequence of the hypercalciura. Again the lag period between application and effect of the diet makes it preferable to combine the low P diet with an early administration of Ca salts. The combined effect of calcium gluconate, citrate and a P-deficient diet on ^{90}Sr retention in mice is presented in Fig. 2. The resulting retention at 6 weeks after injection is 40% of the retention in control mice. Possibly this figure could be further improved if the standard diet had a lower Ca content.

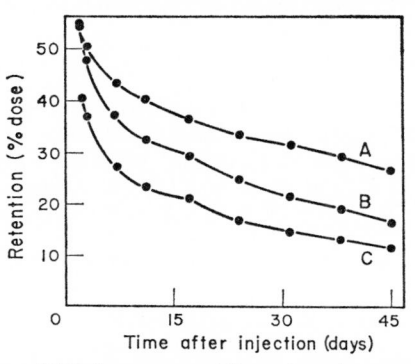

FIG. 2. The retention of ^{85}Sr in groups of five mice: A, control mice; B, mice that received the P-deficient diet, starting 2 days after injection of ^{85}Sr; C, mice that received the P-deficient diet, starting 2 days after injection of ^{85}Sr as well as receiving (four times in the first 2 days) parenteral injections of calcium gluconate (1,250 mg Ca/kg) and of sodium citrate (900 mg/kg).

At present only two alternative approaches can be foreseen that might yield better results: one is the development of substances which can discriminate between Sr and Ca with such selectivity in favour of Sr that in spite of the enormous excess of Ca over Sr at all significant sites, a selective elimination of Sr occurs. Such a specificity is not likely to work in the first step in the elimination process—the exchange of bone mineral with the body fluids. It is not to be expected that selectivity would lead to solubilization of Sr ions from the apatite crystal in excess of the exchange of Ca ions.

The other approach is similarly aimed at a more complete and selective elimination of all circulating Sr ions from the body fluids. This seems to be attainable by treatment with an artificial kidney which can replace the divalent ions completely with Sr-free Ca. Such treatment may be quite effective, perhaps even a factor of two better than the best alternative treatment. However, setting up this treatment will take hours after an accident. If applied after a 6-hour delay it is not likely to result in more effective enhancement of Sr elimination than that obtainable by early injection of Ca salts.

Thus it seems that a low P diet in combination with early Ca and citrate administration provides at present one of the most effective means to enhance the elimination of radioactive Sr.

The capable technical assistance of Misses M. van Doorninck and H. Peet and Mr P. Lelieveld is gratefully acknowledged. Dr H. Reinhold is thanked for reading the radiographs.

REFERENCES

Bacon, J. A., Patrick, H., and Hansard, S. L. (1956). *Proc. Soc. exp. Biol. Med.* **93,** 349–351.

Ito, Y., and Tsurufuji, S. (1958). *Proc. 2nd Int. Conf. peaceful Uses Atom. Energy,* **23,** 443–450.

van Putten, L. M. (1962a). *Int. J. Radiat. Biol.* **5,** 471–476.

van Putten, L. M. (1962b). *Int. J. Radiat. Biol.* **5,** 477–484.

van Putten, L. M., and de Vries, M. J. (1962). *J. nat. Cancer Inst.* **28,** 587–603.

Ray, R. D., Stedman, D. E., and Wolff, N. K. (1956). *J. Bone Jt. Surg.* **38A,** 637–654.

Smith, H., and Bates, T. H. (1965). U.K.A.E.A. P.G. Report 662 (cc). H.M.S.O., London.

Spencer, H., and Samachson, J. (1961). *Clin. Sci.* **20,** 333–343.

Discussion

QUESTION: Have you any evidence of the production of bone tumours in mice that did not receive ^{90}Sr?

ANSWER: No histological studies were made, but there was no evidence from gross inspection and radiography at death.

Studies on The Inhibition of Radiostrontium Uptake from the Human Gastro-intestinal Tract with Sodium Alginate

R. HESP AND B. RAMSBOTTOM

Health and Safety Department, U.K.A.E.A. Windscale and Calder Works, Sellafield, Seascale, Cumberland

SUMMARY

Sodium alginate taken orally before, along with or after ingestion of tracer doses of [85]Sr inhibited the uptake of the tracer from the gastro-intestinal tract in two adult human subjects.

INTRODUCTION

The most probable route by which radiostrontium enters the human body is in the diet, primarily via milk, although after an accidental inhalation of radiostrontium a large proportion of the material may be cleared from the respiratory tract to the gut (ICRP, 1959). About 30% of the radiostrontium deposited in the gastro-intestinal tract is believed to be transferred to the blood stream, and part is deposited in bone. It would be useful in certain circumstances to be able to inhibit the transfer of radiostrontium from the gastro-intestinal tract.

Since milk is one of the body's chief sources of stable Ca, it is desirable that the material used to inhibit the uptake of radiostrontium from the gastro-intestinal tract should, if possible, selectively discriminate against Sr in favour of Ca. The body would then not be deprived of its supply of stable Ca.

It is known that sodium alginate extracted from certain varieties of seaweed has a higher affinity for Sr than for Ca (Haug, 1961; Haug and Larson, 1964), but that the affinity depends upon the form of the alginate, which in turn depends upon the variety of seaweed from which the alginate was derived.

Experiments involving the administration of sodium alginate to humans (Millis and Reed, 1947; Feldman *et al.*, 1952) in amounts of 8–45 g/day, have shown that there was little removal of Ca, sodium and potassium from the gastro-intestinal tract.

Experiments in which sodium alginate was administered to rats (Paul *et al.*, 1964; Skoryna *et al.*, 1965; Waldron-Edward *et al.*, 1964, 1965) have shown that the uptake of radiostrontium from the gastro-intestinal tract was reduced more than that of radiocalcium.

These considerations indicated that sodium alginate might be suitable material for selectively inhibiting the uptake of radiostrontium from the human gastro-intestinal tract.

MATERIALS USED

Sodium alginate derived from *Laminaria hyperborea* stems (Haug, 1961; Haug and Larson, 1964) has been found to have a decreasing affinity for the following divalent metals, in the order $Sr > Cd > Ca$.

The affinity for Sr with respect to cadmium is strongly dependent upon the variety of sodium alginate. Alginic acid is known to include mannuronic and guluronic acids in its constituents (Haug, 1961; Haug and Larson, 1964), and the relative affinity for strontium with respect to cadmium (and presumably Ca) is related to the ratio mannuronic acid/guluronic acid in the variety of sodium alginate. Samples with the lowest values of this ratio were associated with the highest affinity for Sr with respect to cadmium (and presumably Ca). This criterion was used to select Manucol SS/LD2 as the most suitable commercially available form of sodium alginate for our experiments. It also satisfied the requirement of low viscosity.

The sodium alginate solution administered in this series of experiments consisted in each case of 10 g of Manucol SS/LD2 that had been shaken up with 20 ml of absolute alcohol, and then made up to 200 ml with distilled water and flavouring.

A control solution, consisting of 20 ml of absolute alcohol, 180 ml of distilled water, and flavouring was also used.

^{85}Sr was obtained as carrier free strontium chloride in sterile aqueous solution. A total of 0·84 μCi ^{85}Sr was administered (in 10 ml of distilled water) to subject A, and 1·25 μCi to subject B. The resulting dose to whole body and bone was less than 5 millirads.

EXPERIMENTAL PROCEDURE

Details of the oral administrations associated with each experiment are summarized in Table I. No food was consumed during the 12 h before, or 3 h after each administration, although a drink of water was taken about an hour before administration.

Immediately before stages A1 and B1, each subject was examined in a whole-body counter in order to determine the normal body radioactivity due to fallout ^{137}Cs and ^{40}K.

In each stage of the experiment *in vivo* measurements of body radioactivity were made in the Windscale Whole Body Counter in order to determine the patterns of body retention of ^{85}Sr.

TABLE I

Oral administrations associated with each stage of the experiment

Subject	Stage of experiment	Details of administration	Time after first administration (days)
A	A1	Sodium alginate solution drunk 20 min before 0·36 μCi ^{85}Sr	0
	A2	Control solution drunk 20 min before 0·48 μCi ^{85}Sr	26
B	B1	Sodium alginate solution drunk 2 min before 0·43 μCi ^{85}Sr	0
	B2	Sodium alginate solution drunk 20 min after 0·41 μCi ^{85}Sr	28
	B3	Control solution drunk 2 min before 0·41 μCi ^{85}Sr	71

Measurements of the count-rate from ^{85}Sr in the chest and abdomen regions of subject A were made during the first 50 h after each administration. A detector incorporating a NaI (Tl) crystal, 7 cm diam × 12·5 cm high and collimated in lead of thickness 2·5 cm, was used for this purpose.

Daily urine samples were collected. Subject A also gave a series of faecal samples in stages A1 and A2.

MEASUREMENTS OF BODY RADIOACTIVITY

γ-Ray spectra of body radioactivity showed photopeaks due to ^{85}Sr (0·51 MeV), ^{137}Cs (0·662 MeV) and ^{40}K (1·46 MeV). In order to determine the patterns of body retention of ^{85}Sr in each stage of the experiments, the net count-

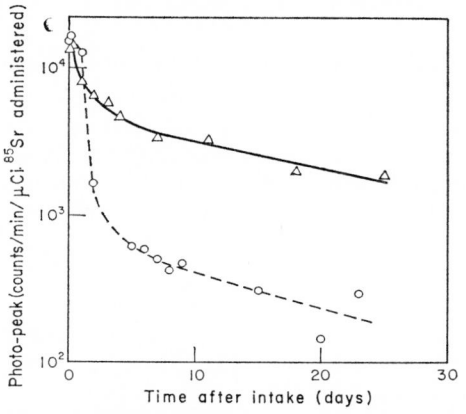

FIG. 1. Body retention of ^{85}Sr in subject A; △, without sodium alginate (A2); ○, with sodium alginate (A1).

M 2   S.M

rate due to ^{85}Sr (corrected for ^{137}Cs and ^{40}K) in the 200 keV band centred on the ^{85}Sr photopeak (0·51 MeV) was calculated for each measurement, and expressed as counts/min per μCi ^{85}Sr administered, corrected for the physical decay of ^{85}Sr (Figs. 1 and 2).

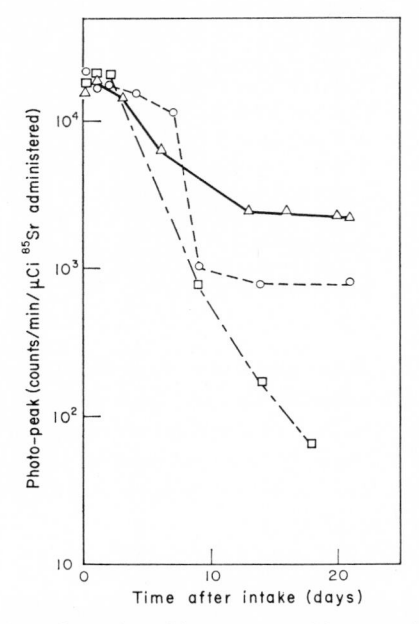

FIG. 2. Body retention of ^{85}Sr in subject B; △, without sodium alginate (B3); ○, with sodium alginate (B2); □, with sodium alginate (B1).

TABLE II

Subject A: whole-body retention of strontium-85

Corrected for physical decay

| Time after administration (days) | Counts/min/μCi ^{85}Sr administered | |
	With sodium alginate (A1)	Without sodium alginate (A2)
18	··	2055 ±40
20	148 ±37	··
23	298 ±38	··
25	··	1870 ±40
Mean value	223 ±27	1963 ±29

Effectiveness of sodium alginate taken 20 min before ^{85}Sr $\dfrac{1963 \pm 29}{223 \pm 27} = 8 \cdot 8 \pm 1 \cdot 1$.

In stages A1 and B1, the net contributions from [85]Sr had dropped to about the limit of detection before stages A2 and B2 started. It was however necessary to make small corrections to the data in stage B3 for the residual [85]Sr from B2.

Estimations of the effectiveness of sodium alginate in inhibiting the uptake of [85]Sr from the gastro-intestinal tract are given in Tables II and III.

The net count-rates per μCi [85]Sr administered to subject A, as measured by means of the collimated detector 49 h after each administration, are given in Table IV.

TABLE III

Subject B: whole-body retention of strontium-85

Corrected for physical decay

Time after administration (days)	Counts/min per μCi [85]Sr administered		
	With sodium alginate (B1)	With sodium alginate (B2)	Without sodium alginate (B3)
13	..	..	2440 ±41
14	166 ±24	780 ±32	..
16	..	..	2390 ±41
18	65 ±28	..	..
20	..	..	2220 ±40
21	..	790 ±33	2170 ±40
Mean	116 ±19	785 ±23	2305 ±20
Effectiveness of sodium alginate taken 2 min before and 20 min after [85]Sr	19·9 ±3·3	2·94 ±0·09	..

TABLE IV

Counting rates over the chest and abdomen of subject A

Counts/min per μCi [85]Sr

	Chest	Abdomen
With sodium alginate (A1)	12·5 ±4·5	50·7 ±5·6
Without sodium alginate (A2)	176 ±7·2	133 ±6·0
Ratio $\left(\dfrac{A2}{A1}\right)$	14·1 ±5·1	2·6 ±0·3

STRONTIUM-85 IN URINE AND FAECES

Each 24-hour-urine sample was analysed for ^{85}Sr, Ca and creatinine. Owing to a considerable day-to-day variation in urine samples given by subject B, the results were all normalized to his average output of 1·32 g creatinine/day in order to smooth out the data. (Figs. 3 and 4, Tables V and VI).

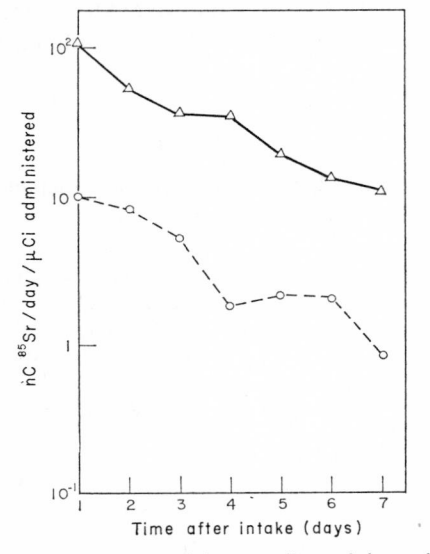

FIG. 3. ^{85}Sr in urine, subject A: △, without sodium alginate (A2); ○, with sodium alginate (A1).

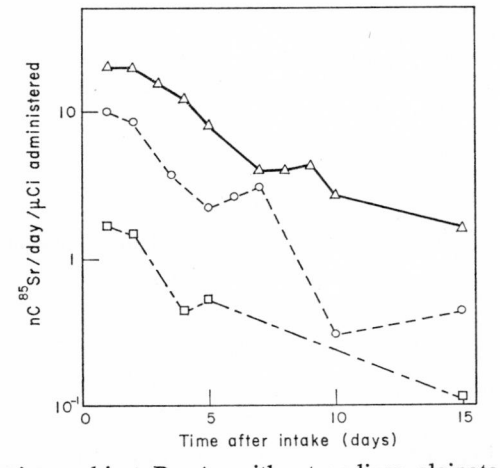

FIG. 4. ^{85}Sr in urine, subject B; △, without sodium alginate (B3); ○, with sodium alginate (B2); □, with sodium alginate (B1).

TABLE V

Subject A: strontium-90, calcium and creatinine in urine

	Stage of experiment	^{85}Sr nC per μCi administered	Ca (mg)	Creatinine (g)
7 days' cumulative excretion	A1	30·6	3443	14·17
7 days' cumulative excretion	A2	280·8	3307	14·17

$$\text{Effectiveness of sodium alginate} \left(\frac{A2}{A1}\right) = \frac{280\cdot8}{30\cdot6} = 9\cdot2 \pm 0\cdot4.$$

TABLE VI

Subject B: strontium-85 in urine, cumulative excretion

Period (days)	Stage of experiment B1	B2	B3	Effectiveness of sodium alginate	
1, 2, 4, 5*	4·1	..	59·81	$\left(\dfrac{B3}{B1}\right) = 14\cdot6$	
1, 2, 4, 5†	25·0	..	508·0	$\left(\dfrac{B3}{B1}\right) = 20\cdot3$	Mean = 17·5 ± 2
1 to 5*	..	28·27	75·31	$\left(\dfrac{B3}{B2}\right) = 2\cdot67$	
1 to 5†	..	168·8	613·5	$\left(\dfrac{B3}{B2}\right) = 3\cdot63$	Mean = 3·15 ± 0·15

* nC/μCi administered.
† nC/μCi administered/g Ca/day.

Faecal samples from subject A were analysed for ^{85}Sr and Ca. Results are presented in Fig. 5. Subject B excreted an average of 175 mg Ca/day in urine.

DISCUSSION

The overall findings of these experiments are summarized in Table VII.

The results of localization measurements on subject A (Table IV) and the amounts of ^{85}Sr in his faecal samples (Fig. 5), support the conclusion that less ^{85}Sr was absorbed from the gastro-intestinal tract when sodium alginate was present. This was also illustrated (Fig. 5) by the higher initial excretion of exogeneous ^{85}Sr in faeces, and the lower excretion of endogenous ^{85}Sr after the fourth day (Hesp and Ramsbottom, 1965a, b).

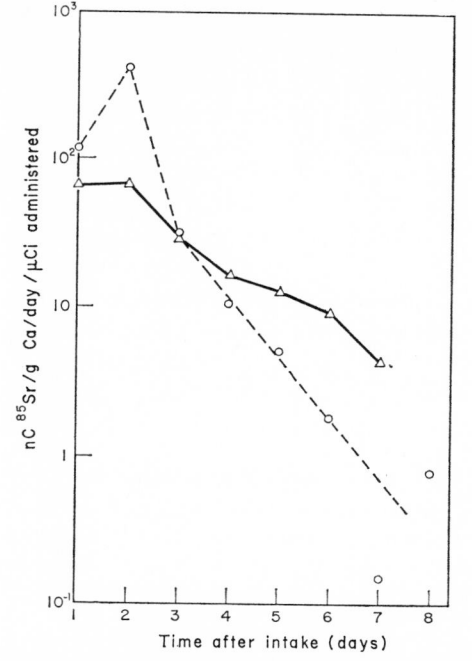

FIG. 5. ^{85}Sr in faeces, subject A; $\triangle$ without sodium alginate (A2); $\bigcirc$, with sodium alginate (A1).

TABLE VII

Approximate values for the factor by which sodium alginate reduced uptake of radiostrontium from the human gastro-intestinal tract

Conditions of administration	Factor by which uptake of ^{85}Sr was reduced
Sodium alginate 20 min before ^{85}Sr	9
Sodium alginate 2 min before ^{85}Sr	19
Sodium alginate 20 min after ^{85}Sr	3

Experiments with rats have also shown that sodium alginate has its maximum influence upon uptake of radiostrontium from the gastro-intestinal tract when both are administered together (Paul *et al.*, 1964).

The actual values presented in Table VII might be slightly different for humans who had not been starved before administration because the ^{85}Sr solution might be expected to mix with the contents of the stomach and remain there longer. Sodium alginate administered 20 min later might then have a greater influence upon uptake of ^{85}Sr from the gastro-intestinal tract.

Having demonstrated that a suitable variety of sodium alginate does reduce the uptake of radiostrontium from the human gastro-intestinal tract it will be necessary to investigate the following points: (*a*) the effect of the contents of the stomach; (*b*) the optimum amount of sodium alginate to administer; and (*c*) the effect of this variety of sodium alginate upon Ca uptake.

Assuming that the most suitable variety of sodium alginate was used in these two experiments, it is worth considering whether an enrichment of the guluronic acid fraction would make the sodium alginate even more effective.

We thank Dr C. T. Blood of Alginate Industries Limited for his advice in the choice of a suitable variety of sodium alginate.

The co-operation of the two volunteers, and the advice of Dr G. W. Dolphin and of our colleagues at Windscale, are gratefully acknowledged.

REFERENCES

Feldman, H. S., *et al.* (1952). *Proc. Soc. exp. Biol. Med.* **79**, 439–441.

Haug, A. (1961). *Acta. chem. scand.* **15**, 1794–1795.

Haug, A., and Larson, B. (1964). *Proc. 4th Int. Seaweed Symp.*

Hesp, R., and Ramsbottom, B. (1965a). 1st International Conference on Medical Physics. Harrogate. [P. G. Report 686 (W)].

Hesp, R., and Ramsbottom B. (1965b). *Nature, Lond.* **208**, 1341–1342.

International Commission on Radiological Protection: Report of Committee II (1959).

Millis, J., and Reed, F. B. (1947). *Biochem. J.* **41**, 273–275.

Paul, T. M. *et al.* (1964). *Can. med. Ass. J.* **91**, 553–557.

Skoryna, S. C. *et al.* (1965). *Can. med. Ass. J.* **93**, 404–407.

Waldron-Edward, D. *et al.* (1964). *Can. med. Ass. J.* **91**, 1006–1010.

Waldron-Edward, D. *et al.* (1965). *Nature, Lond.* **205**, 1117–1118.

Mechanism of Citrate in Influencing the Excretion of Radioactive Strontium

H. SMITH

United Kingdom Atomic Energy Authority,
Chapelcross, Annan

SUMMARY

Carrier-free ^{85}Sr was injected intra-arterially into rabbits. The rate of disappearance from the plasma was followed in animals with abnormally high concentrations of plasma citrate (maintained by continual injection of sodium citrate) and in control animals.

INTRODUCTION

The use of citrate to remove radioactive Sr from animals is not new (Schubert and Wallace, 1950; Cohn and Gong, 1953; Catsch, 1957; Ito and Tsurufiji, 1958; Catsch and Melchinger, 1959; Carlqvist and Nelson, 1960; Smith and Bates, 1965), but to the best of our knowledge, the mechanism of action has not been elucidated. Early administration of citrate, that is, within 30 min of the radioactive isotope would appear to be essential if effective treatment is required. Delay results in a dramatic reduction in efficiency of removal. A typical experiment involving early treatment in rat is shown (Fig. 1); early citrate therapy combined with stable Sr (44 mg/kg) is slightly more effective than this.

Unlike Ca, which is under direct homeostatic control, Sr metabolism is said to be regulated by the total alkaline-earth metal concentration, in particular Ca ion (Comar and Bronner, 1964).

A marked elevation of plasma citrate concentration produces a hypercalcaemia and resultant hypercalcuria. It is thought that the citrate forms an un-ionized calcium citrate complex and that Ca-ion equilibrium in extracellular fluid is maintained by mobilization of Ca from bone. The complex is ultrafilterable, and although much of it is metabolically destroyed in the renal tubular cells, excretion of citrate occurs when the citrate concentration in the tubular fluid is in excess of the maximum rate of catabolism by the kidneys. The stability of the complex is influenced by the hydrogen-ion concentration of the renal tubular fluid; in both man and rat, alkalosis results in increased excretion of Ca and citrate.

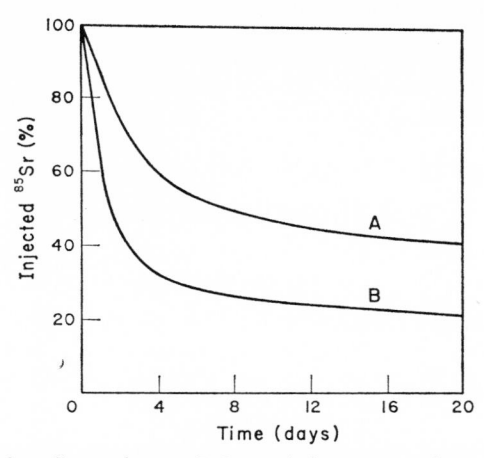

FIG. 1. Effect of sodium citrate (3·5 mM/kg) on retention of ^{85}Sr in rat: A, control rats; B, treated rats.

In the experiments to be described, designed to study the action of citrate specifically upon radioactive Sr, carrier-free ^{85}Sr was injected intra-arterially into rabbits and the rate of disappearance from the plasma followed in the presence or absence of excessive concentrations of citrate.

EXPERIMENTAL DETAILS

Adult female albino rabbits (weight 4·0–4·5 kg) were anaesthetized with nembutal (45 mg/kg/intraperitoneally). Under additional ether anaesthesia, a tracheostomy was performed, the right carotid artery exposed and a polythene tube shunt (capacity, 3 ml; internal bore, 1 mm) introduced. In this way, arterial blood could be withdrawn and citrate injected as desired. The animals were heparinized (15 mg/kg) and 10 μCi of ^{85}Sr injected (0·5 ml, as SrCl$_2$ carrier-free, pH 7). Blood samples (5 ml) were collected in heparinized tubes at frequent intervals, and counted for activity in a well-type plastic phosphor. The distribution of activity in some of the blood samples was also determined and remaining blood was re-injected after counting. Urine samples were obtained by pressure in the suprapubic region and the ejected fluid was collected on absorbent tissue. This was counted as for blood. Rectal temperatures were recorded during experiments and animals developing a hypothermia below 36° C were excluded.

Control animals received no other treatment. Sodium citrate (2·2 mg/kg/min of a 3% solution at 0·3 ml/min) was injected into the cerebral end of the carotid artery in the citrate-treated group. Under these circumstances, it was possible to maintain a plasma citrate concentration of approximately 40 mg/100 ml, and about 20% of the injected dose was excreted in the urine during the period of infusion. High cellular concentrations of citrate can be maintained by this technique without the cardio-toxic effects commonly associated with intravenous injection. Tetany, rarely observed, was alleviated by the injection of calcium gluconate (3 mg/kg in 0·1 ml).

RESULTS

Whole blood activity is expressed as the percentage of injected activity per 5 ml blood. The amount present in the total blood volume can be calculated assuming a blood volume of 180 ml. It is apparent that citrate does not influence the rate of disappearance of ^{85}Sr from blood (Fig. 2), nor does it affect the partition of activity between erythrocytes and plasma (Table I).

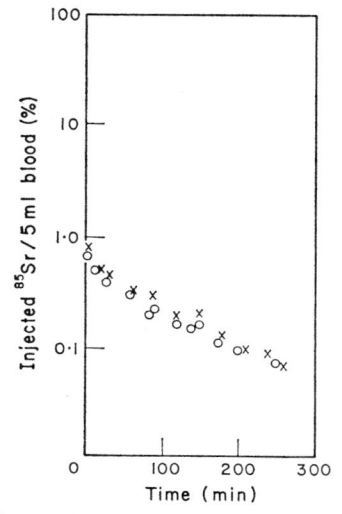

FIG. 2. Influence of citrate on retention of ^{85}Sr in rabbit blood: ○, control; ×, citrate.

Ultrafiltration of plasma, however, revealed that a greater amount of ^{85}Sr was contained in the ultrafilterable fraction of the citrate-treated animals as compared to controls, but this persisted only during the first 60 min. This is consistent with the formation of a citrate complex. The distribution of residual activity in the body was not influenced by the presence of citrate (Table II). Thus the beneficial effects of citrate must depend upon the renal excretion of the complex in the early stages (Table III). Further, the amount excreted can be modified by the pH of the urine. Artificial ultrafiltration or dialysis would obviously eliminate the problem of physiological reabsorption of filtrate and in order to test the hypothesis, an extra-corporal shunt of arterial blood through a dialysis unit was attempted.

TABLE I

Influence of citrate on the distribution of strontium-85 in rabbit blood

	Time (min)						
	5	15	30	60	90	150	210
Control							
% injected dose in:							
(a) Erythrocytes	3·5	4·9	3·8	2·7	3·7	2·0	4·3
(b) Plasma	96·5	95·1	96·2	97·3	96·3	98·0	95·7
% ^{85}Sr in ultrafilterable fraction of plasma	21·0	20·4	21·1	20·9	21·0		
% injected dose per 100 ml of ultrafiltrate	2·7	2·0	1·7	1·2	1·0		
Citrate treated							
% injected dose in:							
(a) Erythrocytes	5·4	3·3	3·1	3·0	4·4	1·8	1·3
(b) Plasma	94·6	96·7	96·9	97·0	95·6	98·2	98·7
% ^{85}Sr in ultrafilterable fraction of plasma	49·0	35·2	24·7	31·4	15·0		
% injected dose per 100 ml of ultrafiltrate	8·0	3·4	2·1	2·0	0·9		

TABLE II

Influence of citrate upon the distribution of strontium-85 in rabbit after 4·5 h

	Control	Citrate treated
Urine	3·5	13·8
Skeleton	86·5	75·4
Soft tissues	0·2	0·6
Heart and lungs	0·1	0·1
Liver	0·1	0·3
Kidneys	0·2	0·3
Stomach and intestines	9·2*	7·9*
Skin	0·2	1·5

* Mainly in large intestine.

TABLE III

**Influence of various parameters upon the renal
excretion of strontium-85 in rabbit**

	% injected dose excreted in 4 h	% injected dose found in dialysate in 4 h
Control	3·46	..
Citrate (urine pH 5)	10·97	..
Citrate (urine pH 6)	13·84	..
Citrate (urine pH 8)	22·55	..
Citrate + dialysis	24·54	2·50
Dialysis	13·44	1·75
NaCl infusion (urine pH 8)	12·85	..

A Technicon dialyser with a Cellophane membrane (pore size 40–60 Å) was used at 38° C. The dialysing fluid was prepared to simulate as closely as possible a protein-free plasma (3,170 mg/l of Na^+; 4,474 mg/l of Cl^-; 156 mg/l of K^+; 1,400 mg/l of HCO_3^-; 80 mg/l of Ca^{2+}; 30 mg/l of Mg^{2+}). Arterial blood circulated through the shunt at the rate of 6 ml/min. Surprisingly little activity appeared in the dialysate (Table III), but this may be due to the relative inefficiency of the unit compared to the kidney. The act of dialysis potentiated the renal excretion of ^{85}Sr, presumably by creating a hypernatraemia, a condition associated with increased calcium clearance (Walser, 1961a, b; Walser and Robinson, 1962). Confirmation of this was obtained by infusing isotonic sodium chloride under alkaline conditions into a rabbit, thereby causing increased renal excretion of ^{85}Sr.

This communication is intended to draw attention to the value of citrate as an early treatment in the removal of radioactive Sr. The optimum conditions of use have still to be determined, but ideally, metabolic blocking of the citrate complex within the renal tubules should lead to maximum excretion. The fact that citrate is ineffective after approximately 1 h is probably due to the fact that the radioactive Sr has moved out of the vascular compartment into bone so that there are negligible amounts of activity available for chelation and subsequent renal clearance.

REFERENCES

Carlqvist, B., and Nelson, A. (1960). *Acta. radiol.* **54,** 305–315.
Catsch, A. (1957). *Naturwissenschaften,* **44,** 94.
Catsch, A., and Melchinger, H. (1959). *Strahlentherapie,* **108,** 63–65.
Cohn, S. H., and Gong, J. K. (1953). *Proc. Soc. exp. Biol. Med.* **83,** 550–554.
Comar, C. L., and Bronner, F. (eds.) (1964). "Mineral Metabolism", vol. 2, part A, p. 529. Academic Press, New York.

Ito, Y., and Tsurufuji, S. (1958). *Proc. 2nd Int. Conf. peaceful Uses Atom. Energy*, **23,** 443–450.

Schubert, J., and Wallace, H. (1950). *J. biol. Chem.* **183,** 157–166.

Smith, H., and Bates, T. H. (1965). *Nature, Lond.* **207,** 799–801.

Walser, M. (1961a). *Am. J. Physiol.* **200,** 1099–1104.

Walser, M. (1961b). *Am. J. Physiol.* **201,** 769–773.

Walser, M., and Robinson, B. H. B. (1962). *Bull. Johns Hopkins Hosp.* **111,** 20–34.

Inhibition of Absorption of Radioactive Strontium by Alginic Acid Derivatives

DEIRDRE WALDRON-EDWARD, T. M. PAUL
AND STANLEY C. SKORYNA

*Gastro-Intestinal Research Laboratory, Donner Building for
Medical Research, McGill University, Montreal*

SUMMARY

Naturally occurring polyuronic acids have been shown to be capable of complexing with Sr ions *in vivo*, rendering them unavailable for transfer across the intestinal mucosa. This finding has an important practical application as a means of removing radioactive Sr from the diet. To extend this application to human subjects poses many problems, including dosage and mode of administration of alginate derivatives, effects of salts of alginic acid other than sodium salts, the intestinal binding of resecreted radioactive Sr and differences in inhibitory ability of different samples of alginate. These problems have been studied with rats as experimental animals.

Under normal laboratory feeding conditions, sodium alginate reduces ^{89}Sr absorption and bone uptake in constant proportions. This was observed at a high level of alginate intake (24% diet), at a low level (1·4%) and at an intermediate level. ^{45}Ca uptake in the bone is affected by long-term feeding with high levels of alginate. Growth-rate and balance studies have shown that interference with Ca metabolism is not sufficient to be detrimental to the health of young growing rats in the period under observation.

Preliminary experiments have shown that calcium alginate can be substituted for sodium alginate in the inhibition studies, in order to boost Ca intake. This finding is interesting because of the implication that alginate acts as a specialized type of ion-exchange material. It also offers new possibilities in the investigation of the complex problem of the kinetics of Ca absorption.

Alginate increases the rate of faecal excretion of endogeneous radiostrontium from rats and from kittens. It is suggested that alginate feeding may be of value in removing previously absorbed ^{90}Sr, introduced by accidental ingestion or by inhalation.

Sodium alginates extracted from various seaweeds as well as material obtained from commercial sources showed wide differences in their ability to inhibit the absorption of radioactive Sr. The materials tested included species obtained from the North American Pacific and Atlantic Coasts and also from locations in the South Pacific (Easter Island and New Zealand). The relationship of these variations to differences in chemistry of the preparations studied are discussed.

INTRODUCTION

In 1957, while studying the induction of malignant bone tumours by radioactive Sr (Skoryna and Kahn, 1959), it occurred to us that inhibition of the carcinogenic action of radioactive Sr might be effected by interference with the intestinal absorption of the isotope. In the subsequent years, a systematic investigation of substances that might inhibit the uptake of radioactivity from the gastro-intestinal tract was undertaken. In spite of the many methods used, the removal of Sr deposits from skeletal tissue by chelating drugs has not been successful. Schubert has discussed the probable reasons for this failure (Schubert, 1958). It has been found possible to remove radiostrontium from milk, the principal dietary source, by means of ion-exchange resins (Migicovsky, 1959; George, 1962; Edmondson et al., 1962; Silverman et al., 1963). A large-scale process for fluid milk has been developed, and tests have shown it to be commercially feasible (Murthy et al., 1961; Fooks et al., 1966). This process, requiring special milk plant equipment, removes about 91 % of the radiostrontium in the milk; other cationic radionuclides are removed with lesser efficiency. This process is of value to infants and children; for adults, consuming a variable diet, some other solution must be sought.

In particular, alginic acid, used extensively in the food industry, has been studied in some detail. Alginic acid derivatives have been found markedly to inhibit intestinal absorption of radioactive Sr if the appropriate alginate–Sr ratio is used (Skoryna, et al., 1964; Paul et al., 1964; Waldron-Edward et al., 1964; Skoryna et al., 1965). On the basis of these studies, a method was described that permits selective suppression of absorption of radioactive Sr from ingested food material permitting the Ca to be available to the body. These results were confirmed by Stara (1965) in cats, by Hesp and Ramsbottom (1965) in a human subject and by van der Borght et al. (1966) in pigs.

The present paper discusses some of the current problems in studies of radiostrontium absorption. This field of investigation is much more complex than it would appear on the surface, because little is known about the basic factors affecting absorption of metal ions, such as pH, electrolytes, transit time, resecretion and also about the mechanisms involved in macromolecule–metal ion binding. Although there is available a commercial product, which can be used effectively in emergency cases to inhibit radioactive strontium absorption, many problems remain, including a survey for suitable resources of phaeophyceae.

NATURALLY OCCURRING POLYURONIC ACIDS: INHIBITORY ACTIVITY

Alginic acid and pectic acid are polymeric uronic acids. These natural polymers are distinguished chemically from other plant gums and mucilages by their ability to precipitate with Ca and Sr. Alginic acid is composed of mannuronic and guluronic acids in varying proportions. It comprises part of the intercellular substance of the brown seaweeds (Baardseth, 1965). Pectic acid is a polygalacturonic acid derived from fruit pectins. Both substances are used commercially in the food and drug industries for many purposes, such as gelling agents, emulsifiers and thickeners. These substances, which are not absorbed from the intestine, could bind the alkaline-earth metals *in vivo* and therefore prevent their absorption.

Both pectic and alginic acids were tested (Skoryna *et al.*, 1964; Waldron-Edward *et al.*, 1965) by injecting aqueous solutions of their sodium salts into ligated segments of the rat's gastro-intestinal tract: the amount of radio-activity in the blood and in bone was measured after 30 min. It was found that the absorption of radioactive Sr was reduced by as much as 80% whereas Ca absorption was affected very little. Quantitatively, alginic and pectic acids acted very similarly.

The method of testing inhibition of absorption by the use of ligated duodenal segments, in which the vascular and lymphatic systems remained intact, has proved very useful in screening-testing large numbers of different batches of alginate as well as other potentially useful materials. With a skilled operator, the results are very consistent and reproducible. However, the method is somewhat tedious even for highly trained personnel, and latterly we have utilized an intubation method for this purpose. The radioactive isotope, mixed with a fixed proportion of carrier, is introduced directly into the stomach of the rat, together with the sample; 24 h later the animal is sacrificed and one femur is assayed for radioactivity (Paul *et al.*, 1964).

Dose response

The addition of a relatively small proportion of alginate to radioactive Sr (2 mg, to Sr^{2+}, 0·5 μl in 0·5 ml) injected into the ligated duodenal segments reduces bone uptake by 60% within the 0·5-h period. Increasing the dose of alginate increases the amount of ^{89}Sr bound and therefore not available for absorption (Table I). The maximum dose given was 16 mg/ml, that is, a ratio of eighty combining units of uronic acid to 1 mmole of metal ion.

TABLE I

Effect of increasing concentrations of sodium alginate on strontium-89 and calcium-45 uptake in bone 30 min after injection into ligated duodenal segment

Each figure represents the mean counts/min in one femur and standard deviation for studies in ten rats

Alginate (mg)	^{89}Sr 5 μCi ^{89}Sr $+0.45$ μM inactive SrCl$_2$	Exp. % control	^{45}Ca 5 μCi ^{45}Ca $+0.55$ μM inactive CaCl$_2$	Exp. % control	OR
Control	31,023 ($\pm$10,300)	..	42,750 ($\pm$9,536)	..	0·72
2 mg	12,401 ($\pm$3,587)	39·9	30,483 ($\pm$5,883)	71·3	0·40
4 mg	9,111 ($\pm$2,736)	29·3	24,966 ($\pm$7,682)	58·4	0·36
8 mg	7,089 ($\pm$928)	22·8	17,707 ($\pm$4,485)	41·6	0·39
12 mg	5,298 ($\pm$2,033)	17·0	13,733 ($\pm$1,861)	32·1	0·38
16 mg	3,672 ($\pm$1,472)	11·8	10,461 ($\pm$1,782)	24·4	0·35

Sodium alginate, dissolved in water to form a stiff jelly, can be fed to animals mixed with their normal laboratory chow. Rats tolerate this diet very well. Alginate can be added up to 24% of the diet in this way. By increasing the amount of alginate progressively from 3% to 24% of the diet, the absorption of radiostrontium from the gut was depressed from 65% to 31% (Fig. 1).

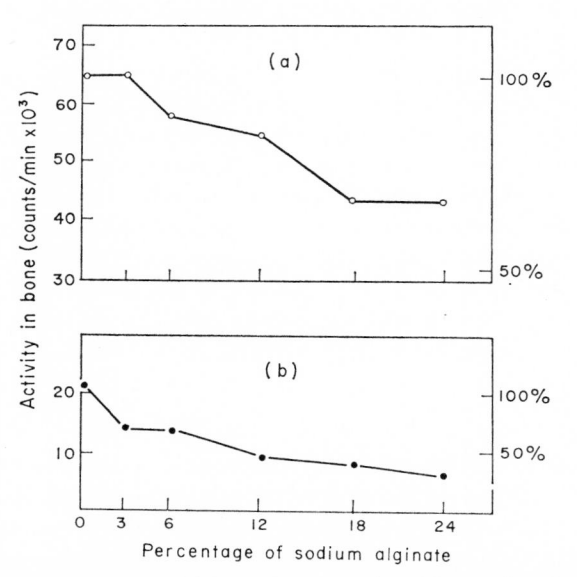

FIG. 1. Effect on (a) ^{45}Ca and (b) ^{89}Sr bone uptake of increasing the percentage of sodium alginate in the diet.

Rats do not consume their normal amount of chow when a larger proportion of alginate is added, and so the animals lose weight.

The absorption of Ca is also affected by sodium alginate, but not to the same extent. The resulting observed ratio (OR) is, therefore, reduced, from the control value of 0·32 to 0·15 when rats of the same age group are placed on the diet containing 24% alginate. This results in an enhancement of the already existing biological discrimination in favour of Ca over Sr.

The actual amount (in μg) of radioactive Sr given in these experiments is very small compared with the inactive Sr and Ca present in the diet. The range of ^{89}Sr to be expected in atmospheric fall-out or as a result of accidental nuclear explosions is probably very wide.

Over a dosage range of from 0·2 to 50 μCi/day, a linear relationship is found between the amount of Sr given and bone uptake (Fig. 2). Addition of sodium alginate to the diet reduced bone uptake in a constant proportion according to the amount of alginate given (Fig. 3).

The importance of this constant proportionality lies in the likelihood that the uptake from a constant low intake of ^{90}Sr over a long period of time could be reduced by the continual intake of a low dosage of sodium alginate. This we have confirmed in rats (Skoryna et al., 1964), in which, over a period of 55 days on a low dosage of ^{89}Sr (0·2 μCi/day), radiostrontium uptake in the bone was reduced by 40% as a result of feeding 140 mg of sodium alginate daily, dissolved in drinking water. The bone uptake of ^{45}Ca in this case remained approximately the same as in the controls so that the OR fell from 0·21 to 0·13.

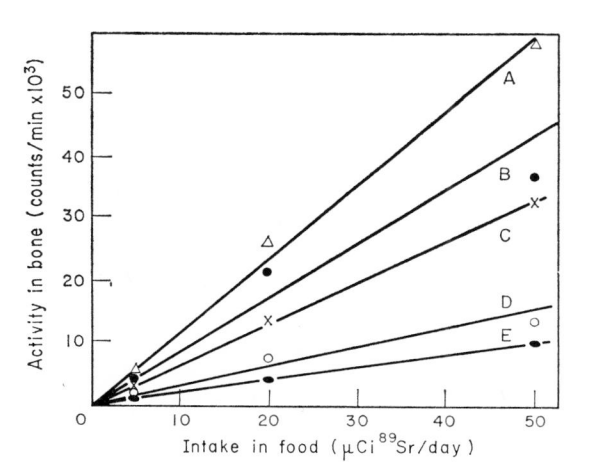

FIG. 2. Relationship between ^{89}Sr activity in bone and in diet: A, control, ^{89}Sr in drinking water; B, control, ^{89}Sr mixed with food; C, as in A, with sodium alginate, 1·4 g/100 ml; D, as in B, with sodium alginate, 12 g/100 g; E, as in D, with sodium alginate, 24 g/100g.

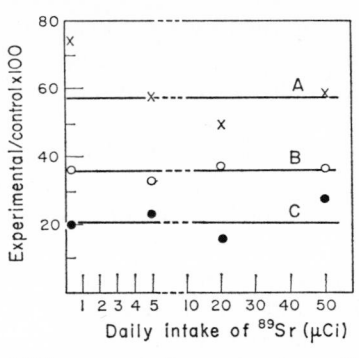

Fig. 3. Bone uptake of ^{89}Sr, expressed as percentage of control value, related to dosage of radiostrontium administered daily for 5 days. Mean values: A, 59%; B, 36%; C, 22%. Sodium alginate given concurrently, as follows: A, 1·4% w/w diet, dissolved in drinking water; B and C, 24% w/w mixed as a jelly with chow.

SODIUM AND CALCIUM ALGINATE

At higher doses of sodium alginate in long-term feeding experiments, the suppressant effect of alginate on Ca absorption becomes apparent in the reduction of ^{45}Ca appearing in the bone.

Standard laboratory chow provides a large pool of available Ca, more than adequate to satisfy the daily requirements of the young, growing rat. The radioactive isotope is diluted considerably with inert Ca when added to the diet so that the specific activity of isotopes is thereby reduced. The turnover of Ca is so large however, that appreciable amounts of ^{45}Ca are deposited in the bone; it is assumed that under normal dietary conditions, ^{45}Ca in the bone represents a predictable proportion of the total Ca absorbed from the intestine.

By feeding large doses of sodium alginate for a long period of time it was anticipated that there should be some gross evidence of Ca deficiency. It was found that the actual amount of Ca consumed with the alginate diet had been reduced to 388 mg from the normal of 450 mg/day. A positive Ca balance was retained, however, and the young rats fed on this diet for 8 weeks maintained a steady growth rate, only slightly less than in control rats of the same age.

The possibility of increasing the intake of Ca by replacing the sodium salt of alginic acid with calcium alginate was next investigated.

Calcium alginate is insoluble in water. It may be obtained commercially as a fine white powder that can be suspended in aqueous solvents. The suspensions can be injected into gut segments with a No. 25 needle, or introduced into the stomach by orogastric tube. It can also be mixed dry with the powdered laboratory chow. This mixture is palatable to the rats; they eat it with no symptoms of distress, such as the constipation observed on feeding dry sodium alginate.

Preliminary screening studies on dose ^{89}Sr/bone uptake response were carried out on rats by means of the ligated duodenal segment method as used for sodium alginate (Table II). [89]Sr absorption was reduced by 82% when the ratio of calcium alginate/Sr (carrier) was 20/1 (μM uronic acid residues/μM Sr) approximately. Increasing the proportion of calcium alginate to 60/1 suppressed absorption of ^{89}Sr to 96%.

TABLE II

Effect of increasing concentrations of calcium alginate on strontium-89 and calcium-45 uptake in bone 30 min after injection into the ligated duodenal segments of rats

Each figure represents mean counts/min in one femur and standard deviation for studies in twelve rats.

Calcium alginate (mg)	[89]Sr 5 μCi [89]Sr + 0·4 μM inactive SrCl$_2$	Expt. % Control	[45]Ca 5 μCi [45]Ca + 0·55 μM inactive CaCl$_2$	Expt. % Control
Control	31,293 ± 3294		37,591 ± 6481	
4 mg	5,707 ± 1486	18·2	17,805 ± 4886	47·3
6 mg	5,129 ± 1379	13·1	10,675 ± 3219	28·3
8 mg	3,700 ± 1304	11·8	11,362 ± 3427	30·2
12 mg	1,362 ± 422	4·3	7,109 ± 1669	18·9

Ca absorption was also altered. The amount of [45]Ca appearing in the femur after injection of 4 mg of alginate was 47% of the control value, compared with 58% when an equal quantity of sodium alginate was used. (No corrections were made in these figures for relative Na and Ca content, nor for the moisture). Increasing the proportion of alginate (60/1) decreased the amount of [45]Ca deposited in the bone to 19% control values at the maximum concentration used.

Similar results were obtained when calcium alginate was introduced in suspension through an orogastric tube into the stomach. After 14 h the count of ^{89}Sr in the femur was 21% of control value, whereas [45]Ca in the bone was reduced to 65% of control. The OR in these experiments was 0·28, comparing very closely with the corresponding figure of 0·25 in the studies with sodium alginate.

These results suggest two interesting features. Firstly that calcium alginate is at least as good as, if not better than, sodium alginate in suppressing the absorption of radioactive Sr. Secondly, that the ratio of ^{89}Sr/[45]Ca bound by the calcium alginate is the same as in the reaction with sodium alginate. This finding supports the view that alginate is, in effect, a specialized type of ion-exchange material. There is a free interchange of the radioactive Sr and Ca in solution with the "bound" Ca.

RESECRETION OF ABSORBED RADIOSTRONTIUM

As previously stated, removal of Sr* from deposits in the bone has not been found possible without causing severe demineralization. Sr is known to be metabolized very slowly by the bone; the released Sr is excreted via the urine, the bile and the gastro-intestinal mucosa, where it is partially re-absorbed. The possibility of binding this endogenous Sr by alginate derivatives has been studied in our laboratories with rats as the experimental animals.

Rats were dosed with 0·25 μCi ^{89}Sr/ml of drinking water for 2 days, providing an average total of 10 μCi to each rat. Sodium alginate was given at four different dose levels for the following 2 weeks. The animals were sacrificed and the total counts/min in one femur from each rat was measured. The average reduction in bone count from the controls was 24%, irrespective of the amount of alginate given (Table III).

TABLE III

Resecretion of radioactive strontium

^{89}Sr was given in drinking water for two days (total 10 μCi). Sodium alginate was added to the laboratory chow of groups of twelve rats as indicated for the following 14 days. One femur was taken from each rat for radioassay.

Sodium alginate (% diet)	Mean (counts/min ±s.d.)	$\dfrac{\text{Experimental}}{\text{Control}} \times 100$
0	12,301 ±4,797	100
5	8,301 ±2,376	67·4
10	10,204 ±3,827	82·9
15	8,731 ±3,821	70·9
20	10,143 ±5,914	82·4

The same batch of sodium alginate was used by Stara in a similar series of experiments carried out on kittens (Stara, 1965). Kittens given a single oral dose of ^{85}Sr, followed by a sodium alginate diet for 30 days (Fig. 4), excreted approximately 14% more of the initial dose in the faeces than the controls. The body burden of ^{85}Sr was reduced in the experimental animals by about 16% of the initial dose.

Following a single intraperitoneal injection of ^{85}Sr, the body-burden curves were determined on two groups of kittens (three each) (Fig. 5). The experimental group was under a similar alginate feeding regime to that in the preceding experiment.

* The decrease in the amount of ^{45}Ca deposited, compared with the corresponding values for sodium alginate, is probably accounted for by the reduction in specific activity of the total Ca administered, rather than a reduction in the actual amount of Ca absorbed.

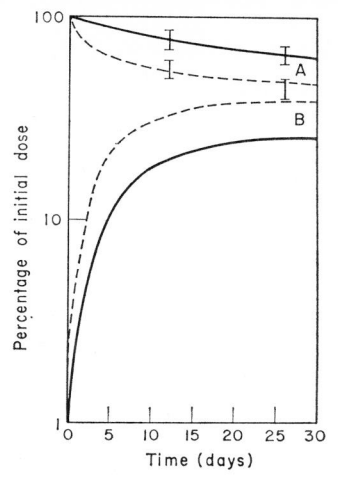

FIG. 4. ⁸⁵Sr retention in kittens after a single oral dose: A, body burden; B, cumulative faecal excretion. Continuous lines, control; broken lines, experimental. Values are expressed as % of initial body burden.

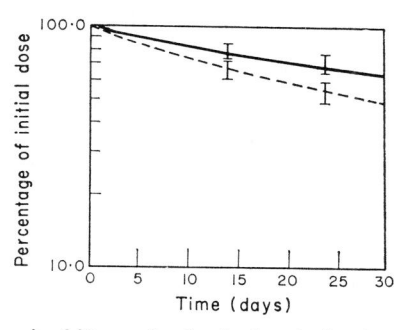

FIG. 5. ⁸⁵Sr retention in kittens (as body burden) after intraperitoneal dose: continuous line, control; broken line, experimental. Values expressed as % of initial body burden.

The rate of endogenous excretion is limited by the rate of exchange at several stages. These occur at the bone–blood and blood–intestinal mucose interfaces. It is also limited by re-absorption. Sodium alginate interferes in this last stage and therefore the increased faecal output of ⁸⁹Sr probably represents a high proportion of the amount normally re-absorbed.

The ability of alginate to bind endogenous radiostrontium in the gut, suggests that alginate may also be used in cases of accidental inhalation of the radioisotope, or in any case where radioactive Sr has already been deposited in the bone.

VARIATION IN THE EFFICIENCY OF ALGINATE SAMPLES

Many samples of sodium alginate from different commercial sources and from several different species and varieties of seaweed have been tested. There have been considerable differences in their capacity to inhibit the absorption of ^{89}Sr (Tables IV and V).

TABLE IV

Effect on strontium-89 uptake in bone of sodium alginate obtained from different commercial sources

5 μCi ^{89}Sr in 0·5 ml with 0·9 μM inactive SrCl$_2$ was injected into ligated duodenal segments of 100–120 g rats together with 2 mg sodium alginate obtained from different commercial sources. After 30 min animals were sacrificed; radioassay was carried out on one femur from each rat

Sample	Mean C.P.M.* ($\pm$s.d.)	$\dfrac{\text{Experimental}}{\text{Control}} \times 100$
Control	31,813 ±11,305	100
Commercial sodium alginate. Lot No.		
130068	24,584 ±9,037	77·2
713849	20,444 ±4,733	77·2
742794	17,141 ±9,018	53·8
741297	18,454 ±6,454	58·0
2002M4471	15,456 ±4,871	48·5
2004M4471	14,460 ±4,487	45·4
AX 450	11,442 ±3,924	35·9
371404	9,805 ±2,372	30·8
192404	7,545 ±2,392	23·7

* Mean for twelve rats.

Variations in the efficiency of these samples may be due to several possible causes. Firstly the purity of the sample; commercial samples are almost invariably prepared from species of phaeophyceae in abundant supply, namely the giant kelp, *Macrocystis pyrifera*, of the California coast, *Laminaria digitata*, *Laminaria hyperborea* and *Ascophyllum nodosum*. The latter three seaweeds contain inert polysaccharide material that tends to be retained as an impurity in the preparation of the alginate by fractional precipitation.

A second factor may be the variability of the selectivity coefficient k of the ion-exchange reaction:

$$\frac{[Sr_{bound}][Na_s]^2}{[Sr_s][Na_{bound}]^2} = k$$

TABLE V

Effect on strontium-89 uptake in bone of sodium alginate obtained from different seaweeds

5 μCi ^{89}Sr plus 0·9 μM inactive $SrCl_2$ with 10 mg of the alginate was administered through an orogastric tube to rats weighing 100–120 g. Animals were sacrificed 24 h later. One femur assayed for radioactivity. (Each figure represents mean counts/min and standard deviation for studies in twelve rats)

Source of alginate	Mean (counts/min ±s.d.)	$\dfrac{\text{Experimental}}{\text{Control}} \times 100$
Control	29,646 ± 5,843	100
Lam. digitata (B. Fundy)	5,886 ± 1,736	19·8
Lam. digitata (Pacific)	5,111 ± 1,359	17·2
Fucus edentatus	9,213 ± 2,321	31·0
Saccorhiza	13,775 ± 4,680	46·4
Hedophyllum sessile	30,110 ± 8,572	101·5
Fucus (Easter Island)*	22,820 ± 4,982	76·9
Durvillia antarctica	26,298 ± 6,567	88·7
Fucus vesiculosus	24,354 ± 7,541	82·1
Ascophyllum nodosum	12,516 ± 3,446	42·2
Lam. cloustonii	12,112 ± 3,446	40·8
Undaria pinnatifuda Commercial	12,575 ± 3,517	42·4

* Variety not identified.

where Na_1 and Sr_1 are the concentration of Na and Sr free in solution, in the reaction $Sr^{2+} + Na$ Alginate $\rightleftharpoons Sr$ Alginate $+ 2Na^+$. Early workers (Specker *et al.*, 1954; Thiele and Hallich, 1957) observed that different divalent metals showed different affinities for alginate. Thiele and Hallich arranged the divalent metals in an "ionotropic series" in the order

$$Pb > Cu > Cd > Ba > Sr > Ca > Zn.$$

Slight variations in this order were found by Schweiger (1962) in studying the stability of different metal–alginate complexes. Müller (1960) and McDowell (1958) observed that the order of the series varied according to the source of the raw material. The affinity series was eventually correlated with variation in the proportion of the two constituent uronic acids by Haug, who with his co-workers (Haug, 1959; Haug and Smidsrød, 1965) has shown that guluronic acid-rich alginate has a higher affinity for divalent metal than the mannuronic acid-rich samples.

Determination of the mannuronic–guluronic ratios have been made in our laboratories on a few alginates extracted from various species of seaweed, and on a few samples of commercial alginate.

N   S.M.

These ratios were compared with the efficiency of inhibiting ^{89}Sr absorption by the intubation method described previously. The mannuronic–guluronic acid ratio fell from 2·63 in a specimen of sodium alginate extracted from a weed obtained in the Southern Hemisphere (*Durvillia antarctica*) with no inhibitory activity, to 1·03 in a commercial sample that had high activity in suppressing Sr absorption.

We thank Dr J. Stara for permission to include his data in this paper. The work described in this paper was supported by a grant from the National Center for Radiological Health, U.S.P.H.S. (Contract PHS 86–65–59).

REFERENCES

Baardseth, E. (1965). *5th Int. Seaweed Symp.* p. 19.

van der Borght, O., Colard, J., van Puymbroeck, S., and Kirchmann, R. (1966). Paper presented at the International Symposium on Radioecological Concentration Processes (Stockholm 25–29 April, 1966).

Edmondson, L. F. *et al.* (1962). *J. Dairy Sci.* **45**, 800–803.

Fooks, J. H., Terrill, J. G., Jr., Heinemann, B. H., Baldi, E. J., and Walter, H. E. (1966). *Hlth Physics*, in press.

George, W. H. S. (1962). *Nature, Lond.* **195**, 155.

Haug, A. (1959). *Acta chem. scand.* **13**, 1250–1251.

Haug, A., and Smidsrød, O. (1965). *Acta chem. scand.* **19**, 341–351.

Hesp, R., and Ramsbottom, B. (1965). *Nature, Lond.* **208**, 1341–1342.

McDowell, R. H. (1958). *Chemy. Ind.* 1401–1402.

Migicovsky, B. B. (1959). *Can. J. Biochem. Physiol.* **37**, 1287–1292.

Müller, B. (1960). Thesis, Technische Hochschule, Karlsruhe.

Murthy, G. K., Masuravsky, E. B., and Campbell, J. E. (1961). *J. Dairy Sc.* **44**, 2158–2170.

Paul, T. M., Waldron-Edward, D., and Skoryna, S. C. (1964). *Can. Med. Assn. J.* **91**, 553–557.

Schubert, J. (1958). *Atompraxis*, **4**, 393–397.

Schweiger, R. G. (1962). *J. org. Chem.* **27**, 1789–1791.

Silverman, J., Ghosh, D., and Belcher, R. L. (1963). *Nature, Lond.* **198**, 780.

Skoryna, S. C., and Kahn, D. S. (1959). *Cancer*, **12**, 306–322.

Skoryna, S. C., Paul, T. M., and Waldron-Edward, D. (1964). *Can. Med. Assn. J.* **91**, 285–288.

Skoryna, S. C., Paul, T. M., and Waldron-Edward, D. (1965). *Can. Med. Assn. J.* **93**, 404–407.

Specker, H., Kuchtner, M., and Hartkamp, H. (1954). *Z. analyt. Chem.* **141**, 33–38.

Stara, J. (1965). Paper presented at the Symposium of Nuclear Medicine (Veterans Administration Hospital, Omaha, January, 1965).

Thiele, H., and Hallich, K. (1957). *Kolloidzeitschrift*, **151**, 1–12.

Waldron-Edward, D., Paul, T. M., and Skoryna, S. C. (1964). *Can. Med. Assn. J.* **91**, 1006–1010.

Waldron-Edward, D., Paul, T. M., and Skoryna, S. C. (1965). *Nature, Lond.* **205**, 1117–1118.

Discussion

QUESTION: Have you tried pure hyaluronic acid?

ANSWER: No. I do not think that it would work, because hyaluronic acids do not precipitate with Ca *in vitro* at these concentrations.

QUESTION: Do you know which variety of seaweed gives the most effective preparation of sodium alginate?

ANSWER: Not yet.

QUESTION: From published work of Haug it seems that the most suitable variety is *Laminaria hyperborea*; I notice that you do not have a sample of that.

ANSWER: A small sample was furnished by Dr. Haug; its activity was about as good as the best of our own samples.

QUESTION: Have you made any experiments *in vitro* to confirm the conclusion that sodium alginate acts as an ion-exchange material?

B. E. C. NORDIN: We have made some studies *in vitro* to establish whether sodium alginate acts as an ion-exchange material or as a weakly dissociated salt. The conclusion is that it behaves as an ion-exchange material; that is why it binds Sr preferentially, whereas weakly dissociated salts, for all practical purposes, bind Sr less effectively than Ca.

Author Index

The numbers in *italics* indicate the pages on which names are mentioned in the reference lists

A

Åberg, B., 243, *245*, 261, 262, *263*
Adams, P. J. V., 139, 146, *148*
Adams, W. S., 124, *128*
Agricultural Research Council Radiobiological Laboratory, 35, 38, *39*, 43, *45*, 85, *90*, 284, *295*
Akiya, S., 273, *278*
Alexander, G. V., 284, *295*
Alexander, L. T., 221, *222*
Andersen, A. C., 183, 185, *194*, 235, *235*
Anderson, R. P., 183, *194*
Andrews, G. A., 113, *128*
Annenkov, B. N., 272, *278*, 304, *305*
Arantani, H., 274, *280*
Archambeau, J. O., 223, *227*
Arden, J. W., 41, *45*
Armitage, C., 274, *280*
Arnold, J. S., 190, *194*
Aubert, J. P., 147, *148*, 150, 158, *159*
Auerbach, H., 208, *212*
Axelrod, D. J., 116, *128*

B

Baardseth, E., 331, *340*
Bacon, J. A., 121, 124, 126, *128*, 271, *278*, 310, *312*
Bailey, N. T. J., 150, 154, *159*
Baldi, E. J., 330, *340*
Baranova, E. F., 237, *245*
Barkhudarov, R. M., 71, 74, *82*
Barnes, D. W. H., 69, *70*, 150, 155, *159*, 161, *166*, 172, *173*
Bartlett, B. O., 34, 35, 36, *39*
Bartley, J. C., 267, *278*
Bates, T. H., 119, 122, 127, *128*, 272, 273, *280*, 310, *312*, 323, *328*
Bauer, G. C. H., 149, 150, 151, 156, 157, 158, *159*, 288, *295*

Bayne, H. R., 271, *279*
Bean, H. W., 43, *45*, 48, 52, *55*, 75, 81, *82*
Belcher, R. L., 330, *340*
Bell, M. C., 223, 224, 226, *227*
Beloborodova, N. L., 237, *245*
Benedict, S. R., 272, *279*
Beninson, D., 24, 25, *30*, 49, 52, *55*
Berger, E., 85, *91*, 123, *129*
Biagi, E., 120, *128*
Biles, C. R., 223, *227*
Bishop, M., 69, *70*, 84, *90*, 132, 134, 135, 138, *138*, 147, *148*, 150, 155, *159*, 161, 165, *166*, 168, 169, 172, *173*
Bjornerstedt, R., 150, 158, *159*
Bligh, P. H., 175, 176, *179*
Blincoe, C., 98, *102*
Bluhm, M., 150, *159*
Bodenlos, L. J., 223, *227*
Bohman, V. R., 98, *102*
Boltz, D. F., 216, *221*
Bond, V. P., 223, *227*
van der Borght, O., 233, 235, *235*, 330, *340*
Boswell, J., 2, *5*
Boyd, J., 150, 158, *159*
Bresnahan, M., 113, 119, *128*
Bronner, F., 147, *148*, 150, 158, *159*, 166, *166*, 323, *327*
Brothers, M., 85, *91*
Browder, A. A., 272, *281*, 304, *305*
Brown, V. M., 104, 108, *109*
Bruce, R. S., 116, *128*, 273, 274, *278*
Brues, A. M., 208, *212*
Bryant, F. J., 41, 43, *45*, 48, 49, 53, *55*, 63, *65*, 76, *81*, 85, 86, *90*, 97, 98, *102*, 148, *148*, 150, 154, *159*, 284, *295*
Budy, A. M., 122, *128*
Bugryshev, P. F., 77, *82*
Buldakov, L. A., 237, *245*
Burton, J. D., 34, 36, *39*
Burykina, L. N., 243, 244, *245*

Bustad, L. K., 185, *194*, 214, 219, 221, *221*, *222*

C

Campbell, J. E., 330, *340*
Carlqvist, B., 119, *128*, 266, 268, 273, 278, 323, *327*
Carlsson, A., 149, 150, 151, 156, 157, 158, *159*, 288, *295*
Carr, T. E. F., 77, 79, *81*, 131, *138*, 139, 146, 147, *148*, 161, 165, 166, *166*, 234, *235*
Case, A. C., 215, *221*
Catsch, A., 119, *128*, 265, 266, 268, 271, 272, 273, 274, 275, 277, *278*, *279*, 304, *305*, 323, *327*
Chamberlain, A. C., 41, 43, *45*, 85, *90*
Chambers, F. W., Jr, 223, *227*
Charles, M. L., 123, *129*, 267, *280*
Chen, P. S., 165, *166*
Clark, C. W., 172, *173*
Clark, I., 270, 272, *278*, *279*
Clarke, W. J., 214, 221, *222*
Cobb, J. R., 150, 158, *159*
Coen, G., 120, *128*
Cohn, S. H., 77, *81*, 84, *90*, 119, *129*, 138, *138*, 267, 268, 270, 272, 275, *278*, *279*, 283, 293, 295, *295*, 323, *327*
Colard, J., 330, *340*
Comar, C. L., 19, 24, 26, *30*, *31*, 63, *65*, 106, *109*, 113, 116, 120, 126, *128*, *129*, 150, 158, *159*, 170, *173*, 185, *194*, 218, 219, 221, *221*, *222*, 223, 224, *227*, 234, *235*, 247, *253*, 266, *281*, 295, *295*, 323, *327*
Congdon, C. C., 226, *227*
Cook, M. J., 214, *222*
Copp, D. H., 116, *128*, 265, 266, 269, 271, *279*
Cotterill, J. C., 41, *45*
Coveart, A. E., 103, *109*
Crabo, B., 261, *263*
Cramer, C., 269, *279*, 294, *295*
Craven, D. L., 150, *159*
Crawford, A., 8, 10, 11, *11*
Czosnowska, W., 27, *30*, 85, *90*

D

Davy, H., 2, 3, 4, 5, *5*, 11, *12*
Dawson, R. M. C., 261, *263*
Day, F. H., 7, *11*

Decker, C. J., 193, *194*
Della Rosa, R. J., 172, *173*, 183, 184, 185, 187, 189, *194*, 235, *235*, 271, 272, *279*
DeRoche, G., 208, *212*
van Dilla, M. A., 190, *194*
Dolphin, G. W., 37, *39*, 83, *90*, 170, *173*
Dooronbekov, Zh., 274, *279*
Dow, E. C., 149, 150, 156, 157, *159*
Dowdle, E. B., 175, *179*
Dowling, J. H., 223, *227*
Downer, R., 247, 250, *253*
Downie, E. D., 237, *245*
Dudley, R. A., 205, *205*
Duggan, M. H., 175, 176, *179*
Dupuis, Y., 175, *179*
Dwyer, L. J., 97, *102*

E

Edmondson, L. F., 330, *340*
Edwards, H. M., Jr, 247, 250, 251, *253*
Eisenberg, E., 138, *138*
Ekman, L., 243, *245*
Elliot, J. R., 271, 277, *279*
Ellis, F. B., 34, 36, *39*
Engfeldt, B., 247, *253*
Enomoto, Y., 70, *70*, 234, 235, *236*, 277, *281*
Evans, R. D., 183, *194*
Eve, I. S., 83, *90*, 170, *173*
Ezmirlian, F., 98, *102*, 113, 116, *128*, 150, *159*

F

Farris, G. C., 94, *102*
Fedorov, N. A., 274, *279*
Feldman, H. S., 313, *321*
Feldstein, A., 77, *81*, 267, 273, 277, *279*, *280*
Figueroa, W. G., 84, *90*
Finkel, A. J., 205, *205*
Fletcher, W., 43, 44, *45*, 57, 58, *61*
Fliedner, T. M., 223, *227*
Flyger, V., 96, 97, *102*
Fooks, J. H., 330, *340*
Ford, M. R., 214, *222*
Fournier, P. L., 175, *179*
Fried, J. F., 274, *279*
Frolen, H., 208, *212*
Frost, H. M., 98, *102*

Fuji, S., 268, 272, 273, 274, *280*
Fujita, M., 258, *260*
Fukuda, R., 268, 271, 274, *280*, 304, *305*

G

Gamachson, J., 77, *81*
Gashwiler, J. S., 94, *102*
George, W. H. S., 330, *340*
Georgi, J., 63, *65*
Ghosh, D., 330, *340*
Gielow, F., 184, 189, *194*
Gillner, M., 261, 262, *263*
Goldman, M., 183, 185, 186, *194*, 235, *235*
Gong, J. K., 275, *279*, 323, *327*
Gordan, G. S., 138, *138*
Gordon, M., 274, *279*
Gorodetzskiy, A. A., 271, *279*
Gottesman, E. D., 123, *129*
Gran, F. C., 139, 146, *148*
Grant, C. L., 247, 250, *253*
Graul, E. H., 277, *279*
Greenberg, D. M., 121, *128*, 172, *173*, 265, 266, 271, *279*
Greenwald, E., 121, *128*
Grimes, J. H., 274, *279*
Gruden, S., 269, *279*
Gusman, E. A., 77, *81*, 119, *129*, 267, 272, *279*

H

Häggröth, S., 97, *102*
Hallich, K., 339, *340*
Hamada, G. H., 219, *222*
Hamilton, J. G., 116, *128*, 269, *279*
Hamilton, T. S., 43, *45*, 48, 52, *55*, 75, 81, *82*
Hansard, S. L., 124, 126, *128*, 271, *278*, 310, *312*
Hardy, E. P., 85, *91*, 221, *222*, 269, 272, *280*
Harley, J. H., 52, *55*, 97, 98, *102*
Harrison, G. E., 25, 29, *30*, 68, 69, *70*, 77, 79, *81*, 84, 85, *90*, 119, *128*, 131, 132, 134, 135, 138, *138*, 139, 143, 147, *148*, 150, 155, 158, *159*, 161, 165, *166*, 168, 169, 170, 172, *173*, 266, 268, 269, *279*
Hartkamp, H., 339, *340*
Hasterlik, R. J., 205, *205*
Hastings, A. B., 172, *173*

Hatano, T., 250, *253*
Haug, A., 313, 314, *321*, 339, *340*
Health and Safety Laboratory, 19, *31*, 74, 76, 77, *81*, *82*
Heaney, R. P., 166, *166*
Hegsted, D. M., 113, 119, *128*
Heinemann, B. H., 330, *340*
Heller, H.-J., 273, *279*
Henderson, E. H., 41, 43, *45*, 63, *65*
Herzog, R. K., 30, *31*
Hesp, R., 319, *321*, 330, *340*
Hjertquist, S. O., 247, *253*
Hodge, H. C., 150, 158, *159*
Hodges, E. J., 86, 89, *90*, *91*
Hodges, R. M., 98, *102*, 150, *159*
Hoglund, G., 97, *102*
Hogue, D. E., 221, *222*
Holgate, W., 195, *206*
Holtzman, R. B., 139, *148*
Honma, T., 268, 272, 273, 274, *280*
Hope, T. C., 8, 9, 10, 11, *12*
Horii, S., 250, *253*
Huggard, A. J., 274, *279*
Hughes, R., 2, *5*
Hundeshagen, H., 277, *279*
Hunt, V. R., 235, *235*

I

Ibsen, K. H., 274, *279*
Ichikawa, R., 70, *70*, 108, *109*, 234, 235, *236*
International Commission on Radiological Protection, 83, 84, 85, *90*, 214, *222*, 313, *321*
Irving, J. T., 85, *90*
Ishibashi, S., 273, *279*
Ishidate, M., 273, *279*
Ishii, K., 274, *280*
Ishimochi, R., 277, *281*
Ito, Y., 269, 273, *279*, 310, *312*, 323, *328*
Itoh, H., 250, *253*
Iwamoto, J., 258, *260*

J

Jeffay, H., 271, *279*
Jensen, J. B., 183, *194*
Jones, D. A., 94, *102*
Jones, D. C., 269, *279*
Jones, H. G., 119, *128*, 266, *279*
Judd, J. M., 103, *109*

K

Kabakow, B., 85, *91*, 122, *128*, 272, 273, *280*
Kahn, D. S., 330, *340*
Kantake, N., 277, *281*
Karniewicz, W., 83, *90*, 285, *295*
Kasatkin, Yu., 274, *279*
Kaspar, L. V., 193, *194*
Kastner, J., 209, *212*
Kawin, B., 119, *128*, 265, 266, 268, *279*, 304, *305*
Keirim-Marcus, I. B., 238, *245*
Kennedy, A., 171, *173*
Kent, A., 8, 9, *12*
Khomutovskiy, O. A., 271, *279*
Kidman, B., 119, *128*
Kirchmann, R., 330, *340*
Knizhnikov, V. A., 71, 73, 74, 77, 79, *82*, 86, *90*
Kollmer, W. E., 267, 269, *279*
Kologrivov, R., 89, *91*
Kornberg, H. A., 113, 116, *128*
Kosterkiewicz, A., 285, *295*
Kostial, K., 269, *279*
Kretchmar, A. L., 226, *227*
Kriegel, H., 267, 269, 274, *279*
Kroll, H., 274, *279*
Kshirsagar, S. G., 158, *159*, 168, 171, *173*, 196, 197, 205, *205*
Kuchtner, M., 339, *340*
Kuikka, A. O., 273, 277, *280*
Kulikova, V. G., 244, *245*
Kulp, J. L., 41, *45*, 48, 49, 51, 53, *55*, 83, 84, 86, 89, *90*, *91*, 98, *102*
Kunde, M. L., 150, *159*
Kurl'andskaya, E. B., 237, *245*
Kusma, J. F., 272, 275, *281*

L

Laake, H., 124, *128*
Lagergren, C., 247, *253*
Larson, B., 313, 314, *321*
Laszlo, D., 25, *31*, 85, *91*, 113, 114, 122, 123, 126, *128*, *129*, 150, 154, *159*, 267, 272, 273, *280*
Lavoisier, A., 10, 11, *12*
Lederer, H., 156, 158, *159*
Lengemann, F. W., 120, *128*, 171, *173*, 218, 219, *222*, 235, *236*

Lewin, I., 116, 119, 125, 126, 127, *128*, *129*
Li, M., 25, *31*, 113, 114, 126, *128*, *129*, 150, 154, *159*, 268, *280*
Likins, R. C., 150, *159*
Lillegraven, A. L., 131, 133, *138*, 139, 142, *148*, 162, 163, *166*
Lindenbaum, A., 274, *279*
Lindquist, B., 149, 150, 151, 156, 157, 158, *159*, 288, *295*
Lindroos, B., 273, 277, *280*
Liniecki, J., 83, *90*, 285, *295*
Lloyd, E., 158, *159*, 168, 171, *173*, 197, 205, *205*
Lloyd, G. D., 41, *45*
Longhurst, W. M., 97, *102*
Lorick, P. C., 113, *128*, 150, 158, *159*, 272, *279*
Lough, S. A., 24, *31*, 185, *194*, 219, 221, *222*
Loutit, J. F., 37, *39*, 41, 43, 44, *45*, 48, 49, 53, *55*, 57, 58, *61*, 76, 77, 79, *81*, 85, 86, *90*, 98, *102*, 131, *138*, 139, 148, *148*, 150, 154, *159*, 161, *166*, 170, *173*, 277, *279*, 284, *295*
Lowenstein, J. M., 274, *280*
Lowrey, R. S., 223, 224, 226, *227*
Lumsden, E., 150, 158, *159*
Lüning, K. G., 208, *212*
Lutkić, A., 269, *279*

M

McArthur, C., 98, *102*, 150, *159*
McArthur, W. H., 226, *227*
McClellan, R. O., 185, *194*, 214, 219, 221, *221*, *222*, 226, *227*, 235, *236*
MacDonald, N. S., 84, *90*, 98, *102*, 113, 116, 124, *128*, 150, 158, *159*, 265, 268, 272, 274, *279*
McDowell, R. H., 339, *340*
MacFadyen, I. J., 151, *159*
MacGregor, J., 150, 154, *159*
McIntyre, I., 64, *65*, 68, *70*
McKenney, J. R., 185, *194*, 214, 215, 221, *221*, *222*
McLean, F. C., 122, *128*, 172, *173*
MacPherson, S., 197, *205*
Mahy, B. W. J., 172, *173*
Marcus, C. S., 23, 30, *31*, 178, 179, *179*, 235, *236*
Marei, A. N., 71, 79, *82*, 86, *90*

Marley, W. G., 37, *39*
Marshall, J. H., 89, 90, *90*, 193, *194*, 293, 294, *295*
Martell, A. E., 274, *280*
Martin, J. H., 97, *102*
Mashimo, T., 277, *281*
Mason, J. I., 131, *138*, 139, *148*, 162, 163, *166*
Masuravsky, E. B., 330, *340*
Matsuo, Y., 277, *281*
Matsushima, Y., 269, 273, *279*
Maxwell, D. C., 274, *279*
Mayneord, W. V., 37, *39*
Mays, C. W., 183, 185, *194*, 235, *235*
Mazzuoli, G., 120, *128*
Medical Research Council, 41, *45*, 86, *90*
Melchinger, H., 265, 266, 268, 271, 272, 273, 274, 275, 277, *278*, *279*, 304, *305*, 323, *327*
Menczel, J., 116, *128*, *129*
Mendel, L. B., 272, *279*
Mercer, E. R., 34, 36, *39*
Migicovsky, B. B., 330, *340*
Migliori, H., 24, 25, *30*
Miller, C. E., 205, *205*
Millis, J., 313, *321*
Mirzoyan, A. A., 226, *227*
Mitchell, H. H., 43, *45*, 48, 52, *55*, 75, 81, *82*
Monroe, R. A., 247, *253*
Morgan, A., 41, *45*, 63, *65*
Morton, A. G., 41, *45*
Moskalev, Yu. I., 237, *245*
Mraz, F. R., 113, 119, *128*, 247, 250, 251, 252, *253*
Müller, B., 339, *340*
Müller, W. A., 268, *279*, 293, 295, *295*
Mulryan, B. J., 150, 158, *159*
Munday, K. A., 172, *173*
Murthy, G. K., 330, *340*
Myttenaere, C., 229, *236*

N

Nakatsuka, M., 274, *280*
Nelson, A., 119, *128*, 208, *212*, 266, 268, 272, 273, 274, *278*, *280*, 304, *305*, 323, *327*
Neuman, M. W., 172, *173*
Neuman, W. F., 150, 158, *159*, 165, *166*, 172, *173*

Nevins, R., 272, *279*
Nicholas, J. A., 150, 158, *159*
Nisbet, J., 139, *148*, 150, 154, *159*
Nix, N., 187, *194*
Nobel, S., 267, 268, *279*, 283, 293, 295, *295*
Nold, M. M., 106, *109*, 113, 120, *128*, *129*, 150, 158, *159*, 218, 219, *222*, 234, *235*
Nordin, B. E. C., 139, *148*, 150, 151, 154, *159*
Norris, W. P., 193, *194*, 209, *212*
Novikova, N. Ya., 73, 79, *82*
Noyes, P., 113, *128*, 150, 158, *159*, 272, *279*
Nozaki, H., 247, 250, *253*
Nusbaum, R., 98, *102*, 150, *159*, 284, *295*

O

Ogawa, E., 268, 271, 272, 273, 274, *280*, 304, *305*
Oldham F., 2, *5*
Ophel, I. L., 103, *109*
Oshima, M., 247, *253*
O'Sullivan, J., 183, *194*
Owen, M., 195, 197, 205, *205*, *206*

P

Palmer, R. F., 113, 116, *128*, 221, *222*, 255, *260*, 268, 269, *280*, *281*
Papadopoulou, D., 113, *129*, 266, *281*
Papworth, D. G., 43, 44, *45*, 57, 58, *61*
Parcher, J. W., 187, *194*
Parfenov, Yu. D., 237, 243, 244, *245*
Parker, A., 63, *65*
Parmley, W. W., 183, *194*
Partington, J. R., 1, *5*, 7, 11, *12*
Patrick, H., 121, 124, 126, *128*, 271, *278*, 310, *312*
Paul, T. M., 273, *280*, *281*, 313, 320, *321*, 330, 331, 333, *340*
Payne, J. W., 272, *281*, 304, *305*
Pecher, C., 218, *222*
Pecher, J., 218, *222*
Peets, E., 267, *280*
Peterson, G., 184, 189, *194*
Petukhova, E. V., 71, 74, *82*
Philips, J. W., 274, *280*
Pond, W. G., 221, *222*
Pool, R., 183, *194*

Posner, A. S., 150, *159*
Powell, T. J., 183, 186, *194*
Prepejchal, W., 209, *212*
van Putten, L. M., 269, 272, 277, *280*, 307, 308, 309, 310, *312*
van Puymbroeck, S., 330, *340*

R

Rae, S. L., 151, *159*
Ramos, E., 24, 25, *30*, 49, 52, *55*
Ramsbottom, B., 319, *321*, 330, *340*
Ray, R. D., 116, *128*, 269, *280*, 310, *312*
Raymond, W. H. A., 29, *30*, 68, *70*, 84, *90*, 132, 134, 135, 138, *138*, 147, *148*, 150, 158, *159*, 161, 165, *166*, 168, 169, 170, *173*
Reber, E. F., 267, *278*
Reed, F. B., 313, *321*
Richards, Y., 274, *280*
Richelle, L. J., 147, *148*, 150, 158, *159*
Ringrose, R. C., 247, 250, *253*
Rivera, J., 24, *31*, 52, *55*, 70, *70*, 84, 85, 86, *90*, *91*, 97, 98, 101, *102*, 185, *194*, 205, *206*, 221, *222*, 269, 272, *280*
Rivera-Cordero, F., 270, 272, *278*
Roberts, D. R., 209, *212*
Robertson, J. S., 84, *90*, 138, *138*
Robinette, W. L., 94, *102*
Robinson, B. H. B., 22, 23, 27, *31*, 327, *328*
Rogacheva, S. A., 277, *280*
Rogers, G., 94, *102*
Rönnbäck, C., 208, *212*, 266, 268, 272, 273, 274, *280*, 304, *305*
Rosén, L., 266, 268, 272, 273, 274, *280*, 304, *305*
Rosenthal, H. L., 108, *109*
Roth, Z., 273, *281*, 297, 298, *305*
Rounds, D. E., 113, 116, *128*
Rowland, R. E., 293, *295*
Rubanovskaya, A. A., 237, *245*, 265, 266, 272, *280*
Rubin, M., 273, *280*
Rundo, J., 131, 132, 133, 134, 135, 138, *138*, 139, 142, 147, *148*, 162, 163, 165, 166, *166*, 168, 169, *173*
Rushton, M., 195, *206*
Russell, R. S., 34, 35, 37, *39*, 221, *222*
Ryabova, E. Z., 271, *279*

S

Sakai, F., 70, *70*
Samachson, J., 25, 27, *31*, 77, 79, *82*, 84, 85, *90*, *91*, 112, 113, 114, 116, 117, 119, 121, 122, 125, 126, 127, *128*, *129*, 138, *138*, 150, 154, 156, 158, *159*, 166, *166*, 267, 268, 269, 272, 273, 277, *279*, *280*, 310, *312*
Saville, P. D., 150, 158, *159*
Schachter, D., 175, *179*
Scheniker, H., 175, *179*
Schenkel, R., 272, *281*
Schilling, G., 230, *236*
Schmid, A., 271, 272, *280*, *281*
Schömer, W., 277, *279*
Schubert, J., 273, 275, *280*, 323, *328*, 330, *340*
Schulert, A. R., 41, *45*, 48, 49, 51, 53, *55*. 63, *65*, 83, 84, 86, *90*, 267, *280*
Schultz, V., 96, 97, *102*
Schweiger, R. G., 339, *340*
Seki, M., 277, *281*
Semenov, D. I., 273, 275, *280*, *281*
Setälä, K., 273, 277, *280*
Shepherd, H., 25, *30*
Shibata, K., 268, 271, 274, *280*
Shikita, M., 269, 273, *279*
Sillén, L. G., 274, *280*
Silverman, J., 330, *340*
Sivachenko, T. P., 271, *279*
Skillman, T. G., 166, *166*
Skoog, W. A., 124, *128*
Skoryna, S. C., 273, *280*, *281*, 313, *321*, 330, 331, 333, *340*
Slavin, W., 94, *102*
Smidsrød, O., 339, *340*
Smith, D. A., 139, *148*, 150, 151, 154, *159*
Smith, F. A., 172, *173*, 271, 272, *279*
Smith, H., 119, 122, 127, *128*, 272, 273, 274, *280*, 310, *312*, 323, *328*
Smith, M. R., 272, *278*
Smith, P. J. A., 274, *279*
Smith, R. L., 274, *280*
Smith, W. H., 172, *173*
Snyder, W. S., 214, *222*,
Sobel, A. E., 267, 268, *279*, 283, 293, 295, *295*
Sowden, E. M., 98, *102*
Spain, P. C., 98, *102*, 113, 116, *128*, 150, *159*
Specker, H., 339, *340*

Speckman, T. W., 209, *212*
Spencer, H., 25, *31*, 77, *81*, 84, 85, *90*, *91*, 113, 114, 116, 119, 121, 122, 123, 125, 126, 127, *128*, *129*, 138, *138*, 150, 154, *159*, 267, 268, 269, 272, 273, 277, *279*, *280*, 310, *312*
Spencer-Laszlo, H., 27, *31*, 77, 79, *82*, 85, *91*, 126, *128*
Spicer, G. S., 41, 43, *45*, 63, *65*, 85, *90*
Spina, G. M., 295, *295*
Spreng, P., 271, *280*
Stanbury, J. B., 149, 150, 156, 157, *159*
Stannard, J. N., 172, *173*, 271, 272, *279*
Stara, J., 330, 336, *340*
Stearns, R., 98, *102*, 150, *159*
Stedman, D. E., 116, *128*, 269, *280*, 310, *312*
Steggerda, F. R., 43, *45*, 48, 52, *55*, 75, 81, *82*
Stepanov, Yu. S., 71, 74, *82*
Stitch, S. R., 98, *102*
Stover, B. J., 185, 190, *194*, 235, *235*
Sutton, A., 25, *30*, 69, *70*, 77, 79, *81*, 119, *128*, 131, 132, 134, 135, 138, *138*, 139, 147, *148*, 150, 155, 158, *159*, 161, 165, *166*, 168, 169, 172, *173*, 266, *279*
Suzuki, S., 268, 271, 272, 273, 274, *280*, 304, *305*

T

Takagi, C., 277, *281*
Takei, Y., 250, *253*
Takeuchi, T., 277, *281*
Takita, H., 273, *279*
Talmage, R. V., 271, 277, *279*
Tamura, Z., 273, *279*
Tanaka, Y., 274, *280*
Taylor, A. N., 30, *31*
Taylor, D. M., 175, 176, *179*
Templeton, W. L., 104, 108, *109*
Teree, T., 119, *129*
Terrill, J. G., Jr, 330, *340*
Tessari, L., 295, *295*
Thiele, H., 339, *340*
Thompson, D. M., 269, *279*
Thompson, J. C., Jr, 26, *31*
Thompson, R. C., 113, 116, *128*, 221, *222*, 255, *260*, 268, 269, *280*, *281*
Thompson, S. P., 3, *5*
Thurber, D. L., 89, *91*
Titterton, E. W., 97, *102*

Tokoki, K., 274, *280*
Touzet, R., 49, 52, *55*
Tregubenko, I. P., 273, 275, *280*, *281*
Trent, D., 94, *102*
Tretheway, H. C., 29, *30*, 150, 158, *159*, 161, *166*, 170, *173*
Tsurufuji, S., 269, 273, *279*, 310, *312*, 323, *328*
Tsuzuki, H., 268, 272, 273, 274, *280*
Turekian, K. K., 98, *102*
Tutt, M., 119, *128*
Twardock, A. R., 219, *222*
Tweedy, W. R., 126, *129*

U

Uchiyama, M., 270, 273, *278*, *281*
Ukita, T., 270, *281*
Ullberg, S., 113, *128*
United Nations, 34, 36, 37, *39*
Updegraaf, H., 172, *173*
Urist, M. R., 84, *90*, 122, *128*, 274, *279*
U.S.A.E.C., 94, *102*
Ushakova, V. F., 237, *245*, 265, 266, 272, *280*

V

Vasington, F. D., 179, *179*
Vaughan, J., 119, *128*, 158, *159*, 168, 171, *173*, 195, 196, 197, 205, *205*, *206*
Vojvodić, S., 269, *279*
Volf, V., 116, *129*, 272, 273, 274, 275, *281*, 297, 298, *305*
Vorozheikina, T. B., 237, *245*
de Vries, M. J., 308, *312*

W

Wade, M. A., 98, *102*
Waldron-Edward, D., 273, *280*, *281*, 313, *321*, 330, 331, 333, *340*
Wall, J., 234, *235*
Wallace, H., 275, *280*, 323, *328*
Walser, M., 22, 23, 27, *31*, 272, *281*, 304, *305*, 327, *328*
Walter, H. E., 330, *340*
Walton, A., 89, *91*
Wasserman, R. H., 19, 23, 30, *30*, *31*, 106, *109*, 113, 116, 120, 126, *128*, *129*, 150, 158, *159*, 175, 178, 179, *179*, 218, 219, 221, *222*, 223, 224, *227*, 234, *235*, 247, *253*, 266, *281*, 295, *295*

Wayne, D. J., 151, *159*
Webb, M. S. W., 41, 43, *45*, 85, *90*
Webber, T. J., 41, *45*
Weber, E., 267, 269, *279*
Whitney, I. B., 218, *222*
Widdowson, E. M., 25, *30*
Wilford, S. P., 274, *279*
Williamson, M., 196, 197, *205*
Wolff, N. K., 116, *128*, 269, *280*, 310, *312*

Y

Yarboro, C. L., 216, *222*
Yartsev, E. I., 71, 79, *82*, 86, *90*
Yoneyama, T., 277, *281*
Young, L. G., 183, 186, *194*

Z

Zander-Principati, G. E., 272, 275, *281*
Zipf, K., 271, 272, *280*, *281*

Subject Index

A

Acute exposure to ^{90}Sr, 190
Adolescent uptake of ^{90}Sr, 185
Age, effect on metabolism, 24, 63, 93, 217, 288
Alginic acid derivatives, 313, 329
Ammonium chloride and enhancement of urinary excretion, 271
Analytical techniques
    activation analysis, 68
    autoradiography, 197
    dialysis, 327
    flame photometry, 68, 103, 183
    atomic absorption spectrometry, 94, 262
    titration, 151
    radiochemical, 63, 94
Atomic absorption spectrometry, 94, 262
Autoradiography, 197

B

Body burden, ^{90}Sr, 44, 83, 116, 216, 229, 243, 308
Bone, accretion of ^{90}Sr, 43
    mineralization rate, 156
    replacement of Sr, 41, 44
    retention of Sr, 292
    models for prediction of ^{90}Sr concentration, 48
    annual turnover of ^{90}Sr, 44, 52
Bone, strontium-90 content
    human, 41, 47, 57, 73, 83
    animal, 93
    predicted and observed, 60
    New York City, 53
    San Francisco, 53
    Britain, 41
    Glasgow, 58
    and diet, 58
    U.S.S.R., 74

Bremsstrahlung counting of ^{90}Sr, 183, 215, 284

C

^{45}Ca, in urine and faeces, 256
Calcium alginate, 329
Calcium, influence on strontium metabolism, 77, 113, 215, 268, 289, 304, 310
    in chromosomes, 30
Calcium measurement, EDTA, 95, 151
    flame spectrophotometer, 68, 103
Carbonate, influence on metabolism, 297
Cereals, ^{90}Sr in, 36
Chelating agents, 273
Chlorides, effect on strontium metabolism, 297
Chromosomes, calcium and strontium in, 30
Chronic exposure, ^{89}Sr, 131
    ^{90}Sr, 83, 181, 273
Citrate, influence on strontium metabolism, 273, 311, 323
Comparative behaviour of Sr, Ca, Ra, 235

D

Deer, strontium metabolism in, 93
Demineralizing agents, 271, 277, 281
Diet, influence on strontium metabolism, 74, 215
Dietary strontium, 33, 37, 58, 72, 84
    and urine, 64
    U.S.S.R., 72
    New York City, 84
    Poland, 85
    Britain, 58, 85
Discrimination factor, definition, 20
    and rate constants, 22

Discrimination between calcium and
  strontium
  body–diet, 24
  bone–diet, 48, 76
  excreta–bone, 259
  fish bone–water, 106
  gastrointestinal, 23
  maternal diet–milk, 218
  renal, 67, 154
  skeletal, 156
  and age, 24
  and dietary calcium, 113
  placental, 183
DNA, 30
Dose commitment, 37

E

Egg production, influence of ^{45}Ca and
  ^{89}Sr, 247
Excretion ratios, urinary/faecal, 161,
  255, 286

F

Flame emission spectrophotometry, 68,
  103
Fluorine in diet, influence on strontium
  uptake, 79
Foetal strontium dynamics, 183
Food chains, ^{90}Sr in, 33

G

Gastrointestinal absorption, 85, 229,
  297, 313, 329

H

High phosphorus diet, effect on strontium metabolism, 116, 215
Hormonal factors, 122

I

Intestinal absorption, availability of
  strontium, 229
Ion-exchange flame photometry, 183

K

Kinetics of ^{85}Sr deposition, 131
  ^{90}Sr deposition, 84, 186
  (short term), 139

L

Lactose, effect on strontium metabolism,
  120
Ligands, strontium binding by, 274
Low level counting, 94
Low phosphorus diet, effect on strontium retention, 269, 289, 307
  effect on ^{90}Sr toxicity, 307
  development of rickets and demineralization of bone, 116
Lysine, effect on strontium metabolism,
  120

M

Magnesium, enhancement of urinary
  excretion of strontium, 272, 302
Manucol, 314
Milk, ^{90}Sr in, 34, 60
Milk, removal of strontium from, 330
Mineralization rates of bone, 156
Milk, contamination, 34, 59
Modification of strontium metabolism
  by
  alginates, 313, 329
  arsenate, 177
  barbital, 177
  carbonates, 297
  changes in thyroid and parathyroid
  function, 125
  citrate, 273, 311, 323
  corticosteroids, 124
  dinitrophenol, 177
  enhancement of urinary excretion, 271
  high calcium diet, 113, 213, 255, 268,
  289
  high phosphorus diet, 116
  high strontium diet (or isotopic
  dilution) 265, 273, 283
  lactose, 120, 177
  lysine, 120
  low calcium diet, 113, 213
  low phosphorus diet, 269, 289, 307
  oligomycin, 177
  parathyroid extract, 126
  phloridzin, 177
  phosphates, 297

Modification of strontium metabolism by—*contd.*
  pilocarpine, 299
  sex hormones, 122
  sulphates, 297
  vitamin D, 122
  whole body irradiation, 223

O

Observed ratio, definition, 19
  urine–diet, relation to urinary calcium, 26
  bone–diet, 52, 108
  bone–water (in fish), 107
  (see also strontium–calcium ratio)
Osteosarcoma, 196, 275
Oxidative phosphorylation, 175

P

Parathyroid extract, effect on strontium absorption, 126
Parathyroid hormone, enhancement of urinary excretion, 271
Phosphate, effect on strontium metabolism, 116, 215, 269, 289, 297, 307
Pilocarpine, effect on strontium metabolism, 273, 299
Placental transfer of ^{90}Sr, 237
Plasma clearance, 258
Plasma, stable strontium and calcium content, 68
  ^{90}Sr in, 168
Pyridoxal, pyridoxine, pyridoxic acid, 272

R

Rate constant, relation to discrimination factor, 21
  tubular reabsorption, 27
γ-Ray spectrometry, 133, 315
X-Ray emission spectrometry, 183
Renal clearance, 69, 161, 269
Renal discrimination between strontium and calcium, 154
Renal tubular reabsorption, 67, 155, 272
Reproductive performance, effect of dietary strontium, 247
Rickets, caused by low phosphorus diet, 116

S

Seaweed, as source of sodium alginate, 313, 330
Skeletal deposition, ^{85}Sr, 13
  ^{90}Sr, 239
Skeletal discrimination, 156
Skeletal retention, 134, 149, 185, 243, 298
  effect of low phosphorus diet on, 116, 269
  effect of high phosphorus diet on, 116
Sodium alginate, 313, 329
Sodium citrate, 323
Sodium salicylate, enhancement of urinary excretion, 272
Sodium sulphate, enhancement of urinary excretion, 272, 302
Soft tissues, distribution of ^{85}Sr, 209
Soft tissue dosage, 207
"Soil factor", 34
Sperm DNA, 261
Squamous carcinoma, 195
Stable strontium, effect on radiostrontium metabolism, 119, 267, 283, 304
  in diet, 73
  measurement, 68, 94, 103, 262
Strontium balance, 143
Strontium/calcium ratio in
  bone and diet, 218
  bone and teeth, 187
  bone, prediction, 83
  bull serum and spermatozoa, 262
  diet and urine, 64
  fish bone, 106
  lake waters, 104
  milk and maternal diet, 218
  newborn bone and maternal diet, 218
  urine and faeces, 260
  urine and plasma, 68
Strontium metabolism in animals
  beagle, 181
  cat, 336
  deer, 93
  dog, 237
  fish, 103
  guinea pig, 229
  hen, 247
  mouse, 207, 307
  rabbit, 167, 195, 323
  ram, 261
  rat, 175, 255, 283, 298, 332

Strontium metabolism in man, three-compartment model, 140
^{85}Sr, in urine and faeces, 256, 116, 318
  measured by whole-body counter, 133, 315
  metabolism, comparison with ^{45}Ca, 149
^{90}Sr, estimation
  autoradiography, 197
  radiochemical, 63, 94
^{90}Sr in human bone, 41, 47, 57, 73
  in human teeth, 71
  in animal bone, 93
  in urine, 63
Sulphate, effect on strontium metabolism, 236, 272, 299, 302
Sutures, uptake of ^{90}Sr and ^{90}Y by, 212

T

Teeth, ^{90}Sr content, 71
Total body monitoring of ^{90}Sr by bremsstrahlung measurement, 183, 215, 284
Total body radioactivity, measurement by γ-spectrometry, 133, 315
Toxicity, ^{90}Sr and ^{226}Ra compared, 181
Tumour induction, 195, 275, 307
Turnover rate, 101, 195
  in deer, 93

U

Urinary calcium, related to OR, 26
  and strontium retention, 268

Urinary clearance, 69, 161, 269
Urinary excretion of radiostrontium, influence of sex hormones, 122
  and diet, 64
  influence of tubular reabsorption, 67
Urinary excretion, enhancement by ammonium chloride, 271
  calcium, 269, 304
  citrate, 323
  corticosteroids, 124
  magnesium, 272, 302
  parathyroid hormone, 126, 271
  pyridoxal etc., 272
  sodium salicylate, 272
  sodium sulphate, 272, 302
  vitamin A, 272
Urinary/faecal excretion ratios, 161, 255, 286

V

Vegetables, ^{90}Sr in, 36
Vitamin A, enhancement of urinary excretion, 271
  as demineralizing agent, 278, 281
Vitamin D, effect on radiostrontium metabolism, 121

W

Whole-body irradiation, effect on ^{89}Sr and ^{45}Ca metabolism, 223
  effect on nitrogen balance, 227
Whole-body radioactivity measurement, 133, 284, 315